TROPICAL PINNIPEDS
Bio-Ecology, Threats and Conservation

TROPICAL PINNIPEDS
Bio-Ecology, Threats and Conservation

Editor

Juan José Alava

Institute for the Oceans and Fisheries
The University of British Columbia
Vancouver
Canada

Coastal Ocean Research Institute
Vancouver Aquarium Marine Science Centre
Vancouver
Canada

and

School of Resource and Environmental Management
Simon Fraser University
Burnaby
Canada

CRC Press is an imprint of the
Taylor & Francis Group, an **informa** business

A SCIENCE PUBLISHERS BOOK

Cover Acknowledgement

- Top-left photo of the Hawaiian monk seal: Reproduced by kind courtesy of M. Sullivan

- Top-right photo of the Galapagos fur seal: Reproduced by kind courtesy of J.J. Alava

- Bottom-left photo of the Juan Fernandez fur seal: Reproduced by kind courtesy of L.P. Osman

- Bottom-right photo of the Mediterranean monk seal: Reproduced by kind courtesy of A.A. Karamanlidis/MOm

CRC Press
Taylor & Francis Group
6000 Broken Sound Parkway NW, Suite 300
Boca Raton, FL 33487-2742

First issued in paperback 2020

© 2017 by Taylor & Francis Group, LLC
CRC Press is an imprint of Taylor & Francis Group, an Informa business

No claim to original U.S. Government works

ISBN-13: 978-1-4987-4139-2 (hbk)
ISBN-13: 978-0-367-78217-7 (pbk)

This book contains information obtained from authentic and highly regarded sources. Reasonable efforts have been made to publish reliable data and information, but the author and publisher cannot assume responsibility for the validity of all materials or the consequences of their use. The authors and publishers have attempted to trace the copyright holders of all material reproduced in this publication and apologize to copyright holders if permission to publish in this form has not been obtained. If any copyright material has not been acknowledged please write and let us know so we may rectify in any future reprint.

Except as permitted under U.S. Copyright Law, no part of this book may be reprinted, reproduced, transmitted, or utilized in any form by any electronic, mechanical, or other means, now known or hereafter invented, including photocopying, microfilming, and recording, or in any information storage or retrieval system, without written permission from the publishers.

For permission to photocopy or use material electronically from this work, please access www.copyright.com (http://www.copyright.com/) or contact the Copyright Clearance Center, Inc. (CCC), 222 Rosewood Drive, Danvers, MA 01923, 978-750-8400. CCC is a not-for-profit organization that provides licenses and registration for a variety of users. For organizations that have been granted a photocopy license by the CCC, a separate system of payment has been arranged.

Trademark Notice: Product or corporate names may be trademarks or registered trademarks, and are used only for identification and explanation without intent to infringe.

Library of Congress Cataloging-in-Publication Data

Names: Alava, Juan José, editor.
Title: Tropical pinnipeds : bio-ecology, threats, and conservation / editor, Juan José Alava, Faculty of Science, Institute for the Oceans and Fisheries, The University of British Columbia, Vancouver, Canada.
Description: Boca Raton, FL : CRC Press, 2017. | "A science publishers book."
| Includes bibliographical references and index.
Identifiers: LCCN 2016054170| ISBN 9781498741392 (hardback : alk. paper) | ISBN 9781498741408 (e-book : alk. paper)
Subjects: LCSH: Seals (Animals)--Tropics. | Pinnipedia.
Classification: LCC QL737.P6 T76 2017 | DDC 599.79--dc23
LC record available at https://lccn.loc.gov/2016054170

Visit the Taylor & Francis Web site at
http://www.taylorandfrancis.com

and the CRC Press Web site at
http://www.crcpress.com

Dedication

*For
Nastenka,
Nastenkita, Juan José
and Joshua*

Foreword

Pinnipeds are a fascinating group of species that are distributed worldwide. They originated in the subarctic region but during their evolution have spread across the globe and all the way into the Antarctic. Even though they crossed the equator in the process, only few species found a way of living in subtropical and tropical areas. The "tropical" species treated in this book have been much less under scientific scrutiny than their more abundant arctic and Antarctic relatives. This is the more surprising since species living at the edge of a group's suitable habitats may offer important insights in limits to adaptation inherent in the amphibious lifestyle of this group.

As nearly all species of pinnipeds, the tropical species have suffered tremendously from hunting in the 19th and 20th century and have slowly recovered or are still in the process of recovering from human persecution. All of these species are comparatively rare (i.e., at least in comparison to the 15 million or so Crabeater seals populating the Antarctic pack ice), not only as a consequence of earlier slaughters but also, because for many of the species the distribution area is limited and their natural population size is consequently low due to geographic restriction alone. Therefore, their conservation and management deserve special attention as evidenced by the fact that 40% of the species fall in the "Endangered" category of the IUCN.

In this context, it is extremely welcome that Juan José Alava and David Aurioles-Gamboa got together and recruited a group of workers that know these species' biology in detail and here report on their phylogenetic relationships, their ecology and threats to their continuing existence. These authors bring together widely differing approaches and report novel data on population status, physiology, life history and health-related problems for the tropical otariids. Moreover these species are all exposed to the El Niño phenomenon and may be the first to suffer from global warming combined with the increasing danger of overfishing: Plenty of good reasons to study them in much more detail. This book offers much of the basic information needed for further studies into the effects of global scale ecological change.

We learn right at the outset that the otariids we find in the region today were late-comers that replaced the earlier phocid inhabitants of tropical and subtropical South America. Several chapters review new knowledge about the more or less tropical seal species and suggest ways to use this knowledge for their conservation. It is particularly helpful that we also hear about the health risks as the spread of diseases with and without human interference, often mediated by introduced domestic animals, may be one of the most dangerous aspects of present anthropogenic effects for the conservation of the often small populations of tropical pinnipeds.

The editors and authors are to be congratulated to have embarked on such a valuable endeavour that will certainly remain a key resource for all pinniped enthusiasts and governmental agencies involved in the management and protection of tropical species.

Fritz Trillmich
University of Bielefeld
33615 Bielefeld
Morgenbreede 45
Germany
Email: fritz.trillmich@uni-bielefeld.de

Preface

In October 2014, I received an invitation from the editorial department of Science Publishers (CRC Press and Taylor & Francis Group) to prepare and submit a proposal for a book on the Biology and Ecology of Seals, Sea Lions, Walruses. While the book's scope as proposed was attractive to me, I realized that it would be difficult to do justice to such a broad subject matter. In the end, I felt it best to focus on a more specific theme and devote the book to the tropical and subtropical pinnipeds of our blue planet. Being originally from a hot tropical country (Ecuador), I considered that a book dedicated to tropical/subtropical sea lions, fur seals and seals inhabiting the tropics (i.e., Tropics of Cancer and Capricorn) and subtropics (i.e., the 38th parallel in each hemisphere) would be highly worthwhile and would coincide with my previous and ongoing research on Galapagos pinnipeds.

Accomplishing this task has been challenging, and would not have been possible without the help of many people. I contacted several esteemed colleagues from tropical and subtropical countries to invite them to contribute chapters on the species for which they have conducted research and expended conservation efforts. The response was positive with about 40 authors/co-authors from Brazil, Canada, Chile, Ecuador and Galapagos Islands, Greece, Mexico, Peru, South Africa, United States of America (e.g., Hawaiian Islands) and Uruguay agreeing to contribute to this book.

After more than two years of substantial work in close collaboration with leading authors and co-authors, the CRC Press editorial and production teams, and the invaluable assistance of invited reviewers, we present *"Tropical Pinnipeds: Bio-ecology, Threats and Conservation."*

This edited book delves into the research and conservation of tropical and subtropical pinnipeds across 16 chapters, including species such as the Hawaiian monk seal, the California sea lion, the Guadalupe fur seal, the Galapagos fur seal, the Galapagos sea lion, the South American sea lion, the South American fur seal, the Juan Fernandez fur seal, the Mediterranean monk seal and the Cape fur seal. Understanding the natural history and threats to these species is critical to support conservation efforts. To this end, the book's topics include current research on evolution, biology, marine ecology, bio-monitoring and state of the art approaches to investigate and assess tropical pinniped populations, behavioral ecology, anthropogenic stressors and health. Special chapters and sections on the conservation of threatened or/and endangered pinnipeds (e.g., Hawaiian monk seal, Galapagos sea lions, Galapagos fur seals, and Mediterranean monk seal) are also included.

I am in deeply indebted to all the authors/co-authors and reviewers who invested their precious time and energies to this book. I am especially thankful to Dr. David Aurioles-Gamboa for co-writing the Introduction for this book, and to Dr. Fritz Trillmich for writing the Foreword. I acknowledge the Institute for the Oceans and Fisheries (University of British Columbia), Coastal Ocean Research Institute (Vancouver Aquarium Marine Science Center) and School of Resource and Environmental Management (Simon Fraser University) for providing resources and platforms to develop this book. I am also grateful to my family for supporting me in this work.

Finally, many thanks to the Galapagos sea lions and fur seals, remarkable marine mammals that continue to inspire me.

Juan José Alava
Institute for the Oceans and Fisheries
University of British Columbia
Vancouver, BC, Canada

Acknowledgements to Reviewers

The editor expresses his profound gratitude to the following colleagues who gently reviewed and provided valuable insights to the chapters submitted to this book.

Dr. Morgan Churchill
Department of Anatomy, New York Institute of Technology, Old Westbury NY 11568-8000, USA.
Email: morgan.churchill@gmail.com

Dr. Patricia Fair
Medical University of South Carolina, Department of Public Health Sciences, 171 Ashley Avenue, Charleston, SC 29425, USA.
Email: fairp@musc.edu

Dr. Frances Gulland
The Marine Mammal Center, 2000 Bunker Road, Fort Cronkhite, Sausalito, CA 94965-2619, USA.
Email: Gullandf@TMMC.org

Dr. Martin Haulena
Marine Mammal Rescue Centre, Vancouver Aquarium Marine Science Centre, P.O. Box 3232, Vancouver, British Columbia, Canada V6B 3X8.
Email: Martin.Haulena@vanaqua.org

Dr. Victoria Otton
School of Resource and Environmental Management, Simon Fraser University, 8888 University Drive, Burnaby, British Columbia, Canada V5A 1S6.
Email: votton@sfu.ca

Dr. David Rosen
Marine Mammal Research Unit, Institute for the Ocean and Fisheries, University of British Columbia, Vancouver, British Columbia, Canada.
Email: rosen@zoology.ubc.ca

Dr. Kristin Sullivan
Freelance Translator, karl8y75.translatorscafe.com, Charlestown, Rhode Island, USA.
Email: karl8y75@yahoo.com

Dr. Fritz Trillmich
University of Bielefeld, 33615 Bielefeld, Morgenbreede 45, Germany.
Email: fritz.trillmich@uni-bielefeld.de

Dr. Stella Villegas-Amtmann
Department of Ecology and Evolutionary Biology, University of California Santa Cruz Long Marine Lab, 100 Shaffer rd, COH, 95060, USA.
Email: stella.villegas@gmail.com

Contents

Dedication   v

Foreword   vii

Preface   ix

Acknowledgements to Reviewers   xi

1. Introduction to Tropical and Subtropical Pinnipeds   1
  Juan José Alava and *David Aurioles-Gamboa*

2. An Overview on the Evolutionary History of Tropical Pinnipeds   12
  Carlos A. Vildoso Morales

3. Variability in the Skull Morphology of Adult Male California Sea Lions and Galapagos Sea Lions   22
  Jimena Bohórquez-Herrera, David Aurioles-Gamboa, Claudia Hernández-Camacho and *Dean C. Adams*

4. Hawaiian Monk Seals: The Biology and Ecology of the World's only Tropical Phocid   50
  Charles Littnan, Michelle Barbieri, Jessica Lopez Bohlander, Tenaya Norris, Stacie Robinson and *Kenady Wilson*

5. Hawaiian Monk Seal Conservation: Past, Present and Future   69
  Charles Littnan, Michelle Barbieri, Jessica Lopez Bohlander, Tenaya Norris and *Stacie Robinson*

6. Guadalupe Fur Seal Population Expansion and its Post-breeding Male Migration to the Gulf of Ulloa, México   91
  David Aurioles-Gamboa, Nereyda Pablo-Rodríguez, M. Patricia Rosas-Hernández and *Claudia J. Hernández-Camacho*

7. Population Status, Anthropogenic Stressors and Conservation of the Galapagos Fur Seal (*Arctocephalus galapagoensis*): An Overview   120
  Juan José Alava, Judith Denkinger, Pedro J. Jiménez, Raúl Carvajal and *Sandie Salazar*

8. Diving Physiology, Foraging and Reproductive Behavior of the Galapagos Sea Lion (*Zalophus wollebaeki*) — 132
Stella Villegas-Amtmann and *Daniel P. Costa*

9. Management Strategies and Conservation Status of Galapagos Sea Lion Populations at San Cristobal Island, Galapagos, Ecuador — 159
Diego Páez-Rosas and *Nataly Guevara*

10. Population Ecology, Trends and Distribution of the Juan Fernandez Fur Seal, *Arctocephalus philippii* (Peters 1866) in Chile — 176
Layla P. Osman and *Carlos A. Moreno*

11. Population Ecology and Conservation Status of the South American Sea Lion in Uruguay — 194
Valentina Franco-Trecu, Massimiliano Drago, Diana Szteren and *Federico G. Riet-Sapriza*

12. Ecology and Conservation Status of the South American Fur Seal in Uruguay — 211
Valentina Franco-Trecu

13. The Uncertain Fate of the Endangered Mediterranean Monk Seal *Monachus monachus* in the 21st Century: Population, Ecology and Conservation Threats — 219
Panagiotis Dendrinos, Styliani Adamantopoulou, Eleni Tounta and *Alexandros A. Karamanlidis*

14. Bioecology and Conservation Threats of the Cape Fur Seal *Arctocephalus pusillus pusillus* — 234
G.J. Greg Hofmeyr

15. Emerging Pathogens and Health Issues in the 21st Century: A Challenge for Tropical and Subtropical Pinnipeds — 245
Karina Acevedo-Whitehouse and *Luis A. Soto-García*

16. Pathologies of Pinnipeds in Brazil — 269
Paula Baldassin, Derek Blaese de Amorim, Max R. Werneck and *Daniela Bueno Mariani*

Index — 287

1

Introduction to Tropical and Subtropical Pinnipeds

Juan José Alava[1,2,3,*] *and David Aurioles-Gamboa*[4]

Pinnipeds are a fascinating and charismatic group of marine mammals playing a crucial role as apex predators and sentinels of the functioning and health of marine ecosystems. Pinnipeds (i.e., Pinnipedia from Latin *pinna* "fin" and *pes, pedis* "foot") belong to the order Carnivora comprising three living families: Otariidae (i.e., sea lions and fur seals), Odobenidae (i.e., walruses) and Phocidae (i.e., seals). They are found in the most extreme environments in Polar Regions to the Earth's equator in the tropics. Pinnipeds comprise about 34 species, of which at least 30% inhabit, distribute and breed permanently in tropical zones from coastal and oceanic regions of the global ocean (see Fig. 1). Defining a pinniped species as tropical or subtropical is a challenging task because several species, including those from temperate regions, may extend their geographic distributions into subtropical or tropical latitudes or are currently expanding their distributional ranges to these regions, and, in some instances, some

---

[1] Institute for the Oceans and Fisheries, University of British Columbia, 2202 Main Mall, Vancouver, BC V6T 1Z4, Canada.
[2] Ocean Pollution Research Program, Coastal Ocean Research Institute, Vancouver Aquarium, Marine Science Centre, P.O. Box 3232, Vancouver, BC V6B 3X8, Canada.
[3] Fundacion Ecuatoriana para el Estudio de Mamíferos Marinos (FEMM), Ecuador.
[4] Laboratorio de Ecología de Pinnípedos, CICIMAR-IPN, La Paz BCS, C.P. 23096, México.
 Email: dgamboa@ipn.mx
* Corresponding author: j.alava@oceans.ubc.ca

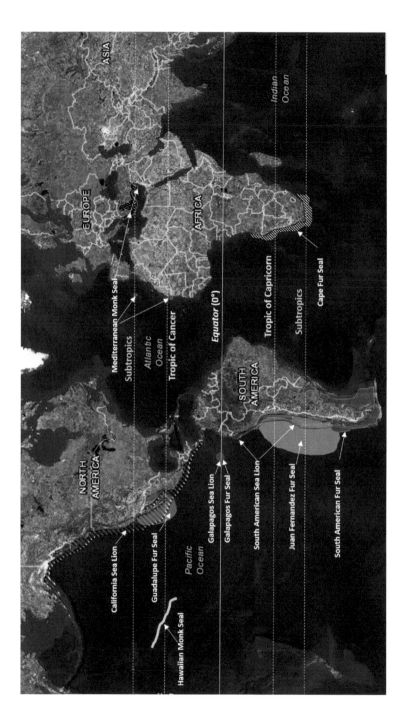

Figure 1. Global distribution of tropical and subtropical pinniped species as defined by the criteria used in this book (i.e., normal distributional ranges found within the Tropics of Cancer and Capricorn and subtropics). Geographical distribution range for each species was explored and approximated based on spatial data from red list maps of the International Union for the Conservation of Nature (IUCN; http://maps.iucnredlist.org/index.html).

colonies are temporally or permanently established (e.g., harbor seals, *Phoca vitulina* in the Mexican Pacific, and hooded seals, *Cystophora cristata*, along the Southeastern coast of USA). For the purpose of this book, the global geographical thresholds used here as the criteria to select both tropical and subtropical species are set up between the Tropic of Cancer in the northern hemisphere at 23° 26' 16" N and the Tropic of Capricorn in the southern hemisphere at 23° 26' 16" S for tropical pinnipeds, as well as the subtropical band (i.e., 38th parallel in each hemisphere) for subtropical species, as shown in Fig. 1. However, some populations of sea lions and seals from the North and South hemispheres are currently expanding their geographical ranges and spreading to new localities.

Following our criteria, living tropical and subtropical species include two phocids and seven to eight otariid species (see Table 1): the Hawaiian monk seal (*Neomonachus schauinslandi*), the California sea lion (*Zalophus californianus*), the Guadalupe fur seal (*Arctocephalus townsendi*), the Galápagos fur seal (*Arctocephalus galapagoensis*), the Galápagos sea lion (*Zalophus wollebaeki*), the South American sea lion (*Otaria flavescens*), the South American fur seal (*Arctocephalus australis*), the Juan Fernandez fur seal (*Arctocephalus philippii*), and the Mediterranean monk seal (*Monachus monachus*). In addition to these species, the Cape (South Africa) Fur Seal (*Arctocephalus pusillus pusillus*) may well fall within the subtropical group as the species is native to and occur along the coast of Namibia and southern Angola (Africa), although this fur seal is considered to be a temperate species (Hofmeyr 2015a, G. Hofmeyr, pers. comm.). Conversely, the recent occurrence of several other species of pinnipeds, including phocid and otariid species predominating in temperate regions, along tropical or subtropical seas and coasts around continents raises new insights to revisit the baseline aimed to categorize a species as either tropical or subtropical.

In the Pacific coast of South America, the recurrent presence of several species of pinnipeds from remote locations and oceanic islands has been usually observed and confirmed in new locations out of their normal distribution ranges, mainly in coastal areas of Colombia, Galápagos Islands, Ecuador, Peru, Chile and Antarctica (Wellington and De Vries 1976, Merlen 1995, Capella et al. 2002, Alava and Salazar 2006, Félix et al. 2007, Acevedo et al. 2011, Torres et al. 2012, Avila et al. 2014, Acevedo et al. 2016). Similarly, both California sea lions (*Z. californianus*) and harbor seals (*P. vitulina*) have colonies found in the Pacific coast of Mexico (Le Boeuf et al. 1983, Gallo-Reynoso and Aurioles-Gamboa 1984), although the California sea lion, recognized as a subtropical species, occupies both sides of the Baja California Peninsula in Mexico (Aurioles-Gamboa and Hernández-Camacho 2015). The recent confirmation of hooded seals (*C. cristata*) observed in subtropical and tropical waters of the Western North Atlantic, where they were misidentified as Caribbean monk seals (*Neomonachus tropicalis*;

Table 1. Current overall population and IUCN listing of extant and extinct tropical and subtropical pinnipeds.

Common name	Scientific name	Population size (individuals)	Population trend	IUCN-red list category	IUCN reference
Hawaiian monk seal	*Neomonachus schauinslandi*	1209	Decreasing	Endangered (EN)	Littnan et al. (2015)
Harbor seal*	*Phoca vitulina*	350,000–500,000	Increasing	Least Concern (LC)	Thompson and Härkönen (2008)
California sea lion	*Zalophus californianus*	387,646	Increasing	Least Concern (LC)	Aurioles-Gamboa and Hernández-Camacho (2015)
Guadalupe fur seal	*Arctocephalus townsendi*	20,000	Increasing	Least Concern (LC)	Aurioles-Gamboa (2015a)
Caribbean monk seal	*Neomonachus tropicalis*	0**	NA	Extinct (EX)	Lowry (2015)
Galápagos fur seal	*Arctocephalus galapagoensis*	15,000	Decreasing	Endangered (EN)	Trillmich (2015a)
Galápagos sea lion	*Zalophus wollebaeki*	14,000–16,000	Decreasing	Endangered (EN)	Alava and Salazar (2006), Trillmich (2015b)
South American sea lion	*Otaria flavescens*	250,000	Stable	Least Concern (LC)	Campagna (2014)
South American fur seal	*Arctocephalus australis*	250,000–300,000	Increasing	Least Concern (LC)***	Campagna (2008)
Juan Fernandez fur seal	*Arctocephalus philippii*	32,278	Increasing	Least Concern (LC)	Aurioles-Gamboa (2015b)
Cape (South African) fur seal	*Arctocephalus pusillus pusillus*	2,000,000	Increasing	Least Concern (LC)	Hofmeyr (2015a)
Mediterranean monk seal	*Monachus monachus*	600–700	Increasing	Endangered (EN)	Karamanlidis and Dendrinos (2015)

* The harbor seal is found throughout coastal waters of the Northern Hemisphere, from temperate to Polar Regions, but this species is included here as a subtropical species given the existence of colonies in the Mexican Pacific, falling within the geographical criteria used in this book to designate a species as tropical or subtropical.
** The population size of the Caribbean monk seal prior to exploitation was estimated as 233,000–338,000 animals (McClenachan and Cooper 2008).
***The Peruvian and northern Chilean populations of South American fur seals are likely to be different stock that may qualify for a threatened status (Campagna 2014).
NA: Not available

Fig. 2) (Mignucci-Giannoni and Haddow 2001, Mignucci-Giannoni and Odell 2001), reflects clear cases of extralimital records in those tropical regions. Vagrants of Sub Antarctic fur seals (*Arctocephalus tropicalis*) have been found in many countries, including Brazil and Chile (Juan Fernandez Islands) and continental coasts of several African nations (Hofmeyr 2015b). Similarly, following previous extinction in the Southeastern Pacific, an increasing number of new sightings and expansion of the distributional range of southern elephant seals (*Mirounga leonina*) have recently been document from the southern tip of South America (Cape Horn Archipelago) as far north as Ecuador (Acevedo et al. 2016), including the Galápagos Islands and Gulf of Guayaquil (Vargas and Steinfurth 2004, Alava and Carvajal 2005).

Moreover, the incidence of tropical pinnipeds arriving from the subtropics to temperate regions have also been documented because of recent tropicalization in the North Pacific which is likely to be driven by the temporal and spatial effects of the more frequent El Niño Southern Oscillation (ENSO) events and general ocean warming (Fukasawa et al. 2004, Wu and Li 2007, Vergés et al. 2014). Interestingly, vagrants and satellite-tracked individuals of Guadalupe fur seals (*A. townsendi*) have been recorded along the Northeastern Pacific (i.e., Pacific Northwest), including the coasts of California, Oregon and Washington State in USA as well as in British Colombia, Canada (Hanni et al. 1997, Etnier 2002; http://www.nmfs.noaa.gov/pr/health/mmume/guadalupefurseals2015.html). These findings obviously influence our framework to assign the tropical or subtropical criterion, causing bias to favor the inclusion or/and exclusion of certain species. In this context, while our quest and searching effort attempted to be systematic to include most species within the scope of the book, this may not be an exhaustive representation of all the species that may exist in or are new to tropical and subtropical zones of the planet.

Figure 2. Artistic illustration of the extinct Caribbean monk seal (*Neomonachus tropicalis*) based on a photograph of an individual held in captivity in the New York Aquarium in 1910 (see original printed photo in Mignucci-Giannoni and Odell 2001). Artwork by N. Alava C.

In the last century, the extinction of the Caribbean monk seal (*N. tropicalis*; Fig. 2), a tropical pinniped that was first discovered during Christopher Columbus' second voyage to America in 1494 (Herrera y Tordesillas 1601, Kerr 1824, Lowry 2015) and that once inhabited the Gulf of Mexico and most of the Caribbean Sea (Lowry 2015), is a poignant reminder of the ongoing and protracted risks of anthropogenic activities jeopardizing the survival and recovering of pinnipeds. The overexploitation of several species of tropical fur seals (e.g., Guadalupe, Galápagos and Juan Fernandez fur seals) during the last two centuries in the Pacific was also a clear example of the massive harvesting that we humans inflicted to satisfy the pelt supply and demand of past and modern societies and that almost brought these species to the brink of extinction. At present times, close to 40% of living tropical/subtropical pinnipeds are categorized as Endangered by the IUCN, while 60% fall within the category of Least Concern, as shown in Table 1. Although the populations of some species are showing increasing or stable trends, other species such as the Hawaiian monk seal, Galápagos fur seal and Galápagos sea lion are readily declining due to natural and anthropogenic factors (Fig. 3). While the population of the endangered Mediterranean monk seal appears to be increasing, this trend should be cautiously interpreted as a continuing declining in population size as recently recommended in the species' IUCN assessment, following the precautionary approach (Karamanlidis and Dendrinos 2015; Fig. 3). Thus, understanding the natural history and threats of these species is of paramount importance to support conservation efforts and management actions aimed to preserve pinnipeds.

To address these tasks, this special book on tropical pinnipeds is focused on reviews and original research fronts on the biology, marine ecology, bio-monitoring, conservation and management, as well as state of the art approaches to investigate and assess tropical pinniped populations, behavior, anthropogenic stressors and health. To accomplish this goal, we envisioned an integration, analysis and discussion of several subjects on new emerging topics in tropical and subtropical pinnipeds and their ecosystems. Based on a combination of original research articles, reviews and short communications by global experts, including academia, marine mammal scientists, graduate students, government and policy-makers, NGOs and private organizations, we hope that this book has brought about new research, baseline data, modeling tools and potential management strategies to provide knowledge and understand the past and current status and threats for the conservation of tropical pinnipeds with the aim of recovering and protecting the marine ecosystems and wellbeing of these species and their habitats.

Along these lines, this book covers several chapters illustrating the recent and newest findings on the bio-ecology, threats and conservation of

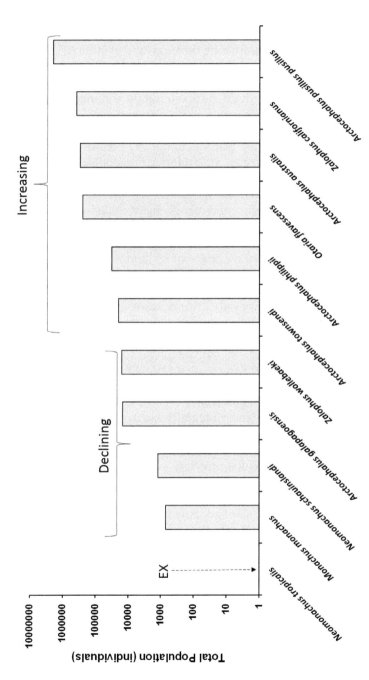

Figure 3. Overall population estimates and trends for tropical and subtropical pinnipeds distributed within the geographical margins of the subtropical band and the Tropics of Cancer and Capricorn. Although the population trend for the Mediterranean monk seal (*M. monachus*) was assessed to be increasing by the last IUCN assessment for the species, here it is considered to be in declining, based on the precautionary principle (Karamanlidis and Dendrinos 2015). EX over the inverted dashed arrow stands for extinct (i.e., the Caribbean monk seal, *N. tropicalis*). Population numbers with associated references are also provide in Table 1.

extant tropical pinnipeds. Briefly, these chapters are described as follows. Chapter 2 (Vildoso 2016) is an overview of the evolutionary history and phylogeny of tropical pinnipeds aimed to illustrate a panorama of the genesis of these species by contemplating their past and evolution, with the recent emergence of present day species from ancient tropical periods and regions to the existing tropics. Chapter 3 (Bohórquez-Herrera et al. 2016) is a remarkable comparative study of the evolutionary and morphological traits of tropical pinniped skulls, using closely related species, both the California and Galápagos sea lions, as a pragmatic examples to assess the variability and diversity of skulls features under the pressure of natural selection and the impact of environmental factors. Chapters 4 and 5 initiates the species-specific chapters, commencing with the bio-ecology, natural and anthropogenic threats of the Hawaiian monk seal (Chapter 4; Littnan et al. 2016), followed by the conservation and management challenges (Chapter 5; Littnan et al. 2016) to preserve this unique tropical phocid of the Pacific ocean. Chapter 6 (Aurioles-Gamboa et al. 2016) is an outstanding study about the population expansion, marine ecology and novel feeding area and habits (i.e., using scat analysis and stable isotope data assessed from vibrissae) of the Guadalupe fur seal, portraying its population increase and recovery. Chapter 7 (Alava et al. 2016) is an overview highlighting the state of the dwindling population, conservation threats (i.e., biological and chemical pollution) and species management recommendations of the smallest tropical otariid, the endangered Galápagos fur seal. This chapter is followed by Chapters 8 and 9, which are dedicated for the endangered Galápagos sea lion, a species living in sympatry with the Galápagos fur seal throughout the Galápagos Islands. Thus, Chapter 8 (Villegas-Amtmann and Costa 2016) is a holistic integrative analysis and review of the physiological plasticity, metabolic rate, foraging strategies and reproductive behavior Galápagos sea lion in the face of climate change. Chapter 9 (Páez-Rosas and Guevara 2016) provides a review on the conservation status and management actions for the protection of the Galápagos sea lions, focused on the largest breeding population inhabiting one of the most semi-urbanized islands of the Galápagos Archipelago. A special article for the Juan Fernandez fur seal is found in Chapter 10 (Osman and Moreno 2016), highlighting the population dynamics and trends, distribution of rookeries and conservation issues following its historical exploitation and recent population recovery around the Juan Fernandez Archipelago off Chile. Chapter 11 and Chapter 12 are new contributions dedicated to the population and reproductive ecology as well as conservation status of Uruguayan pinnipeds, including the South American sea lion (Franco-Trecu et al. 2016) and South American fur seal (Franco-Trecu 2016), respectively.

One of the most important contributions of the book is Chapter 13 (Dendrinos et al. 2016), which is devoted to the population and conservation status, distribution and anthropogenic impacts of the Mediterranean monk seal, a species considered as the rarest and most endangered pinniped in the global ocean. Chapter 14 (Hofmeyr 2016) is a review on the bio-ecology and conservation status offered for the Cape fur seal given its distributional range engulfing subtropical and tropical regions of Africa's southwestern coast.

Finally, the book ends up with special chapters on marine veterinary, conservation medicine and emerging infectious diseases and pathologies of tropical pinnipeds. Chapter 15 (Acevedo-Whitehouse and Soto-García 2016) tailors a novel analysis and the newest review on population health challenges focused on emerging pathogens (e.g., viruses, bacteria and parasites) and several other health topics faced by tropical and subtropical pinnipeds at present times. Chapter 16 (Baldassin et al. 2016) is the first contribution on marine veterinary and pathogens of Brazilian pinnipeds, with particular emphasis on pinnipeds' tuberculosis, clinical pathologies from necropsy examinations and dog attacks in Brazil.

We hope that the book fulfills the expectation of the marine mammal and scientific community, non-government organizations, academia, governments, students and general public, and that the findings and conservation initiatives for tropical pinnipeds raised here can be taken into consideration to support management and educational actions for the preservation of these species.

Acknowledgements

The first author (JJA) acknowledges Fundación Ecuatoriana para el Estudio de Mamíferos Marinos (FEMM) for the continued support and commitment to conduct marine mammals' research during the last three decades in Ecuador. Special thanks to the Institute for the Ocean and Fisheries (IOF, University of British Columbia) and the Coastal Ocean Research Institute (CORI), Vancouver Aquarium Marine Science Centre, for academic support and their continued efforts for ocean conservation. We are in debt with Dr. F. Trillmich for reading this introduction and writing the preface for the book.

Keywords: tropical, pinnipeds, Tropic of Cancer, Tropic of Capricorn, otariids, sea lions, fur seals, phocids, seals

References

Acevedo, J., R. Matus, D. Droguett, A. Vila, A. Aguayo-Lobo and D. Torres. 2011. Vagrant Antarctic fur seal, *Arctocephalus gazella*, in southern channels of Chile. Polar Biol. 34: 939–943.
Acevedo, J., A. Aguayo-Lobo, J.L. Brito, D. Torres, B. Cáceres, A. Vila, M. Cardeña and P. Acuña. 2016. Review of the current distribution of southern elephant seals in the eastern South Pacific. New Zeal. J. Mar. Fresh. Res. doi: 10.1080/00288330.2015.1132746.
Alava, J.J. and R. Carvajal. 2005. First records of elephant seals on the Guayaquil Gulf, Ecuador: On the occurrence of either a *Mirounga leonine* or *M. angustirostris*. Lat. Am. J. Aquat. Mamm. (LAJAM) 4: 195–198.
Alava, J.J. and S. Salazar. 2006. Status and conservation of Otariids in Ecuador and the Galápagos Islands. pp. 495–520. *In*: A.W. Trites, S.K. Atkinson, D.P. De Master, L.W. Fritz, T.S. Gelatt, L.D. Rea and K.M. Wynne [eds.]. Sea Lions of the World. Alaska Sea Grant College Program, Fairbanks, Alaska, USA.
Aurioles-Gamboa, D. and J. Hernández-Camacho. 2015. *Zalophus californianus*. The IUCN Red List of Threatened Species 2015.
Aurioles-Gamboa, D. 2015a. *Arctocephalus townsendi*. The IUCN Red List of Threatened Species 2015.
Aurioles-Gamboa, D. 2015b. *Arctocephalus philippii*. The IUCN Red List of Threatened Species 2015.
Avila, I.C., J.J. Alava and C.A. Galvis-Rizo. 2014. On the presence of a Vagrant Juan Fernandez fur seal (*Arctocephalus philippii*) in the Pacific Coast of Colombia: A new extralimital record. Mastozool. Neotrop. 21(1): 109–114.
Campagna, C. 2014. *Otaria byronia*. The IUCN Red List of Threatened Species 2014.
Campagna, C. (IUCN SSC Pinniped Specialist Group). 2008. *Arctocephalus australis*. The IUCN Red List of Threatened Species 2008.
Capella, J., L. Florez-Gonzalez, P. Falk and D.M. Palacios. 2002. Regular appearance of otariid pinnipeds along the Colombian Pacific coast. Aquat. Mamm. 28: 67–72.
Etnier, M.A. 2002. Occurrence of Guadalupe fur seals (*Arctocephalus townsendi*) on the Washington coast over the past 500 years. Mar. Mam. Sci. 18(2): 551–556.
Félix, F., P. Jiménez, J. Falconí and O. Echeverry. 2007. New cases and first births of the Galápagos fur seal, *Arctocephalus galapagoensis* (Heller, 1904) from the mainland coast of Ecuador. Rev. Biol. Mar. Oceanogr. 42(1): 77–82.
Fukasawa, M., H. Freeland, R. Perkin, T. Watanabe, H. Uchida and A. Nishina. 2004. Bottom water warming in the North Pacific Ocean. Nature 427: 825–827.
Gallo-Reynoso, J.P. and D. Aurioles-Gamboa. 1984. Distribución y estado actual de la población de la foca común (*Phoca vitulina richardsi* Gray, 1864) en la península de Baja California, México. Anales del Instituto de Biología, UNAM. Serie Zoología 55: 323–332.
Hanni, K.D., D.J. Long, R.E. Jones, P. Pyle and L.E. Morgan. 1997. Sightings and strandings of Guadalupe fur seals in central and northern California, 1988–1995. J. Mammal. 78: 684–690.
Herrera y Tordesillas, A.D. 1601. Descripción de las Indias Occidentales. Imprenta Real, Madrid.
Hofmeyr, G.J.G. 2015a. *Arctocephalus pusillus*. The IUCN Red List of Threatened Species 2015.
Hofmeyr, G.J.G. 2015b. *Arctocephalus tropicalis*. The IUCN Red List of Threatened Species 2015.
Karamanlidis, A. and P. Dendrinos. 2015. *Monachus monachus*. The IUCN Red List of Threatened Species 2015.
Kerr, R. 1824. A general history and collection of voyages and travels arranged in systematic order: Forming a complete history of the origin and progress of navigation, discovery, and commerce, by sea and land, from the earliest ages to the present time. William Blackwood, Edinburgh and T. Cadell, London.
Le Boeuf, B.J., D. Aurioles-Gamboa, R. Condit, C. Fox, R. Gisiner, R. Romero and F. Sinsel. 1983. Size and distribution of California sea lion population in México. Proceedings of the California Academy of Sciences 43: 77–85.

Littnan, C., A. Harting and J. Baker. 2015. *Neomonachus schauinslandi*. The IUCN Red List of Threatened Species 2015.
Lowry, L. 2015. *Neomonachus tropicalis*. The IUCN Red List of Threatened Species 2015.
McClenachan, L. and A.B. Cooper. 2008. Extinction rate, historical population structure and ecological role of the Caribbean monk seal. Proc. R. Soc. B 275: 1351–1358.
Merlen, G. 1995. A Field Guide to the Marine Mammals of the Galápagos. Instituto Nacional de Pesca, Guayaquil, Ecuador.
Mignucci-Giannoni, A.A. and P. Haddow. 2001. Caribbean monk seals or hooded seals? The Monachus Guardian 4(2): 8–9.
Mignucci-Giannoni, A.A. and D.K. Odell. 2001. Tropical and subtropical records of hooded seals (*Cystophora cristata*) dispel the myth of extant Caribbean monk seals (*Monachus tropicalis*). Bull. Mar. Sci. 68(1): 47–58.
Torres, D., J. Acevedo, D.E. Torres, R .Vargas and A. Aguayo-Lobo. 2012. Vagrant Subantarctic fur seal at Cape Shirreff, Livingston Island, Antarctica. Polar Biol. 35: 469–473.
Thompson, D. and T. Härkönen. (IUCN SSC Pinniped Specialist Group). 2008. *Phoca vitulina*. The IUCN Red List of Threatened Species 2008.
Trillmich, F. 2015a. *Arctocephalus galapagoensis*. The IUCN Red List of Threatened Species 2015.
Trillmich, F. 2015b. *Zalophus wollebaeki*. The IUCN Red List of Threatened Species 2015.
Vargas, H. and A. Steinfurth. 2004. Primer registro de elefantes marinos en Galápagos. Informe técnico para la Estación Científica Charles Darwin y Parque Nacional Galápagos. Archivos ECCD. Isla Santa Cruz, Galápagos, Ecuador.
Vergés, A., P. D. Steinberg, M.E. Hay, A.G.B. Poore, A.H. Campbell, E. Ballesteros et al. 2014. The tropicalization of temperate marine ecosystems: Climate-mediated changes in herbivory and community phase shifts. Proc. R. Soc. B 281: 20140846.
Wellington, G.M. and T. DeVries. 1976. The South American sea lion (*Otaria byronia*) in Galápagos Islands. J. Mammal. 57: 166–167.
Wu, L. and C. Li. 2007. Warming of the North Pacific Ocean: Local air–sea coupling and remote climatic impacts. J. Climate 20: 2581–2601.

2

An Overview on the Evolutionary History of Tropical Pinnipeds

Carlos A. Vildoso Morales

Introduction

The evolutionary history of pinnipeds in tropical seas is complex, but has experienced noticeable advancements in knowledge during the last three decades. Although our knowledge of pinniped biogeography is improved, several questions still remain due to gaps in our understanding. Especially important are paleontological studies of South American (Muizon 1978, 1981, Cozzuol 2001, Soibelzon and Bond 2013), South African (Hendey and Repenning 1972, Muizon and Hendey 1980) fossils, and molecular phylogenetic analyses (Yonezawa et al. 2008, Churchill et al. 2014). For instance, Churchill et al. (2014) provides a new combined evidence of phylogeny for the Otariidae, and extensively discusses their biogeography and timing of expansion into the southern hemisphere.

It is a striking fact how pinniped evolution is recorded only in a restricted area of the planet, this being noticeable in a large portion of tropical areas, which delivered very scarce or no fossil remains of the group. Even today, the presence of this taxonomic group in waters of tropical regions such as the Indian Ocean Basin, Southeast Asia or Central Pacific Ocean is exceptional,

---

Instituto Peruano de Estudios en Paleovertebrados IPEP - M. A. Fuentes 890, Lima 27, Perú.
Email: ipepaleon@gmail.com

exclusively due to vagrants dispersing very far out from their normal geographical ranges. Only a small number of living and fossil forms can be regarded as truly adapted to life in warm or even warm-temperate waters, in contrast to those taxa adapted to cold or cold-temperate waters. Thus, we can infer that early evolution and main adaptive radiations of pinnipeds developed in temperate to cold climates, and colonization of tropics was possible only under very particular circumstances in the Caribbean Basin and neighboring waters, including coastal areas of South America, and to a lesser extent on western African Coast.

Starting from the generally accepted point of view which places pinniped origins around late Oligocene, the main division of this group can be placed in the latest Oligocene (Deméré et al. 2003, Higdon et al. 2007, Yonezawa et al. 2008, Berta 2009) with two branches: Otariidae plus Odobenidae, and Phocidae with their closely related forms, as shown in the phylogenetic tree of Pinnipedia in Fig. 1. Tropical pinniped evolution continues through the Neogene, showing two stages; the first comprising Miocene to Pliocene, and the second extends from Pliocene to recent times.

Miocene–Pliocene

The two subfamilies of Phocidae (i.e., Phocinae and Monachinae) diverged in the Middle Miocene. The Phocinae seem to be restricted to temperate to cold waters almost exclusively in the Northern Hemisphere, predominantly along circumpolar regions. A possible exception is *Kawas* (Cozzuol 2001), from the Middle Miocene of Patagonia, assigned by the author to Phocinae; this form could be part of a failed attempt of colonization by this subfamily along eastern South American coasts, but always in temperate-cold or cold waters.

In contrast to phocines, monachines are known from both the Northern and Southern hemispheres, and include taxa adapted to tropical waters. The oldest remains of monachines (i.e., *Afrophoca*; Koretsky and Domning 2014) come from Middle Miocene beds in North Africa (Libya). Monk seals (Monachinae) within the tribe Monachini are the only extant pinniped lineage fully adapted to life in warm or temperate-warm waters, in addition to those being strictly temperate. The Mediterranean monk seal (*Monachus monachus*) inhabits temperate waters in the Mediterranean Basin but its range extents to the western African coast; the recently extinct Caribbean monk seal (*Neomonachus tropicalis*) inhabited the warm waters of the Caribbean Basin, and the Hawaiian monk seal (*Neomonachus schauinslandi*) inhabits the Hawaiian archipelago. Known fossil forms seem to cope with this tendency as early as the Tertiary period since we have the Argentinian Late Middle or Late Miocene *Properyptychus argentinus* (Ameghino 1897)

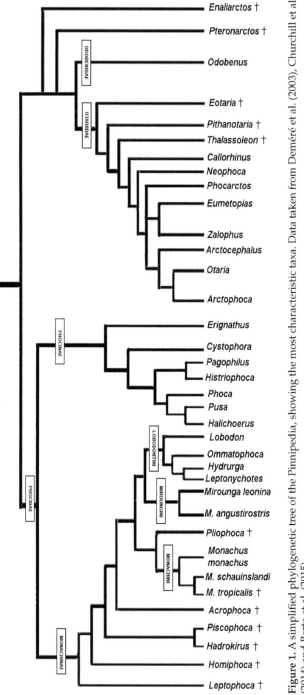

Figure 1. A simplified phylogenetic tree of the Pinnipedia, showing the most characteristic taxa. Data taken from Deméré et al. (2003), Churchill et al. (2014) and Berta et al. (2015).

regarded as a form close to *Monachus* (Soibelzon and Bond 2013), where the bearing sediments and associated fauna indicate deposition under tropical to subtropical climate. Based on these facts, we can speculate that migrations of Monachini have been largely influenced by adaptation to higher-temperature waters, possibly since Middle Miocene.

Muizon (1982) proposes the Monachini originated in European seas, producing several forms including *Monachus*, and later migrated westwards, first reaching South America and the Caribbean where *N. Tropicalis* originated together with related fossil forms (e.g., *Properiptychus*); after this evolutionary process, these species crossed the former seaway between North and South America to finally reach Hawaii where *N. schauinslandi* was differentiated. But recent data suggest instead *Properiptychus* is part of a branch differentiated very early along Eastern South America, and divergence of *Monachus–Neomonachus* became later, may be in Latest Miocene (Scheel et al. 2014).

Another monachine branch also took colonization of tropical or subtropical coastal areas but this process was not accomplished through adaptation to warm or warm-temperate waters, but by taking advantage of cold water currents coming from the southernmost regions. This branch was formerly assigned to Lobodontini (the living Antarctic seals of genus *Lobodon, Hydrurga, Ommatophoca* and *Leptonychotes*, see Fig. 2A) but is

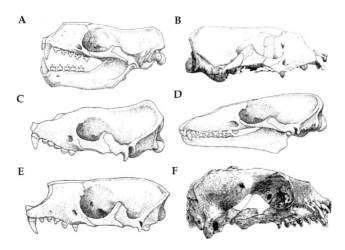

Figure 2. Skulls of representative species of the two main subdivisions of Pinnipedia, including the Phocid and Otariid, which belong to branches living or known to formerly have lived in tropical areas. (2A) skull of a living Monachinae (Lobodontini) phocid, the leopard seal (*Hydrurga leptonyx*); skulls of extinct phocids are described as follows: (2B) skull of *Piscophoca* after Muizon (1981); (2C) skull of *Hadrokirus*, modified from Amson and Muizon (2014); (2D) skull of *Acrophoca*, modified from Muizon (1981); (2E) skull of *Homiphoca*, composite reconstruction based in both Hendey and Repenning (1972) and Muizon and Hendey (1980); and (2F) skull of an extinct otariid, *Hydrarctos*, after Muizon (1978).

currently regarded as a different monachine clade, which shows several convergences with lobodontines. In the western South American coast, these lobodontine-like forms experienced a noticeable diversity in the Late Tertiary (Neogene). Their oldest local record comes from Middle Miocene beds of Ica (Peru) with still undescribed, very generalized forms (C.A. Vildoso-Morales, pers. obs., see Fig. 3), but during the Late Miocene-Early Pliocene, a marked specialization amongst taxa is observed.

The best known forms are those coming from the middle to upper part of the Pisco Formation (Latest Miocene-Early Pliocene) of Arequipa, Peru, some of which are recorded also in the contemporaneous Bahia Inglesa Formation of Northern Chile. The genus *Piscophoca* (Muizon 1981) (Fig. 2B), is a form whose anatomy indicates probable pelagic habits, and regarded by Muizon (1981) as resembling *Monachus*, despite strong similarities with Lobodontini; recent studies (Berta et al. 2015) placed this

Figure 3. Uncovering a skeleton of an undescribed monachine phocid in Middle Miocene Pisco Formation beds, Cerro la Bruja site, Ica Dessert (Peru). Photo credit: C.A. Vildoso Morales/ Archives of Instituto Peruano de Estudios en Paleovertebrados, August 2012.

taxon as a stem monachine. A second form, *Hadrokirus* (Amson and Muizon 2014), apparently belongs to same branch of *Piscophoca*, but shows unusual adaptations in the masticatory apparatus having very robust jaws and teeth (Fig. 2C), indicating a durophagous diet, which may be crustacean or mollusk-based. A third genus, *Acrophoca* (Muizon 1981) is a longirostrine form (Fig. 2D), whose habits have been interpreted as more coastal; his phylogenetic relationships have been formerly suggested (Muizon 1981) to be closer with living *Hydrurga* (leopard seal) and *Lobodon* (crab-eating seal) but now is thought to be a basal monachine close to *Piscophoca*. From Chile, *Piscophoca* and *Acrophoca* are recorded, together with a locally recorded genus *Australophoca* (Valenzuela-Toro et al. 2015a), a dwarf form known only by postcranial remains.

This monachine branch also colonized the western coast of South Africa, where the genus *Homiphoca* (Fig. 2E) is recorded from the Varswater Formation (Lower Pliocene). Following Govender (2015), the phylogenetic relationships of *Homiphoca* link this genus with Peruvian forms *Piscophoca* and *Hadrokirus*; subsequently, its origin could be linked with a migration from South America to Africa, probably in the Late Miocene; this migration is being supported by the occurrence of cold currents in the South Atlantic, and probably the rise of a temporary island chain in this same region under the effect of tectonic events (Govender 2015).

The above mentioned fossil forms share close geographic proximity with extant taxa, living today on Antarctica and surrounding areas, a fact previously remarked by several authors (Muizon 1981, Fyler et al. 2005, Govender 2015). This suggests a probable origin of lobodontines along western South American coast, followed by a gradual displacement towards southernmost regions. Evolution of the other main branch of pinnipeds, the Otariidae, was limited to the North Pacific until very late in the Neogene. In the Early Miocene, Otariidae became separated from Odobenidae and shortly after this family diverged in two well-differentiated assemblages, the first comprising most basal forms of the family, and the second having in one hand *Otaria* and closely related genera, and in the other hand *Arctocephalus* and related forms. It must be mentioned the work of Yonezawa et al. (2009), who supported by DNA analysis, supports the origin of Otariidae in the Northern Hemisphere, followed by a single dispersal into Southern Hemisphere where they started a radiation in the western coast of South America around the Latest Miocene-Early Pliocene.

Contrasting with Phocidae, otariids did not produce a branch showing any adaptation to life in warm or warm-temperate waters; this family colonized, however, extensive areas along tropical coastlines under the influence of cold currents. Figure 4A displays the Miocene-Pliocene biogeographic distributional pattern of tropical pinnipeds as shown by the fossil record.

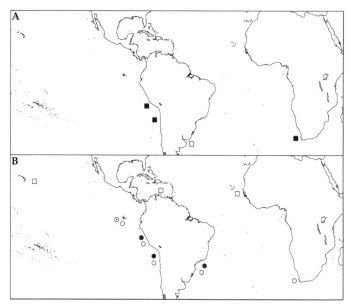

Figure 4. Distribution records for extinct and living tropical pinnipeds. Miocene to lower Pliocene distribution of tropical pinnipeds as indicated by the fossil record (A); recent distribution of tropical pinnipeds (B). White square: Phocidae: Monachini; black square: Phocidae: Lobodontini; white circle: Otariidae (*Arctocephalus*); black circles: Otariidae (*Otaria*); mixed circle: Otariidae (*Zalophus*).

Pliocen–Recent

The fossil record shows that, since the Pliocene, Otariidae began to colonize western South American and western South African coasts, replacing Phocids, which become fully extinct in the tropical areas of both continents. Figure 4B shows the recent distribution of tropical pinnipeds.

In the western South American coast, the Phocid record ends in the Lower Pliocene (Upper Pisco Formation of Peru and Bahia Inglesa Formation of Chile). In younger beds, there are no records of phocids but otarids, with *Hydrarctos* (Muizon 1978), from the Late Pliocene of Peru (Fig. 2F), and the oldest otariid known from South America, followed by a long gap which only ends with the records of indeterminate otariid remains in the Pleistocene of Argentina (Cozzuol 1996) and Late Pleistocene of Chile (Valenzuela-Toro et al. 2013), and *Otaria* (Drehmer and Ribeiro 1998, Rodrigues et al. 2004) and *Arctocephalus* (Oliveira and Drehmer 1997) in the Late Pleistocene of Brazil and Chile (Valenzuela-Toro et al. 2013). Both genera remain as the only ones recorded along tropical South American coasts in the post-Pleistocene. The presence of *Zalophus* in Galapagos Islands (i.e., *Zalophus wollebaeki*) can be connected with a Latest Pleistocene or Early Holocene event of dispersal, from the Eastern North Pacific.

In South Africa, Avery and Klein (2011) mentioned possible coexistence of phocids and otariids in the Lower Pliocene. In the Pleistocene, there are no records of the formerly predominant lobodontine-like taxa, replaced by the otariid *Arctocephalus*. However, Avery and Klein (2011) refer Pleistocene records of monachine phocids of the living Lobodontini genus *Hydrurga* and also the living Miroungini *Mirounga* (elephant seal), which could be part of ephemeral migratory events. Presumably, Miroungini species have had to cross the equator at least once based on their present range, and fossil material has recently been described from Chile (Valenzuela-Toro et al. 2015b, M. Churchill, pers. comm.). It should be remarked that Churchill et al. (2014) using analysis of molecular and morphological evidence, dated the arrival of otariids at the Southern Hemisphere as Latest Miocene. This means that despite the current uncertainty on the subject in the fossil record, there was indeed a period of phocid–otariid coexistence.

The above described process of replacement was probably triggered by climate factors. Climate evolution in the Pliocene shows clear evidence of a tendency to global cooling, beginning the formation and expansion of ice caps; the increase of ice mass, together with diastrophic events, brought a sudden fall of sea level with worldwide regressions affecting those areas occupied by northern Lobodontini-like forms, which suffered a sharp environmental disturbance, with loss, for instance, of hauling habitats (Govender 2015). This view contrasts with that of Valenzuela-Toro et al. (2013) who try to explain the extinction and replacement of tropical lobodontines as an effect of transgressive events.

As aforementioned, Otariidae did not develop forms adapted to higher-temperature waters. Because of this, the occurrence of phocid-otariid replacement was restricted to tropical areas with cold coastal currents; meanwhile, in those areas with warm or warm-temperate waters the local phocids (remarkably, only Monachini) persisted undisturbed until disruption of their habitats under human influence.

Conclusions

While the evolutionary record of tropical pinnipeds is geographically restricted, with fossil forms known only in western African coast, and coastal areas of South America, living forms indicate a still largely unknown history. In the Miocene-Early Pliocene, tropical coastal areas in South America and Southwestern Africa were colonized by Phocidae Monachinae, with Monachini and Miroungini occupying those regions with warm to temperate-warm waters and a distinctive branch formed by *Piscophoca* and closely related forms occupying those with cold or temperate-cold waters.

The present distribution of Monachini in the Pacific and Caribbean Basin probably goes back to this same interval. In the Pliocene, Otariidae begin to replace formerly dominant phocids in tropical areas of South America and South Africa. Under this premise, the current distributional pattern of tropical pinnipeds was probably in place since Late Pleistocene.

Acknowledgements

The author would like to thank Dr. J.N. Martínez and Lic. J. Apolin Meza for his kind review and useful comments which helped to improve the manuscript, and also M.P. Sciammaro for her continuous support and comments. Special acknowledgements are expressed to Dr. M. Churchill for providing observations and edits which greatly improved this work. Undescribed monachine materials referred here were examined from Instituto Peruano de Estudios en Paleovertebrados (IPEP) collections.

Keywords: pinniped evolution, paleobiogeographical distribution, Pliocene, Miocene, South American pinnipeds, Phocidae, Monachinae, Monachini, Lobodontini, Otariidae

References

Amson, E. and C. Muizon. 2014. A new durophagous phocid (Mammalia: Carnivora) from the late Neogene of Peru and considerations on monachine seals phylogeny. Jour. Syst. Paleont. 12(5): 523–548.

Ameghino, F. 1897. Mammiferes crétacés de l´Argentine. Deuxiéme contribution á la connaisance de la faune mammalogique des couches á *Pyrotherium*. Boletín del Instituto Geográfico Argentino 18: 1–117.

Avery, G. and R.G. Klein. 2011. Review of fossil phocid and otariid seals from the southern and western coasts of South Africa. Trans. R. Soc. S. Afr. 66(1): 14–24.

Berta, A. 2009. Pinniped evolution. pp. 861–868. *In*: W.F. Perrin, B. Wursig and J.G.M. Thewissen [eds.]. Encyclopedia of Marine Mammals, 2nd Ed. Academic Press, San Diego and London.

Berta, A., S. Kienle, G. Bianucci and S. Sorbi. 2015. A reevaluation of *Pliophoca etrusca* (Pinnipedia, Phocidae) from the Pliocene of Italy: Phylogenetic and biogeographic implications. J. Vert. Paleontol. 35(1). e889144, DOI: 10.1080/02724634.2014.889144.

Churchill, M., R. Boessenecker and M.T. Clementz. 2014. Colonization of the Southern Hemisphere by fur seals and sea lions (Carnivora: Otariidae) revealed by combined evidence phylogenetic and Bayesian biogeographical analysis. Zool Jour. Linn. Soc. 172: 200–225.

Cozzuol, M.A. 1996. The record of aquatic mammals in Southern South America. pp. 321–342. *In*: G. Arratia [ed.]. Contributions from Southern South America to Vertebrate Paleontology. Münchner Geowisssenschaftliche Abhandlungen, Reihe A, Geologie und Palaöntologie. 30.

Cozzuol, M.A. 2001. A "northern" seal from the Miocene of Argentina: Implications for phocid phylogeny and biogeography. J. Vert. Paleontol. 21(3): 415–421.

Deméré, T.A., A. Berta and P.J. Adam. 2003. Pinnipedimorph evolutionary biogeography. Bull. Am. Mus. Nat. Hist. 279: 32–76.

Drehmer, C.J. and A.M. Ribeiro. 1998. A temporal bone of an Otariidae (Mammalia, Pinnipedia) from the Late Pleistocene of Rio Grande do Sul State, Brazil. Rev. Univ. Guarulhos, Geociencias 3: 39–44.

Fyler, C.A., T.W. Reeder, A. Berta, G. Antonelis, A. Aguilar and E. Androukaki. 2005. Historical biogeography and phylogeny of monachine seals (Pinnipedia: Phocidae) based on mitochondrial and nuclear DNA data. J. Biogeogr. 32: 1267–1279.

Govender, R. 2015. Preliminary phylogenetics and biogeographic history of the Pliocene seal, *Homiphoca capensis* from Langebaanweg, South Africa. Trans. R. Soc. S. Afr. 70(1): 25–39.

Hendey, Q.B. and C.A. Repenning. 1972. A pliocene phocid from South Africa. Ann. S. Afr. Mus. 59: 71–98.

Higdon, J.W., O.R.P. Bininda-Emonds, R.M.D. Beck and S.H. Ferguson. 2007. Phylogeny and divergence of the pinnipeds (Carnivora: Mammalia) assessed using a multigene dataset. BMC Evol. Biol. 7.216. DOI: 10.1186/1471-2148-7-216.

Koretsky, A.I. and D.P. Domning. 2014. One of the oldest seals (Carnivora, Phocidae) from the Old World. J. Vert. Paleontol. 34(1): 224–229.

Muizon, C. 1978. *Arctocephalus* (*Hydrarctos*) *lomasiensis subgen. nov. et nov.* sp. Un nouvel Otaridé du Miopliocéne de Sacaco (Pérou). Bull. Inst. Fr. Et. And. 7(3-4): 169–188.

Muizon, C. and Q.B. Hendey. 1980. Late tertiary seals of the South Atlantic Ocean. Ann. S. Afr. Mus. 82: 91–128.

Muizon, C. 1981. Les Vertébrés Fossiles de la Formation Pisco (Pérou). Premiere partie: Deux nouveaux Monachinae (Phocidae, Mammalia) du Pliocéne de Sud-Sacaco. Mém. 6, Institut Francais d´Etudes Andines, Paris.

Muizon, C. 1982. Phocid phylogeny and dispersal. Ann. S. Afr. Mus. 89(2): 175–213.

Oliveira, E.V. and C.J. Drehmer. 1997. Sobre algunos restos de Pinnipedia (Mammalia, Carnivora) do Quaternario do Estado do Rio Grande do Sul, Brasil. Rev. Univ. Guarulhos, Geociencias 2: 19–22.

Rodrigues, P., F. Prevosti, J. Ferigolo and A. Ribeiro. 2004. Novos materiais de Carnivora para o Pleistoceno do Estado do Rio Grande Do Sul, Brasil. Rev. Brasil. de Paleont. 7: 77–86.

Scheel, D.M., G.J. Slater, S.O. Kolokotronis, C.W. Potter, D.S. Rotstein, K. Tsangaras et al. 2014. Biogeography and taxonomy of extinct and endangered monk seals illuminated by ancient DNA and skull morphology. Zookeys 409: 1–33.

Soibelzon, L.H. and M. Bond. 2013. Revisión de los carnívoros (Mammalia, Carnivora) acuáticos y continentales del Mioceno de la Mesopotamia argentina. pp. 170–178. *In*: D. Brandoni and J.L. Noriega [eds.]. El Neógeno de la Mesoptamia argentina. Asociación Paleontológica Argentina, Publicación Especial, 14.

Valenzuela-Toro, A.M., C.S. Gutstein, R.M. Varas-Malca, M.E. Suarez and N.D. Pyenson. 2013. Pinniped turnover in the South Pacific Ocean: New evidence from the Plio-Pleistocene of the Atacama Desert, Chile. J. Vert. Paleontol. 33(1): 216–223.

Valenzuela-Toro, A.M., N.D. Pyenson, C.S. Gutstein and M.E. Suarez. 2015a. A new dwarf seal from the Late Neogene of South America and the evolution of pinnipeds in the southern hemisphere. Papers Paleontol. 2: 101–115.

Valenzuela-Toro, A.M., C.S. Gutstein, M.E. Suárez, R. Otero and N.D. Pyenson. 2015b. Elephant seal (*Mirounga* sp.) from the Pleistocene of the Antofagasta Region, northern Chile. J. Vert. Paleontol. 35: 3, e918883, doi:10.1080/02724634.2014.918883.

Yonezawa, T.N., N. Kohno and M. Hasegawa. 2008. The monophyletic origin of sea lions and fur seals (Carnivora: Otariidae) in the Southern Hemisphere. Gene 441: 89–99.

3

Variability in the Skull Morphology of Adult Male California Sea Lions and Galapagos Sea Lions

Jimena Bohórquez-Herrera,[1,]* *David Aurioles-Gamboa,*[2] *Claudia Hernández-Camacho*[2] *and Dean C. Adams*[3]

Introduction

The skull undergoes a complex developmental process during which strong evolutionary forces shape its morphology. In vertebrates, the skull is important because it encompasses several vital organs such us the brain, the senses, the feeding structures, and a large portion of the respiratory system (Webster and Webster 1974, Kardong 2001, Chai and Maxson 2006). The skull facilitates the gathering of information about the surrounding

---

[1] Unidad de Investigación, Universidad Santo Tomás, Seccional Tunja. Cll. 19 # 11 - 64, Tunja, Boyacá, Colombia.
[2] Laboratorio de Ecología de Pinnípedos "Burney J. Le Boeuf", Centro Interdisciplinario de Ciencias Marinas, Instituto Politécnico Nacional. Av. IPN s/n, Col. Playa Palo de Santa Rita, C.P. 23096, La Paz, Baja California South, México.
 Email: dgamboa@ipn.mx; jcamacho@ipn.mx
[3] Department of Ecology, Evolution, and Organismal Biology. Iowa State University. Ames, IA 50011, United States.
 Email: dcadams@iastate.edu
* Corresponding author: jimena.bohorquez@usantoto.edu.co, jimenabh@gmail.com

environment, allowing individuals to respond to the particular benefits or threats presented to them. Thus, the skull must be strong enough to protect these vulnerable organs, while also facilitating the stimulation of the eyes, ears, and nose, and permitting food and oxygen intake, brain thermoregulation, and the amplification or reduction of sound (Webster and Webster 1974, Kardong 2001, Chai and Maxson 2006).

In order to understand the phenotypic changes that occurred during skull evolution, we must first obtain information on the sequential interactions between genes, cells, tissues, and organs. Important factors include the distinct patterns of tissue growth and differentiation and the various biomechanical interactions occurring during epigenetic development as well as environmental factors (Chai and Maxson 2006, Hallgrímsson et al. 2007a, Richtsmeier and DeLeon 2009).

Species evolution typically occurs as a result of the mutation rate or selection intensity (Barton and Partridge 2000). The intensity of natural selection responds to the rate of change in environmental conditions (Barton and Partridge 2000) and is determined by the interaction of each organism with its environment (Odling-Smee et al. 2003) in the form of behavior (Duckworth 2009). Thus, environmental changes and behavioral variability between individuals are potential causes of evolutionary change as they alter selection pressures (Badyaev 2005). Natural selection pressures mold distinct morphological traits in different environments. Thus, variability in the anatomical, morphological, and functional characteristics of the skull are a good reflection of evolutionary forces (Webster and Webster 1974, Kardong 2001, Chai and Maxson 2006).

Other researchers have used linear morphometrics to assess California sea lion skull morphology (Orr et al. 1970, Zavaleta 2003), identifying differences between colonies on either coast of the Baja California Peninsula and suggesting that the skull plays an important role in male reproductive success with the skull being used as a defense and attack structure (Zavaleta 2003). Linear morphometrics also revealed differences in skull morphology between the western and central colonies of the Galapagos Islands (Wolf et al. 2008).

Behavior and reproduction

California sea lions *Zalophus californianus* (Lesson, 1828) and Galapagos sea lions *Zalophus wollebaeki* (Sivertsen, 1953) are ideal species for evaluating patterns of phenotypic divergence for several reasons, including differences in their terrestrial habitats and the distribution of females at different colonies. The latter appears to be associated with the territorial behavior of reproductive males (Heath 1989, Boness 1991, Bohórquez-Herrera et al. 2014). These species exhibit a lek-like polygynous mating system where males fight each other in front of females in order to acquire territories and mate (Heath 1989, Boness 1991, Robertson et al. 2008, Bohórquez-Herrera et al. 2014).

Recent research has shown that differences in the environmental conditions of California sea lion habitats affect the behavioral plasticity of this species' mating system, with males inhabiting high-temperature environments defending their territories while immersed in water as a thermoregulatory strategy. Such a strategy might result in decreased polygyny in the Gulf of California (GC) colonies relative to those on the Pacific coast (PC) of the Baja California Peninsula (Bohórquez-Herrera et al. 2014).

On the other hand, Galapagos sea lions display distinct territorial behavior, with the breeding season being unaffected by male dominance. Combined with the region's drastic environmental conditions, this characteristic has resulted in an extended reproductive season that lasts five to six months during which females choose males based on their permanence in the area rather than dominance *per se*. Thus, male success in aggressive displays does not play an important role in the breeding system for this species (Pörschmann et al. 2010).

Habitat and distribution

The California sea lion is the only pinniped that permanently inhabits the GC; occasionally, other species also have been spotted there, including *Phoca vitulina* Linnaeus, 1758, *Mirounga angustirostris* (Gill, 1866), and *Arctocephalus townsendi* Merriam, 1897 (Aurioles-Gamboa and Zavala-González 1994). This species' habitat range extends from British Colombia on the southwest coast of Canada to the Islas Marias on the southwest coast of Mexico, including both the western and eastern coasts of the Baja California Peninsula (Lowry and Forney 2005).

California sea lions inhabit 29 colonies in the GC (13 reproductive colonies and 16 resting colonies) (Aurioles-Gamboa and Zavala-González 1994). More than 80% of the population in this area is located on small islands with large rocks, clear substrates, and convex coastlines, as part of a strategy used to avoid high temperatures (Aurioles-Gamboa and Zavala-González 1994, González-Suárez and Gerber 2008, Robertson et al. 2008). On the west coast of the Baja California Peninsula, California sea lions inhabit ten colonies, with pups being most abundant on Cedros Island and the San Benito Islands (Lowry and Maravilla-Chavez 2005).

Galapagos sea lions are distributed throughout the archipelago, primarily in colonies on the eastern islands (San Cristobal and Española); the species coexists with the fur seal (*Arctocephalus galapagoensis* Heller, 1904), which exclusively inhabits the western islands (Isabela and Fernandina). Despite the relatively reduced habitat of this species, population tendencies reflect high variability, with an increase in the fur seal population and a considerable reduction in Galapagos sea lions on the western side of the archipelago in recent years (Páez-Rosas et al. 2014).

Population structure

As is common among pinnipeds, these species exploit extensive feeding areas and are able to travel long distances, allowing them to aggregate and breed at different colonies (Campagna et al. 2001, Westlake and O'Corry-Growe 2002). However, this high level of dispersion is skewed by sex since females exhibit breeding philopatry; thus, males are nearly seven times more dispersed than females (Lawson and Perrin 2007, González-Suárez et al. 2009). The high degree of male dispersion suggests low spatial segregation in the feeding habitats exploited by individuals from the same colony, or a high interconnection between colonies (González-Suárez et al. 2009, Szteren and Aurioles-Gamboa 2011). Clear separation by ecological zones, which cannot be explained by abundance or population growth alone, has been observed for both sea lion species, a pattern confirmed by intraspecific variation in DNA sequences (Aurioles-Gamboa and Zavala-González 1994, Maldonado et al. 1995, Wolf et al. 2008, González-Suárez et al. 2009).

For California sea lions, variability in mitochondrial DNA (maternal inheritance) and nuclear DNA (both maternal and paternal inheritance) reveals a pattern of isolation due to the distance between reproductive colonies. When combined, these methods can be used to distinguish between individuals from the Benitos Islands (Pacific coast) and GC colonies. Within the Gulf, colonies in the CG-South (Los Islotes) can also be distinguished from those in the GC-Central and GC-North. Nuclear DNA does not distinguish between the GC-Central and GC-North colonies (San Jorge, Los Lobos, Granito, and San Esteban); however, mitochondrial DNA does distinguish these populations (Maldonado et al. 1995, González-Suárez et al. 2009, Schramm et al. 2009).

Based on abundance and degree of territory occupation, the GC population includes three distinct regions: north, central, and south. The GC-North colonies include islands with the highest occupation per unit area, representing > 80% of the population; the GC-Central population includes > 15% of the total population, while the GC-South population includes < 3% of the total population (Aurioles-Gamboa and Zavala-González 1994). Other studies have also grouped this species' colonies based on variables like population size, diet, microbial parasites, and skull diseases (Ward et al. 2010, Szteren and Aurioles-Gamboa 2011).

Microsatellite data as well as mitochondrial DNA suggest isolation due to the distance between Galapagos sea lion colonies, including early signs of genetic differentiation (Wolf et al. 2008).

Foraging ecology

Studies analyzing differences in the foraging ecology behavior exhibited by individuals from different California sea lion colonies have also revealed important information regarding their population structure. Based on more than 40 different studies on the species' diet, Espinosa (2007) identified six different groups in USA and Mexico. These six groups included: (1) San Miguel Island (*Merluccius productus, Loligo opalescens*, and *Engraulis mordax*), (2) the northwest coast of the Baja California Peninsula to Punta Eugenia (*E. mordax, Sardinops sagax caeruleus, Trachurus symmetricus*, and *M. productus*), (3) the southwest coast of the Baja California Peninsula from Punta Eugenia to Magdalena Bay (*L. opalescens, M. angustimanus*, and *Kathestotoma averruncus*), (4) Los Islotes in GC-South (*Serranus aequidens, Aulopus bajacali*, and *Pronotogrammus multifasciatus*), (5) Farallón de San Ignacio, San Pedro Nolasco, San Pedro Martir, Granito, and Los Cantiles Islands in GC-Central (*E. mordax*), and (6) San Esteban, El Pardito, El Rasito, and Los Machos in GC-Central and GC-North (*S. sagax caeruleus*).

Wolf et al. (2008) identified significant differences in niche partitioning among Galapagos sea lions, with animals in the western region using different food sources than those in the central area. Although there appear to be spatio-temporal differences in foraging strategies in terms of which foraging grounds are used (Páez-Rosas and Aurioles-Gamboa 2013), this species feeds exclusively on fish in both regions of the archipelago (Páez-Rosas et al. 2014).

The goal of this study was to assess the genetic, population, ecological, behavioral, and environmental variability between California sea lions and Galapagos sea lions. We assessed patterns of phenotypic divergence in the skulls of adult males from different California sea lion populations and compared them with the Galapagos sea lion population, as the latter have modified their behavior and habitat use in response to environmental conditions (Wolf et al. 2005).

Materials and Methods

Study area

The study area includes California sea lion reproductive colonies along the west coast of California, USA, and both sides of the Baja California

Peninsula, Mexico (Fig. 1). The GC is located on the east coast of Baja California Peninsula (22° and 32° N and 105°–107° W), measuring 1,200 km by 100–150 km and separating the peninsula from the Mexican mainland (Allison 1964, Lluch-Cota et al. 2007). The largest freshwater discharges in the GC derive from rivers in the mountain regions of Sonora and Sinaloa, while the largest sediment discharge comes from the Colorado River (USA) to the north of the GC (Moore and Curray 1982).

The GC is a high evaporation basin that communicates with the Pacific Ocean to the south. The water surface temperature ranges from 14°C to 21°C in February, and from 28°C to 31°C in August; however, the salinity concentration does not vary drastically (35–35.8 UPS) (Roden 1964).

Unlike the GC, the Pacific coast of the peninsula (PCP) has only a few fluvial systems; however, considerable sediments derive from the Magdalena-Santo Domingo coastal plain (Pedrín-Avilés and Padilla-Arredondo 1999). The PCP is a dynamic zone exposed to important oceanographic events like currents, gyres, and blooms. Specifically, the PCP experiences intense coastal upwellings produced by northwesterly winds that intensify in spring (April to May). Thus, deep offshore waters emerge

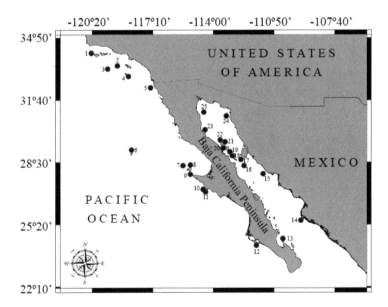

Figure 1. Map showing the location of *Zalophus californianus* reproductive colonies throughout the species' distribution in USA and Mexico. This study compared six population units: PCP-North (1. San Miguel, 2. Guadalupe Island, 7. San Benito Islands, 8. Cedros Island, 9. Natividad Island, 10. San Roque, 11. Asunción Island), PCP-South (12. Santa Margarita Island), GC-South (13. Los Islotes, 14. Farallón de San Ignacio), GC-Central (15. San Pedro Nolasco, 16. San Pedro Mártir, 17. San Esteban, 18. El Rasito, 19. El Partido, 20. Los Machos, 21. Los Cantiles, 22. Granito Island), and GC-North (23. Lobos Island, 24. San Jorge Island, 25. Rocas Consag).

carrying cold waters with a high nutrient concentration to the surface, increasing the phytoplankton population and productivity in the area. The intensity of this phenomenon depends on wind strength, topography, and bathymetry, as well as other oceanographic events like "El Niño" or "La Niña", which can drastically affect productivity (Espinosa-Carreon et al. 2004, Zavala et al. 2006).

We also analyzed skull samples from Galapagos sea lions *Z. wollebaeki*. The Galapagos Islands are located in the equatorial Pacific, forming an archipelago (Fig. 2) at the confluence of major ocean currents. Cold waters from the South Equatorial Current circulate from east to west; meanwhile, the Equatorial Counter Current brings subsurface waters from the opposite direction. Moreover, the Panama Current provides warm waters from the northeast; its intensity varies seasonally, and it indirectly influences atmospheric pressure differentials (i.e., "El Niño) from the Pacific Basin. The affluents of different bodies of water also generate distinct points of upwellings along the continental coast, mostly around the islands located on the western end of the archipelago (Feingold and Glynn 2014).

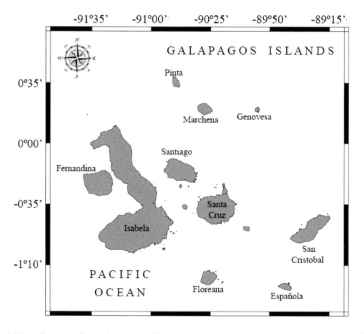

Figure 2. Map showing the *Zalophus wollebaeki* reproductive colonies on the Galapagos Islands (Equatorial Pacific).

Skull morphology

Photographic collection

For the morphological characterization, photographs were taken from the dorsal, ventral, and lateral views of the adult male skulls. The PCP-North California sea lion skulls (see Fig. 1) are from the Mammalogy Department at the Natural History Museum of Los Angeles County in California, USA. Skulls from the populations inhabiting Mexican coasts are from the Laboratorio de Ecología de Pinipedos "Burney J. Le Boeuf" at the Centro Interdisciplinario de Ciencias Marinas (CICIMAR, Interdisciplinary Center for Marine Sciences) of the Instituto Politécnico Nacional (IPN, National Polytechnic Institute) located in La Paz, Baja California Sur, Mexico. We obtained photographs of Galapagos sea lion skulls from the Ornithology and Mastozoology Collection at the California Academy of Sciences in San Francisco, California, USA. We only performed the morphological analysis on skulls with complete structures in order to ensure that the landmarks could be located; thus, the sample number for each view varied depending on how damaged the visible structures of a particular view were (Table 1).

Table 1. Skull sample size for each view of each population analyzed.

Region	Dorsal view	Ventral view	Lateral view
PCP-North	51	50	52
PCP-Central	8	7	10
PCP-South	54	39	54
GC-North	4	0	5
GC-Central	65	56	68
GC-South	16	6	15
Z. wollebaeki	8	8	7
TOTAL	206	166	211

Geometric morphometrics

Skull morphology was assessed using geometric morphometrics (GM), one of the newest methods for studying shape variation; it is also considered the most suitable methodology for cephalometrics studies (Bookstein 1991, Rohlf and Marcus 1993, McIntyre and Mossey 2003). This technique evaluates morphological information using landmarks on anatomical structures that adequately describe the design of those structures; the locations of these landmarks are then translated to Cartesian coordinates in two- or three-dimensional space (Van der Molen et al. 2007).

The use of GM eliminates differences in the size, position, and orientation of each object; thus, the resulting information refers only to its geometric configuration (Rohlf and Slice 1990). This permits the use of a wide range of statistical tools that not only ensure the correct interpretation of morphological variation but also facilitate identification of its causes (Adams et al. 2013b). Thus, the combination of GM and multivariate statistics (the essence of this technique) allows us to characterize structural morphology completely and without redundancy (Klingenberg 2013a).

GM takes into account the spatial relationships between landmarks without the need to predefine the set of features to sample (Klingenberg 2013a). Thus, GM facilitates analysis of biological forms, preserving physical structure, maintaining geometric relationships, and permitting analysis of morphological variation from a global perspective. Moreover, this technique enables us to make statistical comparisons of morphology, generating graphical representations that facilitate visualization of the direction, location, and magnitude of each morphological variation (Adams and Funk 1997, Van der Molen et al. 2007, Adams et al. 2013a, Klingenberg 2013b).

Two types of marks are used for this procedure: landmarks and sliding landmarks (Fig. 3). Landmarks are homologous variables on particular structures with common ancestral origin that can be easily identified on the phenotype (Bookstein 1991, Adams et al. 2004, Polly 2008). Sliding landmarks are homology-free variables applied to the phenotype using a mathematic algorithm "homologous" in its orientation relative to other landmarks. This permits us to include in morphological analysis those areas of each structure that are defined by curves or surfaces that cannot be delimited with landmarks, permitting comparison of morphological variations that have been gained or lost as part of evolutionary processes (Adams et al. 2004, Slice 2007, Polly 2008).

We used TPS software for image processing. Digitized landmarks were translated to two-dimensional coordinates, then a Generalized Procrustes Analysis (GPA) was applied, using geomorph package using R statistical computing environment (Adams et al. 2013a). The GPA makes rotational adjustments, overlapping one landmark over another, and optimizes a goodness of fit measure where the specimens are translated, scaled, and rotated until all position, size, and orientation data have been removed and the only variable remaining is morphology (Rohlf and Slice 1990, Van der Molen et al. 2007).

Sliding landmarks go through an additional process that "slides" them over a curve edge until they fits as closely as possible the position of a reference specimen (Green 1996, Bookstein 1997, Adams et al. 2004). Thus,

Skull Morphology of Sea Lions 31

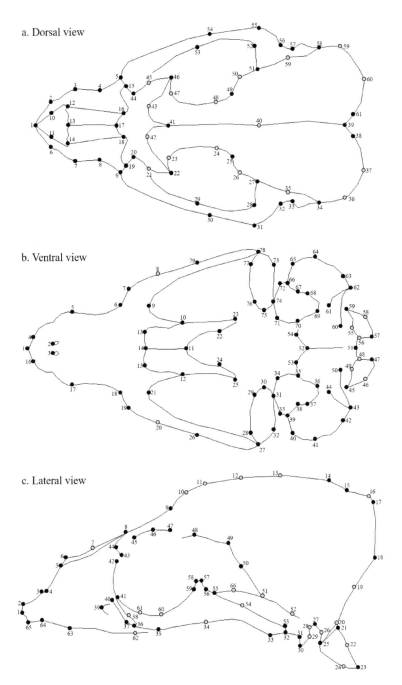

Figure 3. Schema of the (a) dorsal, (b) ventral, and (c) lateral views of the adult male skulls, showing the location of landmarks (black dots) and sliding landmarks (gray dots).

a sliding landmarks can slide along the local direction parallel to the same edge in order to minimize the bending energy needed to produce the relative change of the curve area relative to the reference specimen (Bookstein 1997, Adams et al. 2004, Pérez et al. 2006).

Statistical analyses

We assessed the variation in the GPA data using Principal Component Analysis (PCA). PCA does not take into account the global structure, which is why it does not neccesarily reveal groups even when they are present. However, PCA facilitates the identification of changes in morphology associated with the greatest variation; thus, it identifies which aspects of morphology are particularly variable, or particularly constant.

We used multivariate analysis of variance (MANOVA) with posthoc paired tests in order to assess whether there were significant differences between geographically neighboring population. To graphically represent the relationships among population means, we performed an additional PCA of the mean skull shape for each population. Morphological differences were then represented graphically with thin plate splines, which express changes in landmark configuration (morphological changes) as deformations of the tissues in which the landmarks are embedded (Bookstein 1991, Slice et al. 1998, Slice 2007). These procedures were done using geomorph package using R statistical computing environment (Adams et al. 2013a).

Results

We carried out a PCA of all specimens analyzed in order to identify general variation between the California and Galapagos sea lion populations. Because PCA does not consider group structure, none of the resulting figures reflects different population groups (Figs. 4–6). However, these graphic representations enable us to identify changes associated with the greatest variation in order to identify the morphological features that are particularly variable.

The morphological variables responsible for the greatest variation (first component) include the anteroposterior elongation of the dorsal zone of the upper jaw and the frontal zone of the skull (Figs. 4 and 6), the lateral elongation of the zygomatic arches and mastoid process (Figs. 5 and 6), the height of the sagittal crest, the anterior position of the zygomatic arch, and the dorsal process of the temporal apophysis of the zygomatic bone (Fig. 6).

In addition, some other features made a lesser contribution (second principal component), explaining a considerable amount of variation. These included the posterior elongation of the braincase, the posterolateral

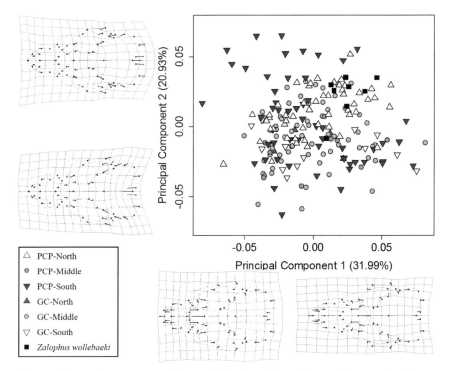

Figure 4. Principal component analysis and thin plate splines of the maximum and minimum values for each component; *dorsal view* of the skulls of all individuals analyzed.

elongation of the zygomatic apophysis of the temporal bone (Fig. 4), the anterolateral elongation of the mastoid apophysis and the tympanic bullae, the anterior elongation of the upper jaw, the anterior position of the occipital condyles (Fig. 5), and the elevation of the zygomatic arches (Fig. 6).

In contrast, some morphological features show more stability, with little variability between sea lion populations. These include the morphology of the nasal bones (Fig. 4), the width of the upper jaw (Figs. 4 and 5), the morphology of the dorsal lamina of the palatine bone, the base of the occipital bone (Fig. 5), the dorsal end of the orbital edge above the temporal apophysis of the zygomatic bone, and the zygomatic apophysis of the frontal bone (Fig. 6).

The MANOVA revealed significant differences between California sea lions and Galapagos sea lions for all of the skull views analyzed (Table 2). The paired comparisons for each view (dorsal, ventral and lateral) showed similar tendencies, with specific differences being explained by the variability of different skull modules. Fewer significant differences were observed between populations based on the ventral view of the skull, while the dorsal and lateral views more clearly separate nearly all neighboring

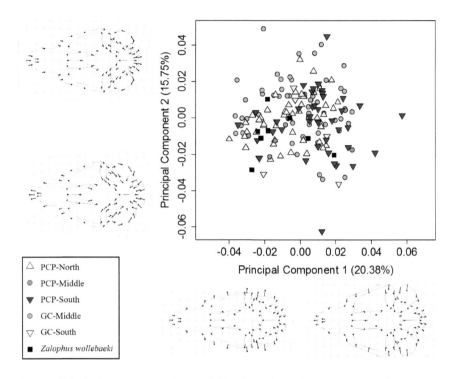

Figure 5. Principal component analysis and thin plate splines of the maximum and minimum values for each component; *ventral view* of the skulls of all individuals analyzed.

populations. The PCP-Central and PCP-South populations were not significantly different, nor were the GC-North and GC-Central populations (Table 2). It is important to note that PCP-Central and GC-North are the populations with the smallest sample size, which may explain the lack of significant differences.

The PCA of the mean skull shapes for each population allowed us to identify morphological variations between sea lion populations. For the dorsal and ventral views (Figs. 7–8), the greatest variation was explained by a strong separation between the *Z. wollebaeki* and *Z. californianus* populations. This difference derives from the posterior elongation of the frontal bone and the braincase in Galapagos sea lions (Fig. 7) and the fact that the smaller occipital condyles and the lateral contraction in the zygomatic apophysis of the temporal bone, the mandibular fossa, and the mastoid process together produce a much thinner skull compared to California sea lions. The dorsal and ventral views also reveal that *Z. californianus* incisors are closer to the tip of the face, and that the bases of the canine teeth are larger than in *Z. wollebaeki* (Figs. 7 and 8).

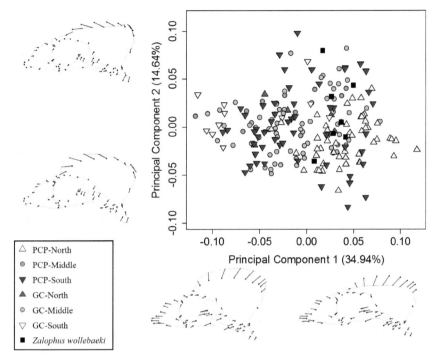

Figure 6. Principal component analysis and thin plate splines of the maximum and minimum values for each component; *lateral view* of the skulls of all individuals analyzed.

Table 2. MANOVA results and paired comparison (Euclidean distances) for neighboring populations of *Z. californianus* and the *Z. wollebaeki* population for each of the skull views analyzed. Values in bold correspond indicate populations with no significant differences between them.

Paired comparison		Dorsal view		Ventral view		Lateral view	
		$F_{(6,205)} = 3.030$ $p = 0.001$		$F_{(5,165)} = 3.012$ $p = 0.001$		$F_{(6,210)} = 6.283$ $p = 0.001$	
		d	p	d	p	d	p
PCP-North	PCP-Central	0.036	0.013	**0.019**	**0.323**	0.069	0.001
PCP-Central	PCP-South	**0.030**	**0.068**	**0.022**	**0.095**	**0.029**	**0.246**
PCP-South	GC-South	0.025	0.029	**0.021**	**0.273**	0.040	0.007
GC-South	GC-Central	0.024	0.039	0.027	0.021	0.037	0.015
GC-Central	GC-North	**0.024**	**0.768**	-	-	**0.031**	**0.543**
PCP-North	Z. wollebaeki	0.043	0.001	0.025	0.013	**0.035**	**0.213**
PCP-Central	Z. wollebaeki	0.059	0.002	**0.029**	**0.091**	0.067	0.012
PCP-South	Z. wollebaeki	0.049	0.001	0.032	0.001	0.059	0.009
GC-North	Z. wollebaeki	0.062	0.006	-	-	0.070	0.029
GC-Central	Z. wollebaeki	0.053	0.001	0.030	0.001	0.062	0.002
GC-South	Z. wollebaeki	0.055	0.001	**0.030**	**0.089**	0.082	0.002

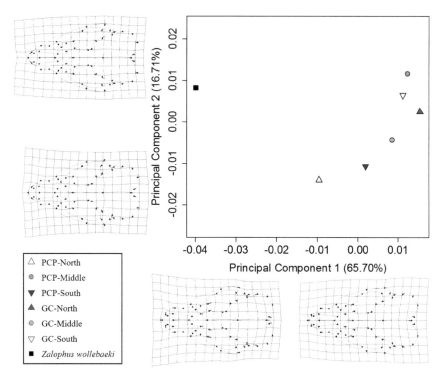

Figure 7. Principal component analysis and thin plate splines of the maximum and minimum values for each component; *dorsal view* of the mean skull shape for each region.

The paired comparisons of the dorsal view showed significant differences between neighboring populations of California sea lions and the Galapagos sea lion population (Table 2). The exceptions are the PCP-Central and PCP-South ($d = 0.030$, $P = 0.068$), and the GC-North and GC-Central ($d = 0.024$, $P = 0.768$) populations, as no significant differences were identified between them. As mentioned above, these results may be influenced by small sample sizes.

The second component of the dorsal view PCA reveals that the variation between PCP populations is much greater than that between GC populations (Fig. 7). This variation is explained by the posterior elongation of the nuchal crest, the lateral elongation of the zygomatic apophysis of the temporal bone, and the space between incisive teeth being defined by a lateral external bulge (on each side) (Fig. 7). The GC populations are concentrated in the center of this component; meanwhile, the PCP populations show greater variation with PCP-North and PCP-South being located at the negative end of the axis, while PCP-Central is located at the positive end along with Galapagos sea lion population.

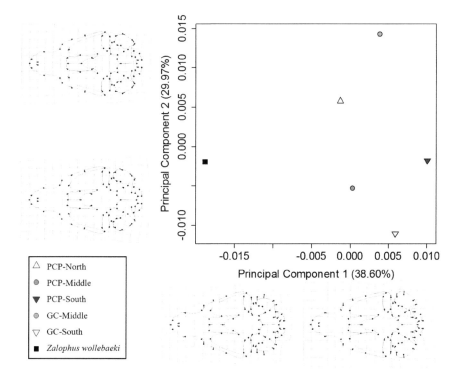

Figure 8. Principal component analysis and thin plate splines of the maximum and minimum values for each component; *ventral view* of the mean skull shape for each region.

The paired comparisons of the ventral view did not show significant differences between PCP populations (PCP-North vs. PCP-Central, PCP-Central vs. PCP-South), or between PCP-South populations and the GC. Likewise, the Galapagos sea lion population did not differ significantly from the PCP-Central or GC-South populations (Table 2). The Galapagos sea lion population and the PCP populations are located in the center of the axis, indicating less variation in their ventral view features, while the GC-Central and GC-South populations are located on opposing ends of the axis, suggesting considerable variation (Fig. 8). The variability explained by this second component is due to the anterolateral elongation of the mastoid process, the position of the palate fissures, and the lateral elongation of the zygomatic process of the temporal bone (Fig. 8).

Based on the results of the lateral view, the PCP-Central and PCP-South populations were not significantly different from one another nor were they significantly different from the GC-North and GC-Central populations. The Galapagos sea lion population also was not significantly different from the PCP-North population (Table 2). These similarities are represented graphically in the PCA of the lateral view (Fig. 9), where the

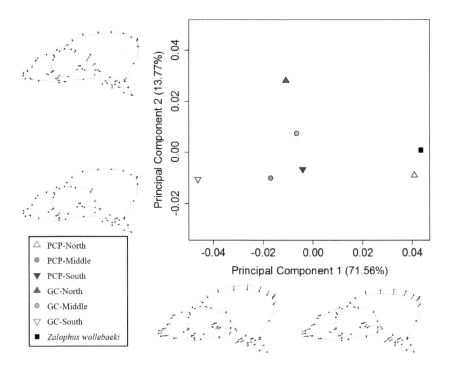

Figure 9. Principal component analysis and thin plate splines of the maximum and minimum values for each component; *lateral view* of the mean skull shape for each region.

first axis of variation includes the GC-South population at its negative end, while the PCP-North population and the Galapagos sea lion population are located at the positive end. The variability summarized by this axis is associated with the height of the sagittal crest, the anterior elongation of the nasal, incisive, and maxilar bones, and the anteroventral elongation of the zygomatic arches (Fig. 9). The second component explains the variability in GC-North individuals as deriving from the ventral inflection point of the frontal bone where it joins the parietal bone; as well as the position of the infraorbital foramen, and the inflection point of the occipital bone where it joins the mastoid process in the posterior area of the skull (Fig. 9).

Discussion

The skull is composed of several integrated components, the development, functioning, and evolution of which occur simultaneously (Webster and Webster 1974, Kardong 2001, Chai and Maxson 2006, Klingenberg 2013a).

However, this integration is not absolute as the skull is composed of different modules that are relatively independent (Klingenberg 2008, 2010, 2013a).

The cranium, jaws, and teeth are structures that respond differently to environmental and natural selection processes as they are subjected to different genetic, developmental, and functional controls (Caumul and Polly 2005, Rychlik et al. 2006, Hallgrímsson et al. 2007a, b). The amount of time required for the ontogenetic development of each structure is one factor that contributes to morphological variation (Hillson 1986, Kardong 2001, Hallgrímsson et al. 2007a, b). Thus, we divided the skull into three partially independent and embryologically distinct modules: 1. The basicranium, which derives from the chondrocranium; 2. The neurocranium, which is composed of the dermal bones of the cranium; and 3. The face, which derives initially from the splanchnocranium with a posterior development of elements from the dermal bones of the cranium (Hallgrímsson et al. 2007a, b) (Fig. 10).

The cranial base is considered the central integrative structure of the skull (Lieberman et al. 2000, Hallgrímsson et al. 2007a, b) as it is in the center of the skull structure, located under the brain and neurocranium and below and behind the face. As such, the cranial base indirectly acts as a transmitter between the face and the neurocranium (Hallgrímsson et al. 2007a, b). Likewise, the cranial base is the first zone of the skull to reach adult size and shape, primarily growing by endochondral ossification, in which bone tissue is produced from cartilage. The neurocranium fully develops next, followed by the face; unlike the cranial base, these structures develop through intramembranous ossification of sutures, wherein a membrane facilitates bone formation (periosteum). This bone can shape and reabsorb bone tissue, undergoing continuous self-remodeling (Hallgrímsson et al. 2007a, b).

Skull module integration has been the subject of several evolutionary biology studies involving these species and taxonomical groups. Other researchers have employed a variety of statistical techniques to address fundamental biological questions regarding evolutionary, genetic, and functional integration (Klingenberg 2013a). Although we did not evaluate skull structure modularity here, the PCAs provide us with general information on skull structure integration and modularity (Klingenberg 2013a).

Thus, we can infer integration between the described modules (Fig. 10) based on general morphological variations (reflected in the first four principal components) for each skull view (Figs. 4–6). In all cases, there was greater similarity within each module of morphological variation (measured by the direction and magnitude of the vectors) than between them. However, the direction and magnitude of the change varied for different skull modules (Fig. 10). Thus, we infer that each of the

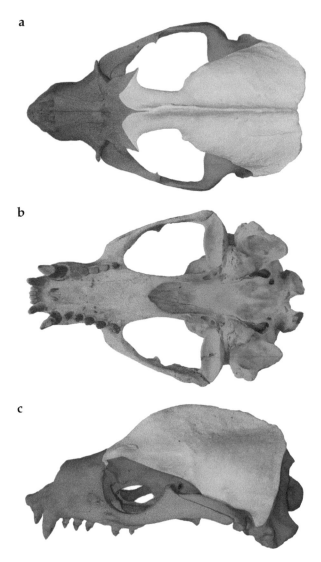

Figure 10. Developmental modules shown on the (a) dorsal, (b) ventral, and (c) lateral views of a *Zalophus californianus* skull (blue: face, green: neurocranium [braincase], red: cranium base).

skull modules develops independently. This has been reported for other mammal species, with modularity reflecting complex integration in skull development (Hallgrímsson et al. 2007a, b). However, our interpretations are based on general patterns of variability in skull integration and modularity. Additional and more specific analyses should be undertaken in order to answer these important biological questions at established levels of confidence (Klingenberg 2013a).

Overall, the greatest morphological variations between California and Galapagos sea lions occurred in the dorsal zone of the upper jaw, the frontal area of the skull, the zygomatic arches, the mastoid process, and the sagittal crest. The braincase reaches adult size relatively early because it serves as a supportive and protective structure for the brain; however, its growth continues as mechanical forces act on it and regulate transcription factors in the sutures (Moss and Young 1960, Opperman et al. 2005). The sagittal crest is a simple bone extension in the central line of the braincase that serves as the adherence point for the temporal muscles; thus, its development is directly associated with the growth of this muscle (Washburn 1947) and the higher sagittal crest, the more developed the temporal muscle. The sagittal crest is a feature with high phenotypic plasticity due to its importance in mastication, defining bite pressure as the movement and force of the mandible is defined largely by the size of the temporal muscles and sagittal crest (Searfoss 1995, Holbrook 2002, Cameron and Groves 2004).

On the other hand, the frontal zone of the face, particularly the jaws and the zygomatic bone, reach adult size after the base and braincase have done so; thus, their somatic growth patterns continue in accordance with the muscles inserted into those bones (Moss and Young 1960, Cheverud 1995, Opperman et al. 2005, Hallgrímsson et al. 2007a, b). As bones that extend on each side of the skull, the zygomatic arches are of great importance as their size and distance from the central skull body are indicators of the relative size of the temporal bones passing through them. They also connect the posterior part of the mandible to the cranium; thus, the larger the aperture of the zygomatic arches, the stronger and larger the temporal muscles, as is the case for most carnivores (Searfoss 1995).

These structures also offer support to the temporal muscles, anchoring the masseter muscles, which are responsible for raising the lower jaw (Searfoss 1995). Several studies have shown that zygomatic arch morphology varies during growth, adjusting to feeding changes that occur during different life stages (Rafferti et al. 2000, Usami and Itoh 2006). The frontal zone of the skull, which includes the zygomatic arches and the dorsal zone of the upper jaw, exhibited greater phenotypic plasticity, reflecting the influence of the environment on the organisms' ontogeny (Cheverud 1995, Opperman et al. 2005). The morphological variability of the skulls of different sea lion populations reflects the diversity and availability of feeding resources throughout their distribution.

In contrast, the cranial base is less affected by the environment during ontogenetic development as it is the first region of the skull to reach adult size, it acts as a base that supports the brain structure, and it is the confluence of many nerves and blood vessels (Cheverud 1995, Lieberman et al. 2000, Opperman et al. 2005). The morphological variations observed on the ventral view were associated with the components previously mentioned (zygomatic arches and upper jaw), as well as the mastoid process. Considering the early development of the cranial base, the variability present in the mastoid process may reflect directional selection rather than phenotypic plasticity.

The mastoid apophysis is a rounded projection of the temporal bone located behind the ear canal. It is important as an area of adherence for the muscles associated with head movements like the splenius capitis, longissimus capitis, and sternocleidomastoid. The sternocleidomastoid muscle originates in the manubrium extending from the medial third of the clavicle to the mastoid process and superior nuchal line of the occipital bone; it is responsible for controlling lateral flexion (nodding) and rotation (turning the head left and right). The splenius capitis originates in the first dorsal vertebrae in the dorsal cervical ligament of the neck; inserted in the occipital bone and the mastoid process, it permits unilateral tilting and rotation of the head in the same direction. Finally, the longissimus capitis refers to the "minor complex" (semispinalis) and originates in the sixth thoracic dorsal vertebra and extends to the occipital bone and the mastoid process; its function is focused on extension, tilt, and rotation movements (Tate 2009).

Recently, Bohórquez-Herrera et al. (2014) reported a significant change in the behavior of individuals in terms of where they spend most of their time. This study highlighted how the *Z. californianus* mating system was affected by changing environmental conditions, with males inhabiting high temperature environments in the GC spending most of their time immersed in the water as a thermoregulatory strategy. This change in mating behavior was exhibited even when males were defending their territories, displaying aggression toward male competitors by chasing (swimming) them, biting them, and destroying their fins. This behavior is a complete behavioral transformation of the mating system originally described for this species (Heath 1989, Boness 1991), which is still present in the PCP populations where males defend their territories on land (Bohórquez-Herrera et al. 2014).

The lateral elongation of the mastoid process in the sea lions populations analyzed here implies that there is an important variation in the adhesion areas of these muscles; thus, when the area of articulation increases, the length of the lower arm of the head muscles also increase (Antón and Galobart 1999), providing greater contraction power (Salesa et al. 2005). Following Bohórquez-Herrera et al. (2014), we argue that morphological variation in the mastoid process is related to variation in neck movements, and differences in the force of gravity acting on individuals who spend most of their time in the water versus those that remain on land.

We also observed variation in the development of different skull modules between the different populations of California sea lions and the Galapagos sea lion population. The ventral view, which is mainly composed of the cranial base module (Fig. 10), includes the fewest significant differences between populations (Table 2). This view only allowed us to discriminate between the GC-Central and GC-South populations, as well as between the Galapagos sea lion population and the PCP-North, PCP-South, and GC-Central populations.

Despite the small size of the Galapagos sea lion population (Table 1), the cranial base is significantly different from the California sea lion populations. The lack of significant differences between the Galapagos population and PCP-Central and GC-South California sea lion populations might be related to the small sample size.

The different skull views analyzed allowed us to differentiate each California sea lion population from the Galapagos sea lion population based on the first PCA component of variation. These analyses showed that *Z. wollebaeki* is characterized by a thin skull due to the lateral contraction of the zygomatic apophysis of the temporal bone, the mandibular fossa, and the mastoid process. This equatorial population also exhibited a reduced area at the base of canine teeth and smaller occipital condyles. The front area and braincase are posterior elongated, while the nasal, incisive, and maxillary bones are anterior elongated. The species also presents a low sagittal crest and thin, dorsal elongated zygomatic arches.

The separation between the Galapagos sea lion population and the California sea lion populations is based on variation in the development of three skull modules. The cranial base is responsible for many of these differences (zygomatic process of the temporal bone, mandibular fossa, mastoid process, and occipital condyles), reflecting the fact that skull features separate the two species at the evolutionary level. Moreover, variations in the modules of the face and the braincase likely represent the influence of environmental factors.

Using linear morphometrics, Zavaleta (2003) analyzed the skull morphology of adult males of these two species, finding that the principal differences were in the face area (condyle-canine length, upper jaw length, palate length, incisor foramens length, and interorbital minimum width), with only one difference being associated with the cranial base (length of the posterior lacerate foramen). This is in contrast to our study, in which the cranial base included many significant differences that distinguish these two species.

The morphological variations identified in the face module are associated with the variability of the Galapagos population; a low, but significant genetic and morphological differences have been reported between populations on western versus central islands (Wolf et al. 2008). Morphological variations in the dorsal and lateral zones of the mandible and face separate these populations (Wolf et al. 2008). Finally, the morphological variability in the braincase is likely due to environmental factors.

Since the first axis of variation of the PCA clearly separated the two species, the morphological variability within California sea lion populations was observed mostly on the second axis of variation. In the ventral view, the separation between GC populations was based mostly on variation in the mastoid apophysis; in the GC-Central they are posterior elongated, while in the GC-South they are anterior elongated (PC2, Fig. 8). The cranial base is the first module that reaches adult size and shape; thus, morphological variations are generally associated with processes of natural selection, rather than phenotypic plasticity. As a result, the minimal morphological variability in the cranial base of PCP populations may be due to a normalizing selection. In contrast, selection does not seem to favor mean shapes among the GC populations; thus, we argue that disruptive or directional selection processes are at work in this region.

In contrast, for the dorsal view we observed greater variability between the PCP and CG populations (shorter distance between the third incisor and the tip of the face, posterior elongation of the nuchal crest, and lateral elongation of the temporal zygomatic process) (CP2, Fig. 7). Variability in this view mainly derived from two modules that reach adult size and shape late in individual ontogeny (face and braincase); thus, variability is likely a reflection of environmental factors.

Despite the fact that face and braincase variability of California sea lion populations was relatively low in the dorsal view, the lateral view revealed considerable variation in these two modules associated with environmental factors. In the lateral view, the PCP-Central and -South populations did not show significant differences as they have little morphological variability (defined by the first two components of variation in the PCA) relative to other populations. Likewise, the GC-North and GC-Central populations showed little variation in the main axis of morphological variation, and both are separated from the GC-South population as the latter presents skulls with high sagittal crests, posterior compressed frontal zones, and wide zygomatic arches. The PCP-North population presented opposing features, and was more similar to Galapagos sea lions.

Thus, the lateral view of the skull showed a high variability in the three skull modules for the different California sea lion populations; the GC-North and GC-Central populations in particular stand out as they exhibit anteroventral elongated braincases. The high morphological variability indicates that the lateral view of the skull best reflects the influence of environmental variability during ontogenic development. This is observed in sea lion skull morphology in the form of higher sagittal crests.

The skull morphology analysis previously carried out by Zavaleta (2003) identified one PCP population of California sea lions and two GC populations. In the present study, we identified a clear segregation of the PCP-North population from the PCP-Central and PCP-South populations. Likewise, the GC-South population we separated from the GC-North and GC-Central populations present in the Gulf of California.

Acknowledgements

This research was partially funded by Mexico's National Council of Science and Technology (Consejo Nacional de Ciencia y Tecnología, CONACyT) as part of the project "Comprehensive study on breastfeeding in the California sea lion: critical period in the survival of pups" (CONACYT No. 132415). The National Polytechnic Institute (Instituto Politécnico Nacional, IPN) also sponsored this research as part of the following projects: "Trophic level estimation of marine organisms inside and outside the lagoon complex of Magdalena-Almejas Bay based on isotopic analysis" (SIP 20140053, SIP 20150340) and "Diet, trophic level and accumulation of trace metals monitoring in a top predator in the region of the lagoon complex of Magdalena-Almejas Bay" (SIP 20150347, SIP 20140111). The Saint Thomas University (Universidad Santo Tomás) (Boyacá, Colombia) also funded this study as part of the project "Impact assessment and vulnerability of the Siscunsí (Boyacá) moor in the Martinera and Las Cintas Basins." J.B.H.'s doctoral research was funded by scholarships from the Consejo Nacional de Ciencia y Tecnología (CONACyT; National Council for Science and Technology), and the Programa Integral de Fortalecimiento Institucional (PIFI; Comprehensive Institute Building Program) from the IPN. We also thank the access granted to skull collections by Dr. Jim Dines and Dr. David Janiger of the National History Museum of Los Angeles County, and Dr. Maureen Flanery of the California Academy of Sciences. Special thanks to the researchers, students, and volunteer staff from the Laboratorio de Ecología de Pinnípedos "Burney J. Le Boeuf" at the Centro Interdisciplinario de Ciencias Marinas del Instituto Politécnico Nacional (CICIMAR-IPN; Interdisciplinary Center for Marine Sciences at the National Polytechnic Institute), and its skull collection where most of the samples were taken and the study was carried out. The Armada de México Fuerza Naval del Pacífico (Pacific Naval Force Division of the Mexican Navy) stationed at Cabo Cortés, Baja California, provided invaluable logistical support in the field. We also thank Dr. K. Sullivan for revising the English text.

Keywords: pinniped skull, morphology, sagittal crest, basicranium, zygomatic arches, evolution, California sea lion, Galapagos sea lion

References

Adams, D.C. and D.J. Funk. 1997. Morphometric inferences on sibling species and sexual dimorphism in *Neochlamisus bebbianae* leaf beetles: Multivariate applications of the thin-plate spline. Syst. Biol. 46(1): 180–194.

Adams, D.C., F.J. Rohlf and D.E. Slice. 2004. Geometric morphometrics: Ten years of progress following the "revolution". Ital. J. Zool. 71: 5–16.

Adams, D.C., E. Otarola-Castillo and E. Sherratt. 2013a. Geomorph: Geometric morphometric analysis of 2d/3d landmark data. R package version 1.1-5. http://cran.r-project.org/web/packages/geomorph/.

Adams, D.C., F.J. Rohlf and D.E. Slice. 2013b. A field comes of age: Geometric morphometrics in the 21st century. Hystrix 24(1): 7–14. doi: 10.4404/hystrix-24.1-6283.

Allison, E.C. 1964. Geology of areas bordering Gulf of California. pp. 3–29. *In*: T.H. van Andel and G.G. Shor, Jr. [eds.]. Marine Geology of the Gulf of California. The American Association of Petroleum Geologists, Wisconsin.

Antón, M. and A. Galobart. 1999. Neck function and predatory behaviour in the scimitar-toothed cat *Homotherium latidens* (Owen). J. Vertebr. Paleontol. 19(4): 771–784.

Aurioles-Gamboa, D. and A. Zavala-González. 1994. Algunos factores ecológicos que determinan la distribución y abundancia del lobo marino *Zalophus californianus*, en el Golfo de California. Cienc. Mar. 20(4): 535–553.

Badyaev, A.V. 2005. Stress induced variation in evolution: From behavioural plasticity to genetic assimilation. P. R. Soc. B. 272: 877–886.

Barton, N. and L. Partridge. 2000. Limits to natural selection. BioEssays 22: 1075–1084.

Bohórquez-Herrera, J., C.J. Hernández-Camacho, D. Aurioles-Gamboa and V- Cruz-Escalona. 2014. Plasticity in the agonistic behavior of male California sea lions, *Zalophus californianus*. Anim. Behav. 89: 31–38.

Boness, D.J. 1991. Determinants of mating systems in the Otariidae (Pinnipedia). pp. 1–44. *In*: D. Renouf [ed.]. The Behavior of Pinnipeds, Chapman and Hall, Cambridge.

Bookstein, F.L. 1991. Morphometric Tools for Landmark Data: Geometry and Biology. Cambridge University Press, USA. 435 p.

Bookstein, F.L. 1997. Landmark methods for forms without landmarks: Localizing group differences in outline shape. Med. Image Anal. 1: 225–243.

Cameron, D.W. and C.P. Groves. 2004. Bones, Stones and Molecules: "Out of Africa" and Human Origins. Academic Press, USA.

Campagna, C., R. Werner, W. Karesh, M.R. Marín, F. Koontz, R. Cook et al. 2001. Movements and location at sea of South American sea lions (*Otaria flavescens*). J. Zool. 257: 205–220.

Caumul, R. and P.D. Polly. 2005. Phylogenetic and environmental components of morphological variation: Skull mandible and molar shape in marmots (Marmot, Rodentia). Evolution 59(11): 2460–2472.

Chai, Y. and R.E. Maxson, Jr. 2006. Recent advances in craniofacial morphogenesis. Dev. Dynam. 235: 2353–2375.

Cheverud, J.M. 1995. Morphological integration in the saddle-back tamarin (*Saguinus fuscicollis*) cranium. Am. Nat. 145: 63–89.

Duckworth, R.A. 2009. The role of behavior in evolution: A search for mechanism. Evol. Ecol. 23: 513–531.

Espinosa, M. 2007. Variabilidad especial de la dieta del lobo marino de California (*Zalophus californianus californianus*, Lesson 1828). M. Sc. Thesis, Centro de Investigación Científica y de Educación Superior de Ensenada, Baja California, Mexico. 175 p.

Espinosa-Carreon, T.L., P.T. Strub, E. Beier, F. Ocampo-Torres and G. Gaxiola-Castro. 2004. Seasonal and interannual variability of satellite-derived chlorophyll pigment, Southface height, and temperature off Baja California. J. Geophys. Res.-Oceans 109(C03039).

Feingold, J.S. and P.W. Glynn. 2014. Coral research in the Galápagos Islands, Ecuador. pp. 3–22. *In*: J. Denkinger and L. Vinueza [eds.]. The Galápagos Marine Reserve: A Dynamic Social-Ecological System. Springer, New York.

González-Suárez, M. and L.R. Gerber. 2008. Habitat preferences of California sea lions: Implications for conservation. J. Mammal. 89(6): 1521–1528.
González-Suárez, M., R. Flatz, D. Aurioles-Gamboa, P.W. Hendrick and L.R. Gerber. 2009. Isolation by distance among California sea lion populations in Mexico: Redefining management stocks. Mol. Ecol. 18: 1088–1099.
Green, W.D.K. 1996. The thin-plate spline and images with curving features. pp. 19–87. In: K.V. Mardia, C.A. Gill and I.L. Dryden [eds.]. Image Fusion and Shape Variability. University of Leeds Press, Leeds.
Hallgrímsson, B., D.E. Lieberman, W. Liu, A.F. Ford-Hutchinson and F.R. Jirik. 2007a. Epigenetic interactions and the structure of phenotypic variation in the cranium. Evol. Dev. 9: 76–91.
Hallgrímsson, B., D.E. Lieberman, N.M. Young, T. Parsons and S. Wat. 2007b. Evolution of covariance in the mammalian skull. Novart. Fdn. Symp. 284: 164–185.
Heath, C.B. 1989. The behavioral ecology of the California sea lion, *Zalophus californianus* Tesis de doctorado. Santa Cruz: University of California, Santa Cruz. 255 p.
Hillson, S. 1986. Teeth. Cambridge University Press, Cambridge.
Holbrook, L.T. 2002. The unusual development of the sagittal crest in Brazilian tapir (*Tapirus terrestris*). J. Zool. 256: 215–219.
Kardong, K.V. 2001. Vertebrados: Anatomía comparada, función, evolución. Spain, McGraw-Hill/Interamericana. 732 p.
Klingenberg, C.P. 2008. Morphological integration and developmental modularity. Annu. Rev. Ecol. Evol. S. 39: 115–132.
Klingenberg, C.P. 2010. Evolution and development of shape: Integrating quantitative approaches. Nat. Rev. 11: 623–635.
Klingenberg, C.P. 2013a. Cranial integration and modularity: Insights into evolution and development from morphometric data. Hystrix 24(1): 43–58.
Klingenberg, C.P. 2013b. Visualization in geometric morphometrcis: How to read and how to make graphs showing shape changes. Hystrix 24(1): 15–24.
Lawson, L.J. and N. Perrin. 2007. Advances in our understanding of mammalian sex-biased dispersal. Mol. Ecol. 16: 1559–1578.
Lieberman, D.E., O.M. Pearson and K.M. Mowbray. 2000. Basicranial influence on overall cranial shape. J. Human Evol. 38: 291–315.
Lluch-Cota, S.E., E.A. Aragón-Noriega, F. Arreguín-Sánchez, D. Aurioles-Gamboa, J.J. Bautista-Romero, R.C. Brusca et al. 2007. The Gulf of California: Review of ecosystem status and sustainability challenges. Prog. Oceanogr. 73: 1–26.
Lowry, M.S. and K.A. Forney. 2005. Abundance and distribution of California sea lions (*Zalophus californianus*) in central and northern California during 1998 and summer 1999. Fish. Bull. 103: 331–343.
Lowry, M.S. and O. Maravilla-Chavez. 2005. Recent abundance of California sea lions in western Baja California, Mexico and the United States. pp. 485–497. In: D.K. Garcelon and C.A. Schwemm [eds.]. Proceedings of the Sixth California Islas Symposium. National Park Service Technical Publication CHIS-05-01, Institute for Wildlife Studies, Arcata, California.
Maldonado, J.E., F. Orta, B.S. Stewart, E. Geffen and R.K. Wayne. 1995. Intraspecific genetic differentiation in California sea lions (*Zalophus californianus*) from southern California and the Gulf of California. Mar. Mammal Sci. 11(1): 46–58.
McIntyre, G.T. and P.A. Mossey. 2003. Size and shape measurement in contemporary cephalometrics. Eur. J. Orthodont. 25(3): 231–242.
Moore, D.G. and J.R. Curray. 1982. Geologic and tectonic history of the Gulf of California. Deep Sea Drilling Project 14: 1279–1294.
Moss, M.L. and R.W. Young. 1960. A functional approach to craniology. Am. J. Phys. Anthropol. 18: 281–292.
Odling-Smee, F.J., K.N. Laland and M.W. Feldman. 2003. Niche Construction: The Neglected Process in Evolution. Princeton University Press, Princeton.

Opperman, L.A., P.T. Gakunga and D.S. Carlson. 2005. Genetic factors influencing morphogenesis and growth of sutures and synchondroses in the craniofacial complex. Seminars in Orthodontics 11: 199–208.
Orr, R.T., J. Schonewald and K.W. Kenyon. 1970. The California sea lion: Skull growth and a comparison of two populations. P. Natl. Acad. Sci. 37: 381–394.
Páez-Rosas, D. and D. Aurioles-Gamboa. 2013. Spatial variation in the foraging behaviour of the Galápagos sea lions (*Zalophus wollebaeki*) assessed using scat collections and stable istope analysis. J. Mar. Biol. Assoc. UK, 1–9. doi: 10.1017/S002531541300163X.
Páez-Rosas, D., M. Rodríguez-Pérez and M. Riofrío-Lazo. 2014. Competition influence in the segregation of the trophic niche of otariids: A case study using isotopic Bayesian mixing models in Galapagos pinnipeds. Rapid Commun. Mass Spectrom. 28: 2550–2558. doi: 10.1002/rcm.7047.
Pedrín-Avilés, S. and G. Padilla-Arredondo. 1999. Morfología y sedimentología de la plataforma continental del Southoeste de la Península de Baja California, México. Rev. Mex. Cienc. Geol. 16(2): 132–146.
Pérez, S.I., V. Bernal and P.N. González. 2006. Differences between sliding semi-landmark methods in geometric morphometrics, with an application to human craniofacial and dental variation. J. Anat. 208: 769–784.
Polly, P.D. 2008. Developmental dynamics and G-matrices: Can morphometric spaces be used to model evolution and development? Evol. Biol. 35: 1–20.
Pörschmann, U., F. Trillmich, B. Mueller and J. Wolf. 2010. Male reproductive success and its behavioural correlates in a polygynous mammal, the Galápagos sea lion (*Zalophus wollebaeki*). Mol. Ecol. 19: 2574–2586. Doi: 10.1111/j.365-294X.2010.04665.x.
Rafferti, K.L., S.W. Herring and F. Artese. 2000. Three dimensional loading and growth of the zygomatic arch. J. Exp. Biol. 203: 2093–3004.
Richtsmeier, J.T. and V.B. DeLeon. 2009. Morphological integration of the skull in craniofacial anomalies. Orthod. Craniofac. Res. 12(3): 149–158.
Robertson, K.L., C.W. Runcorn, J.K. Young and L.R. Gerber. 2008. Spatial and temporal patterns of territory use of male California sea lions (*Zalophus californianus*) in the Gulf of California, Mexico. Can. J. Zoolog 86: 237–244.
Roden, G.I. 1964. Oceanographic aspects of the Gulf of California. pp. 30–58. *In*: T.H. van Andel and G.G. Shor, Jr. [eds.]. Marine Geology of the Gulf of California. The American Association of Petroleum Geologists, Wisconsin.
Rohlf, F.J. and D.E. Slice. 1990. Extensions of the Procrustes method for the optimal superimposition of landmarks. Systematic Zool. 39: 40–59.
Rohlf, F.J. and L.F. Marcus. 1993. A revolution in morphometrics. Trends Ecol. Evol. 8: 129–132.
Rychlik, L., G. Ramalhinho and D. Polly. 2006. Response to environmental factors and competition: Skull, mandible and tooth shapes in Polish water shrews (Neomys, Soricidae, Mammalia). J. Zool. Syst. Evol. Res. 44(4): 339–351.
Salesa, M.J., M. Antón, A. Turner and J. Morales. 2005. Aspects of the functional morphology in the cranial and cervical skeleton of the sabre-toothed cat *Paramachairodus ogygia* (Kaup, 1832) (Felidae, Machairodontinae) from the Late Miocene of Spain: Implications for the origins of the machairodont killing bite. Zool. J. Linn. Soc-Lond. 144: 363–377.
Schramm, Y., S.L. Mesnick, J. de la Rosa, D.M. Palacios, M.S. Lowry, D. Aurioles-Gamboa et al. 2009. Phylogeography of California and Galápagos sea lions and population structure within the California sea lion. Mar. Biol. 156(7): 1375–1387. doi: 10.1007/s00227-009-1178-1.
Searfoss, G. 1995. Skulls and Bones: A Guide to the Skeletal Structures and Behavior of North American Mammals. Stackpole Books, USA.
Slice, D.E., F.L. Bookstein, L.F. Marcus and F.J. Rohlf. 1998. A glossary for geometric morphometrics. Publication number 944 of the postgraduate studies on ecology and evolution, State University of New York at Stony Brook. Available online at: [http://life.bio.sunysb.edu/morph/glossary/gloss1.html].
Slice, D.E. 2007. Geometric morphometrics. Annu. Rev. Anthropol. 36: 261–281.

Szteren, D. and D. Aurioles-Gamboa. 2011. Ecological regionalization of *Zalophus californianus* rookeries, as a tool for conservation in the Gulf of California. Cienc. Mar. 37(3): 349–368.
Tate, P. 2009. Gross anatomy and functions of skeletal muscles. pp. 227–264. *In*: P. Tate [ed.]. Seeley's principles of Anatomy and Physiology. McGraw Hill, USA.
Usami, A. and I. Itoh. 2006. Morphological changes in the zygomatic arches during growth. Pediatric Dental Journal 16(2): 179–183.
Van der Molen, S., N. Martínez and R. González. 2007. Introducción a la morfometría geométrica: Theoric-practice course. Centro Nacional Patagónico, Universitat de Barcelona. 82 p.
Ward, E.J., H. Chirakkal, M. González-Suárez, D. Aurioles-Gamboa, E.E. Holmes and L. Gerber. 2010. Inferring spatial structure from time-series data: Using multivariate state-space models to detect metapopulation structure of California sea lions in the Gulf of California, Mexico. J. Appl. Ecol. 47(1): 47–56.
Washburn, S.L. 1947. The relation of the temporal muscle to the form of the skull. Anat. Record. 99: 239–248.
Webster, D. and M. Webster. 1974. Comparative Vertebrate Morphology. Academic Press, New York. 516 p.
Westlake, R.L. and G.M. O'Corry-Growe. 2002. Macrogeographic structure and patterns of genetic diversity in Harbor seals (*Phoca vitulina*) from Alaska to Japan. J. Mammal. 83(4): 1111–1126.
Wolf, J.B., G. Kauermann and F. Trillmich. 2005. Males in the shade: habitat use and sexual segregation in the Galápagos sea lion (*Zalophus californianus wollebaeki*). Behav. Ecol. Sociobiol. 59(2): 293–302. doi: 10.1007/s00265-005-0042-7.
Wolf, J.B.W., C. Harrod, S. Brunner, S. Salazar, F. Trillmich and D. Tautz. 2008. Tracing early stages of species differentiation: Ecological, morphological and genetic divergence of Galápagos sea lion populations. BMC Evol. Biol. 8: 150.
Zavala, J., O. Salmerón, V. Aguilar, S. Cerdeira and M. Kolb. 2006. Caracterización y regionalización de los procesos oceanográficos de los mares mexicanos. Available online at: http://www.conabio.gob.mx/gap/index.php/Procesos_oceanográficos.
Zavaleta, L. 2003. Variaciones geográficas en morfometría craneal en machos adultos de lobo marino de California (*Zalophus californianus*) en México. M.Sc. Thesis. Centro Interdisciplinario de Ciencias Marinas, Instituto Politécnico Nacional. La Paz, B.C.S., Mexico.

Hawaiian Monk Seals
The Biology and Ecology of the World's only Tropical Phocid

Charles Littnan,[1,]* Michelle Barbieri,[1,a]
Jessica Lopez Bohlander,[1,b] Tenaya Norris,[2]
Stacie Robinson[1,c] and Kenady Wilson[3]

Introduction

The pinniped group known as monk seals include two genera and three species of geographically separated seals: the Mediterranean monk seal (*Monachus monachus*), the Caribbean monk seal (*Neomonachus tropicalis*; now extinct), and the Hawaiian monk seal (*Neomonachus schauinslandi*; Scheel et al. 2014; Fig. 1). Monk seals are considered the most primitive of all living phocid species with anatomical features that are little changed from ancestors that lived more than 14 million years ago (mya; Berta et al. 2015).

---

[1] NOAA Fisheries Service, Hawaiian Monk Seal Research Program, 1845 WASP Blvd., Building 176, Honolulu, HI 96818.
[a] Email: michelle.barbieri@noaa.gov
[b] Email: jessica.bohlander@noaa.gov
[c] Email: stacie.robinson@noaa.gov
[2] The Marine Mammal Center, 2000 Bunker Road, Sausalito, CA 94965.
Email: norrist@tmmc.org
[3] Alaska Fisheries Science Center, NOAA 7600 Sand Point Way N.E., Seattle, Washington 98115.
Email: kenady.wilson@noaa.gov
* Corresponding author: charles.littnan@noaa.gov

Figure 1. Images of Hawaiian monk seals at various life stages, including weaned pup (A), subadult (B) and adult (C) age classes. The thick blubber layer and thin covering of hair similar to more temperate phocids are still found on this tropical pinniped.

Monk seals represent a unique evolutionary branch as a group of seals that established themselves in more temperate and tropical climes than those colonized by other phocids. However, this unique lineage is in jeopardy. The Caribbean monk seal became extinct during the last 60 years (Kenyon 1977, McClenachan and Cooper 2008), and the Mediterranean monk seal population currently stands far below historical levels, at an estimated 600 animals spread across an extensive range and numerous national boundaries (Karamanlidis et al. 2016). The Hawaiian monk seal is the sole remaining member of the genus *Neomonachus* and could be the best hope for the continuation of an evolutionarily important lineage of monk seals (Scheel et al. 2014).

The endangered status of Hawaiian monk seals has made them the focus of intensive research and recovery efforts over the last several decades. In this chapter we discuss the general biology, ecology, and life history of the species with an emphasis on aspects that are affected by its tropical distribution. In the following chapter, we will demonstrate how a strong foundation of research has evolved into the world's most proactive marine mammal recovery program to prevent the Hawaiian monk seal from going extinct.

Species Description, Distribution and Status

Pups of both sexes weigh approximately 14–17 kg at birth and are weaned when they weigh 50–100 kg (Kenyon and Rice 1959, Wirtz 1968). Substantial post-weaning mass loss in monk seal pups (NMFS unpublished data), and other phocid pups (Reiter et al. 1978, Worthy and Lavigne 1983), is a normal part of their life history. Unlike other monachine seals there is no sexual dimorphism in monk seals (Baker et al. 2014). Few adults have been weighed or measured, so a complete growth curve is not available, but the maximum weight and length of monk seals is estimated to be 205 kg and 2.3 m long (Reif et al. 2004).

Similar to their temperate and polar phocid counterparts, monk seals have a relatively thin covering of hair and a thick layer of blubber. Hawaiian monk seals, except pups, undergo a catastrophic molt, shedding the pelage along with the outer layers of skin (Kenyon and Rice 1959). Monk seal pups are born with black lanugo (fetal hair) that is shed around the time of weaning (Fig. 1A), at which time pups become silvery gray with darker pelage on their backs relative to their stomachs (Johanos et al. 1994). Juveniles, subadults (Fig. 1B), and adults (Fig. 1C) have a silvery gray coat that is molted annually and slowly becomes a light brown over the course of the year, with lighter ventral pelage.

Distribution and status

The Hawaiian monk seal is the only truly tropical seal species in the world today. It is not clear when monk seals reached the Hawaiian Islands (Repenning and Ray 1977), but monk seals entered the Pacific Ocean and diverged from their closest relative, the Caribbean monk seal, no less than 3.5 mya, before the closing of the Isthmus of Panama. Hawaiian monk seals are endemic to the Hawaiian Islands and can be found throughout the archipelago from Kure Atoll in the west to the island of Hawaii in the east, with records of occasional sightings at Johnston Atoll (Fig. 2).

There is limited data available prior to the 1950s to estimate the overall Hawaiian monk seal population size, but it is assumed that monk seals had a broad distribution across the Hawaiian Archipelago prior to the arrival of humans. There is substantial archaeological and paleo-ecological evidence suggesting Hawaiian settlement by humans no later than 1200 AD (Kirch 2011). Hawaiian monk seal remains have been found in Hawaiian middens (domestic waste piles) dating to 1400–1750 AD (Watson et al. 2011), but monk seals do not have a large presence in native Hawaiian culture as other marine animals do, such as whales, turtles, and sharks. It is thought that monk seals were extirpated from the main Hawaiian Islands (MHI; east of

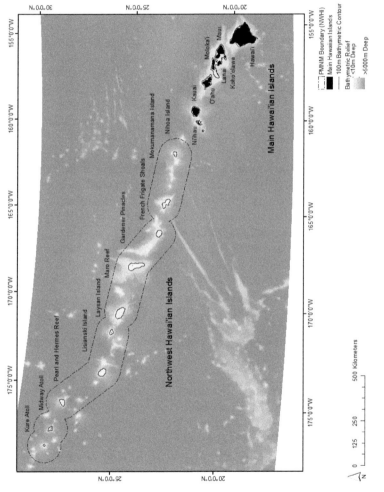

Figure 2. Map of the Hawaiian Archipelago. Hawaiian monk seals are found across the entire chain of Hawaiian Islands. The largest portion of the population lives in the remote Northwestern Hawaiian Islands (NWHI), a large marine protected area (PMNM: Papahānaumokuākea Marine National Monument). A smaller but growing population of seals are found in the human-populated main Hawaiian Islands.

Kauai/Niihau) early in human settlement and relegated to the more remote Northwestern Hawaiian Islands (NWHI), thus limiting their interactions with early cultures in the MHI.

Genetic work indicates that a few thousand monk seals persisted for many generations, with a recent bottleneck caused by hunting in the 19th century shortly after the first Europeans came to Hawaii in 1778 (Schultz et al. 2009, 2010). In 1805, the Russian explorer Lisianski provided the first written account of monk seals in the NWHI. What rapidly followed was a period of intense commercial and subsistence hunting with thousands of seals killed for their meat, skins, and oil by sealers and shipwrecked sailors. The largest collection recorded was for 1500 monk seal skins collected by *The Gambia* in 1859, though this account has been questioned (Kenyon and Rice 1959). Much of this harvest ended by 1886 with few seals observed in the islands by the turn of the century. In the early 1900s, after a period of partial recovery, a large-scale depletion of the species was likely caused by human disturbance or ecological shifts in the NWHI (Kenyon 1972, Gerrodette and Gilmartin 1990, Ragen 1999).

Efforts to measure the monk seal population in the 1950s, based on beach counts (numbers of individuals on the beach in one day), counted a maximum of 1540 individuals. This corresponded to a potential population of roughly 3000, the majority of which was located in the NWHI (NMFS 2007). Only seven sightings of Hawaiian monk seals were documented in the MHI between 1928 and 1956 (Kenyon and Rice 1959). Monitoring in the NWHI after 1958 showed a population in steep decline. In fact, it is estimated that the population underwent a nearly 70% decrease between 1958 and 2012 (Carretta et al. 2015).

A tale of two subpopulations: The main and the Northwestern Hawaiian Islands

It is estimated that there were approximately 1,300 Hawaiian monk seals in existence in 2015 (NMFS, unpublished data). The majority of Hawaiian monk seals are found in the NWHI with the most recent estimate of 810 individuals for the six NWHI breeding colonies decreasing at a rate of 3.3% annually Fig. 3 (Caretta et al. 2014). A smaller number of seals reside in the MHI with a population growth rate of 6.5% per year and a minimum abundance of 185 individuals (Caretta et al. 2015).

It is not just monk seal population size and trajectory that differentiate the remote NWHI and more developed MHI. The NWHI are part of the Papahānaumokuākea Marine National Monument (PMNM), one of the largest marine protected areas in world. In this region, there are eight HMS breeding colonies (listed from west to east): Kure Atoll, Midway Atoll, Pearl and Hermes Reef (PHR), Lisianski Island, Laysan Island, French Frigate

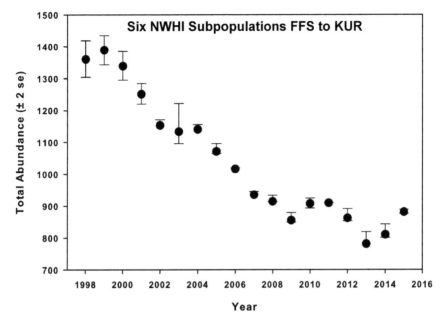

Figure 3. Decline of Hawaiian monk seals in the Northwestern Hawaiian Islands (NWHI), from French Frigate Shoals (FFS) to Kure Atoll (KUR), based on beach counts of the primary breeding sub-populations. Error bars are standard errors (SE).

Shoals (FFS), Necker Island, and Nihoa Island. Most of these islands are low-lying, with the most extreme examples at PHR and FFS where virtually all land is less than 2 m above sea level, and consequently highly vulnerable to storm surge, erosion, and sea level rise (Baker et al. 2006). Necker and Nihoa are basalt islands lifting hundreds of meters above the ocean.

The primary driver for the NWHI population decrease is chronic poor juvenile survival, which has degraded the age class structure at most of the NWHI breeding sites (Fig. 4). Recently it was found that fewer than one in five monk seal pups born in the NWHI survived to reproductive age (Harting et al. 2014). The numerous threats to monk seals in the NWHI are spatially and temporally variable, with the greatest threat being insufficient prey resources that limit the survival of young seals. Descriptions of these threats and the strategies to mitigate them are presented in the following chapter.

The MHI are geologically the youngest and largest islands of the archipelago and consist of: Niihau, Kauai, Oahu, Molokai, Lanai, Maui, Kahoolawe, and Hawaii. Small uninhabited volcanic islets (Kaula Rock, Lehua, Manana, and Molokini Crater) sit offshore of the main islands. These larger, higher elevation islands provide more protected terrestrial habitat for monk seals to haul-out than the low-lying sandy islets of the NWHI.

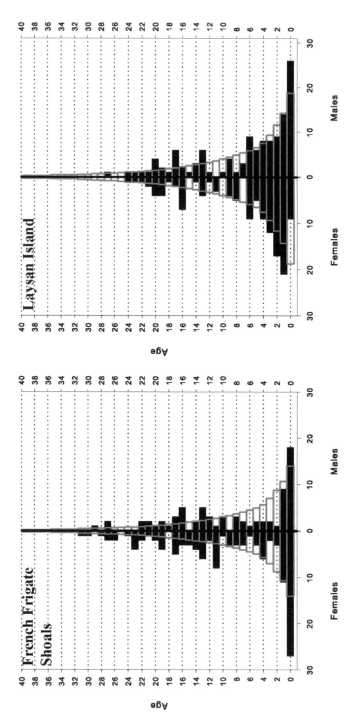

Figure 4. Examples of the population age-sex structure for a subset of monk seal breeding sub-populations (black bars) versus an "optimal" age-sex structure (red overlay). Chronic poor juvenile survival has created top-heavy structures with older females being the primary reproductive contributors and few females recruiting into the breeding class. Recent improved survival at Laysan Island has shifted the population closer to the ideal structure relative to French Frigate Shoals.

In comparison to the monk seals in the NWHI, MHI seals are generally in excellent physical condition, likely indicative of more accessible or abundant food resources in the MHI (Baker and Johanos 2004). This increased prey availability is likely a by-product of limited competition from other seals (sparse population in early stages of re-establishment) and apex predators (such as sharks and other large fish whose populations have been limited by human activity such as fishing). The result is greater juvenile survival, earlier onset of sexual maturity, and more regular pupping by adult females compared with seals in the NWHI. The excellent body condition of the MHI seals is a favorable indicator for continued population growth in the region. Yet there are unique concerns and management challenges associated with an increasing MHI monk seal population that are discussed in the following chapter.

Life History, Ecology and Health

Hawaiian monk seals spend most of their lives at sea and come to shore only to pup, nurse, molt, and rest. Seals haul-out primarily on sandy beaches, but are observed hauled-out on virtually all substrates, including emergent reef, rocky beaches, and vegetated areas. The majority of seals use the terrestrial and marine habitat around the islands or atolls on which they were born (natal sites), but some seals travel between sites. The likelihood of a seal moving to a new site increases with age, and the proportion of seals observed at non-natal sites varies with location (Johanos et al. 2014). Movements between the NWHI and MHI, while observed, are rare; however, movements within the NWHI and within the MHI are more common.

Sociobiology and reproduction

Asynchronous breeding is one of the most striking aspects of monk seal life history that differs from most pinnipeds and is likely a result of their tropical distribution. Monk seal births have been documented in all months of the year (NMFS, unpublished data), although they are most common between February and August and peak in April and May (Johanos et al. 1994). Pregnant females usually select the same pupping site each year, which often is the site of their own weaning (Westlake and Gilmartin 1990), and give birth to a single offspring. Occasional births of twins have been documented, but these are known to have lower survival than single born pups (Schultz et al. 2011).

As with many phocids (Kovacs and Lavigne 1986), female monk seals usually fast and remain with their pups throughout the nursing period.

On average, pups nurse for 5–6 weeks (Fig. 5; Kenyon and Rice 1959, Johnson and Johnson 1984, Boness 1990, Johanos et al. 1994). Nursing monk seal mothers are generally intolerant of other adult seals, including other mothers with pups (Kenyon and Rice 1959, Boness 1990). Yet, occasional pup-switching does occur, especially when mother-pup pairs occur close together, suggesting that they do not consistently distinguish their own offspring from others' pups (Boness 1990, Job et al. 1995, Boness et al. 1998). A mother also may foster another pup if her own becomes lost or dies. Switching or fostering of pups appears to have minimal effects on first-year survival in cases where the pups are of comparable size (Boness 1990).

Figure 5. An adult female monk seal nurses her newly born pup at Laysan Island. Mothers suckle their pups for approximately 6 weeks before weaning them.

Weaning occurs abruptly when the mother leaves her pup and returns to sea to resume feeding. Over the next few months, the considerable amount of weight the mother lost during lactation will be regained. Approximately 3–4 weeks after weaning a pup, postpartum females mate, and 5–6 weeks later, they haul out for 10–14 days, or possibly longer, to molt. Generally, females that do not give birth in a given year will molt a month earlier (Johanos et al. 1994).

Females give birth for the first time between the ages of four and ten years of age and reproductive parameters vary substantially among breeding sites (Harting et al. 2007). The most striking difference in fecundity in the NWHI is that sexual maturity at Laysan Island occurs up to 4 years earlier than other sites, with females at FFS having the latest onset of reproduction and lowest mean fecundity. Since the onset of sexual maturity in pinnipeds usually coincides with the attainment of some percentage of final body size (Laws 1956), the observed difference at FFS is consistent with the smaller weaning sizes observed for this site (Craig and Ragen 1999, NMFS, unpublished data) and is indicative of poorer nutritional condition for adult and immature seals when compared to Laysan. This is further supported by the correlation of much better body condition and earlier age at primiparity for seals in the MHI where all 6-yr-old seals are adult sized compared with 80% at Laysan (Baker et al. 2011).

Genetics

Initial studies of genetic variation based on mitochondrial DNA sequences and nuclear DNA fingerprints suggest that the species currently is characterized by low genetic variability, minimal genetic differentiation among breeding colonies and, perhaps, some naturally occurring local inbreeding (Kretzmann et al. 1997). More recent microsatellite DNA analysis tested additional genetic markers from more individuals (over 2,000 to date), confirming the lack of genetic variation in the Hawaiian monk seal (Kretzmann et al. 2001, Shultz et al. 2009, 2010). The long-term evolutionary history of the species as well as recent human impacts are both possible sources for this extremely low genetic variation. The potential for genetic drift should have increased when seal numbers were reduced by European harvest in the 19th century, but any tendency for genetic divergence among breeding colonies is probably mitigated by the inter-island movements of seals. High genetic connectivity (similarity in genetic markers among samples) also suggests that there is no subdivision or differentiation in the population, either between cohorts (temporal population structure) or between sites (spatial population structure; Schultz et al. 2010). Both high genetic connectivity and field observations suggest that as many as 10% of animals move to sites away from their natal site (Shultz et al. 2011, Johanos et al. 2014).

Diet

As generalist predators, monk seals have a diverse diet that includes fishes (e.g., eel, wrasses, squirrelfish, soldierfish, triggerfish, parrotfish), cephalopods (i.e., squid and octopus), and crustaceans (i.e., crab, shrimp, and lobster; Goodman-Lowe 1998; Parrish et al. 2005, Piché et al. 2010, Iverson et al. 2011, Cahoon et al. 2013). The diets of seals in the NWHI and MHI were compared by looking through identifiable hard parts of prey found in seal scat and regurgitates from monk seals. While there were some small differences the general diet was similar and comprised of fish (78–97%), followed by cephalopods (11–16%) and crustaceans (1–6%; Goodman-Lowe 1998, Cahoon et al.

2013, NMFS, unpublished data). Monk seals prefer solitary, cryptic prey that hide in rock crevices, the sand, or under rocks on the seafloor. Seal-mounted video cameras showed that seals ignore prey known to occur in their diet if the prey is encountered swimming in the water column. Monk seals appear to only pursue prey at deeper depths that were hiding on or near the seafloor (Parrish et al. 2005).

Foraging behavior and habitat use

As benthic foragers, monk seals search for food in coral reef habitat and substrate composed of sand and talus (loose eroded carbonate fragments) on marine terraces. They forage primarily in depths from 18–90 m (Parrish et al. 2000, 2005, Stewart et al. 2006, NMFS, unpublished data). The depth of the seafloor is only this shallow in the Hawaiian Archipelago around atolls, islands, submerged reefs and banks, and seamounts. Monk seals in the NWHI have been observed foraging in deep-water coral beds at depths of 500 m (Parrish et al. 2002), suggesting that seals may use a wide range of habitats, perhaps to avoid competition while foraging (Parrish et al. 2000, Stewart et al. 2006) or to access different types of prey (Stewart et al. 2006). There is a great degree of individual variability in the foraging behavior of monk seals, and it is therefore difficult to identify specific behavioral traits that are common throughout the species, across age and sex classes.

Interspecific competition

Seal-mounted video cameras were deployed on seals in the NWHI in the late 1990s and revealed a much greater encounter rate between seals and other apex predators on the summits of neighboring bank and at depths of 60–80 m (Parrish et al. 2008) than was later found in the MHI. Monk seals observed foraging in the NWHI average more than four encounters per minute with predatory fishes that followed closely, waiting for a chance to take the prey uncovered by the seals (Parrish et al. 2008). This intense competition may explain why some seals forage in deeper water and also provides insight as to why MHI monk seals are more successful at acquiring sufficient food resources. Video studies in the MHI found that apex predator escorts were far less common, and if present were rarely more than 1 or 2 individuals (NMFS, unpublished data). Fishing in the MHI has dramatically reduced apex predator biomass relative to the NWHI (Friedlander and DeMartini 2002), likely benefitting monk seals by decreasing interspecific competition.

Oceanography and impacts on foraging success and survival

In marine ecosystems, large-bodied top predators often show a noticeable response to large-scale oceanographic processes (Voigt et al. 2003). In Hawaiian monk seals, studies have shown survival changes in response to the latitudinal position of the Transition Zone Chlorophyll Front (TZCF; Baker et al. 2007), El Niño events (Antonelis et al. 2003), and the Pacific Decadal Oscillation (PDO; Polovina et al. 1994, 1995, Baker et al. 2012). For example, Baker et al. (2012) showed that the positive PDO phase, generally characterized by cooler more productive waters around the Hawaiian Archipelago, is associated with more favorable foraging conditions for monk seals. This relationship was stronger in the northern portion of the NWHI, where the islands are closer to the TZCF, which is a convergence zone with increased primary productivity that migrates annually over 1000 km in latitude. This compilation of data suggests that monk seal survival is affected by a combination of long- and short-term fluctuations in ocean conditions.

Health and disease

Disease in Hawaiian monk seals can be discussed in two general groupings: existing and emerging. There is evidence of disease in the monk seal population, but the greatest disease threats to the species are pathogens with the potential to be introduced into this geographically isolated, and therefore relatively immunologically naïve population. This section will focus on the existing diseases in the population, and disease threats and mitigation are discussed in the next chapter.

Parasites

Endoparasitism is common in Hawaiian monk seals, as in most free-ranging marine mammals (Dailey 2001). Gastrointestinal helminths are the most commonly identified parasites and can cause mild to moderate gastric ulceration (Banish and Gilmartin 1992) and reduced nutritional condition (Reif et al. 2006). The efficacy of topical anti-helminthics as an intervention strategy to treat some of these parasites and promote weight gain remains under investigation (Gobush et al. 2011).

Antibodies to coccidian protozoa, including *Toxoplasma gondii*, *Sarcocystis* spp. and *Neospora caninum*, have been detected in Hawaiian monk seals (Aguirre et al. 2007, Littnan et al. 2007). These results are difficult to interpret because rising serum titers are not always documented due to temporal availability of samples. Single point titers alone suggest exposure, but cross-reactions are possible (Honnold et al. 2005, Littnan et al. 2007). The relative contributions of highly pathogenic *T. gondii* genotypes and mixed protozoal infections to the occurrence of fulminant disease remain under investigation for Hawaiian monk seals.

The first case of disseminated toxoplasmosis and mortality in a Hawaiian monk seal was reported in 2005 (Honnold et al. 2005). Subsequently, several more mortalities from disseminated infections have been documented in the MHI. The diverse diet of monk seals makes identifying risk factors for oral exposure to *T. gondii* challenging. The increase in infection rates in the past decade has made toxoplasmosis the primary health concern for monk seals may be attributed to the growing human population in Hawaii, as human presence is correlated with that of domestic and feral cats, the definitive host of *T. gondii*. Improvements in surveillance effort and diagnostic capacity also may play a role.

Bacteria

Leptospirosis is the disease caused by spirochete *Leptospira* bacteria that are shed in the urine of terrestrial mammals and typically cause kidney failure in susceptible hosts. Although leptospirosis is a well-known threat to human health in Hawaii (Katz et al. 2002), serologic surveys of Hawaiian monk seals across the archipelago since 1997 yield only sporadic evidence of exposure (Aguirre et al. 2007, Littnan et al. 2007, NMFS unpublished data). Mortalities from this disease are limited to two cases, and leptospirosis is not considered as having a population-level impact at this time.

There are a number of bacteria such as, *Brucella* and *Chlamydophila abortus* that are known to affect reproductive success in many mammalian species. Sero-surveys and analysis of placental tissues demonstrated a low prevalence of these reproductive pathogens in the species and no indication of population level impacts.

Fecal cultures of MHI monk seals have documented the presence of several organisms, including common commensal flora and those that may indicate contamination from human sources (e.g., *Salmonella*). Other organisms cultured from Hawaiian monk seal feces include *Campylobacter jejuni, E. coli, Edwardsiella tarda, Klebsiella, Pleisomonas shigelloides, Vibrio alginolyticus* and *Vibrio cholerae* (Littnan et al. 2007). Though detected in feces both of apparently healthy and malnourished individuals, systemic disease from pathogenic enteric bacteria has not been identified.

Viruses

Exposure to several viruses has been documented, though none have been found to impact survival or their role in clinical disease is poorly understood. Herpesvirus exposure is documented in Hawaiian monk seals in the wild and under human care (Goldstein et al. 2006, Aguirre et al. 2007). A previously unidentified gamma herpesvirus was isolated from nasal swabs collected from Hawaiian monk seals and named Hawaiian monk seal herpesvirus, though it was not associated with clinical signs typical of herpesvirus infections in other pinnipeds (e.g., respiratory infections, reproductive impairment, leukocytosis). It was detected at an estimated prevalence of 20% across the NWHI and was more commonly detected in seals in rehabilitation compared with wild seals (Goldstein et al. 2006). Its role in clinical disease, if any, is not well described in this species but it may (similar to other herpes virus infections) recrudesce in stressed or otherwise compromised animals.

Evidence of calicivirus exposure also has been documented serologically in specimens from seals collected during the 1978 Unusual Mortality Event (UME) at Laysan and onward (Gilmartin et al. 1980, Aguirre et al. 1999). Calicivirus (San Miguel sea lion virus) typically causes raised skin vesicles on the extremities, though these are uncommonly observed in Hawaiian monk seals. Given the lack of clinical signs associated with positive titers in the 1978 UME, it is thought that the virus is endemic within the population. A more comprehensive discussion of viral threats (e.g., morbilliviruses, influenza viruses, West Nile virus) that may impact the species is provided in Chapter 5.

Biotoxins

Hawaii lacks many of the marine biotoxins that can have measurable impacts on marine mammal populations in other climates. However, herbivorous coral reef fish in this tropical region accumulate ciguatoxin, the causative agent of ciguatera fish poisoning in humans. In other species, it leads to gastrointestinal or neurologic illness. Exposure of Hawaiian monk seals to this potent neurotoxin was first documented in 1978 during the investigation of a series of mortalities that led to the declaration of an Unusual Mortality Event (UME) at Laysan Island (Johnson and Johnson 1978, Gilmartin et al. 1980). Since then, diagnostic methods have been expanded and exposure to ciguatoxin has been documented in several additional seals (Bottein et al. 2011), though the risk factors for exposure are difficult to evaluate. The potential impacts of this toxin on monk seals remain unclear but studies are underway to better evaluate the potential cumulative impact of this biotoxin on monk seal survival, especially in resource-limited habitats.

Contaminants

Persistent Organic Pollutants (POPs) are a classification of compounds used for many applications, including transformer insulation, flame retardants, and pesticides and include compounds such as Polychlorinated biphenyls (PCBs), polybrominated diphenyl ethers (PBDEs), and dichlorodiphenyltricholroethanes (DDTs). Many of these compounds have been outlawed in the U.S. and Europe but still persist in the environment. POPs can be transported long distances, through oceanic or atmospheric mechanisms, from areas in which they are still being produced and used to remote areas and areas where they are banned (Bard 1999). These compounds are lipophilic, toxic, bioaccumulate in organisms, and biomagnify through the food chain.

Many POPs have been detected in the blood and blubber of Hawaiian monk seals from the MHI and NWHI (Wilcox et al. 2004, Ylitalo et al. 2008, Lopez et al. 2012). As is commonly observed in other mammalian species, adult female monk seals were found to have the lowest levels of most POP compounds due to the ability to offload these molecules through nursing. Adult males had the highest levels, with juvenile seal concentrations being intermediate. Of notable interest, PCBs were significantly greater in adult males and juveniles at Midway compared with other locations in the NWHI and MHI.

Many effects of POPs are sub-lethal but can disrupt the function of various organ systems or body processes. POPs can have reproductive and immunosuppressive effects (i.e., endocrine disruption and immunotoxicity) and also have been associated with cancer in marine mammals (DeLong et al. 1973, de Swart et al. 1994, Ylitalo et al. 2005). It can be difficult to assess these sub-lethal effects without well-planned controlled experiments, which can be difficult, if not impossible, to conduct on free-ranging marine mammals due to logistical and ethical constraints. Therefore, little is known about the species-specific effects of POPs on Hawaiian monk seals. However, mean levels measured were below those seen to cause effects in other marine mammal species.

Conclusion

Hawaiian monk seals are a unique lineage; they are the last of their genus and the only completely tropical phocid in the world. This exceptional distribution has resulted in physiological, ecological, and behavioral adaptions not observed in other phocids. However, this unique species faces the very real possibility of extinction. A host of natural and anthropogenic threats have driven and continue to push Hawaiian monk seals to the brink of oblivion. The fragile tropical marine ecosystems and low-lying coral islets that monk seals inhabit are under great threat from climate change, in particular ocean warming and sea level rise, further heightening the risk to the species. The species and its decline has been the focus of intensive research over the last 30 years as scientists and managers struggle to pull the species back from the brink. The following chapter discusses the evolution of the Hawaiian monk seal recovery effort and the current cutting edge program working to conserve the world's only tropical seal.

Acknowledgements

The authors would like to thank the staff of NOAA's Hawaiian Monk Seal Research Program for providing data, images, and input into the drafting of this chapter.

Keywords: Hawaiian monk seal, *Neomonachus schauinslandi*, endangered, marine mammal, life-history, ecology

References

Aguirre, A.A., J.S. Reif and G.A. Antonelis. 1999. Hawaiian monk seal epidemiology plan: Health assessment and disease status studies. NOAA Technical Memorandum NOAA-TM-NMFS-SWFSC-280.
Aguirre, A.A., T.J. Keefe, J.S. Reif, L. Kashinsky, P.K. Yochem, J.T. Saliki et al. 2007. Infectious disease monitoring of the endangered Hawaiian monk seal. J. Wildl. Dis. 43(2): 229–241.
Antonelis, G.A., J.D. Baker and J.J. Polovina. 2003. Improved body condition of weaned Hawaiian monk seal pups associated with El Niño events: Potential benefits to an endangered species. Mar. Mamm. Sci. 19(3): 590–598.
Baker, J.D. and T.C. Johanos. 2004. Abundance of the Hawaiian monk seal in the main Hawaiian Islands. Biol. Conserv. 116(1): 103–110.
Baker, J.D., C.L. Littnan and D.W. Johnston. 2006. Potential effects of sea level rise on the terrestrial habitats of endangered and endemic megafauna in the Northwestern Hawaiian Islands. Endang. Species Res. 3(3): 21–30.
Baker, J.D., J.J. Polovina and E.A. Howell. 2007. Effect of variable oceanic productivity on the survival of an upper trophic predator, the Hawaiian monk seal *Monachus schauinslandi*. Mar. Ecol. Prog. Ser. 346: 277–283.
Baker, J.D., E.A. Howell and J.J. Polovina. 2012. Relative influence of climate variability and direct anthropogenic impact on a sub-tropical Pacific top predator, the Hawaiian monk seal. Mar. Ecol. Prog. Ser. 469: 175–189.
Baker, J.D., T.C. Johanos, T.A. Wurth and C.L. Littnan. 2014. Body growth in Hawaiian monk seals. Mar. Mam. Sci. 30: 259–271. doi: 10.1111/mms.12035.
Banish, L.D. and W.G. Gilmartin. 1992. Pathological findings in the Hawaiian monk seal. J. Wildl. Dis. 28(3): 428–434.
Bard, S.M. 1999. Global transport of anthropogenic contaminants and the consequences for the Arctic marine ecosystem. Mar. Pollut. Bull. 38(5): 356–379.
Berta, A., J.L. Sumich and K.M. Kovacs. 2015. Marine Mammals: Evolution Biology, Third Edition, Academic Press. San Diego, CA 738 pp.
Boness, D.J. 1990. Fostering behavior in Hawaiian monk seals: Is there a reproductive cost? Behav. Ecol. Sociobiol. 27(2): 113–122.
Boness, D.J., M.P. Craig, L. Honigman and S. Austin. 1998. Fostering behavior and the effect of female density in Hawaiian monk seals, *Monachus schauinslandi*. J. Mammal. 79(3): 1060–1069.

Bottein, M.Y.D., L. Kashinsky, Z. Wang, C. Littnan and J.S. Ramsdell. 2011. Identification of ciguatoxins in Hawaiian monk seals *Monachus schauinslandi* from the northwestern and main Hawaiian Islands. Environ. Sci. Technol. 45(12): 5403–5409.

Cahoon, M.K., C.L. Littnan, K. Longenecker and J.R. Carpenter. 2013. Dietary comparison of two Hawaiian monk seal populations: The role of diet as a driver of divergent population trends. Endang. Species Res. 20: 137–146.

Carretta, J.V., E.M. Oleson, D.W. Weller, A.R. Lang, K.A. Forney, J. Baker et al. 2015. U.S. Pacific Marine Mammal Stock Assessments: 2014. U.S. Department of Commerce, NOAA Technical Memorandum, NOAA - TM - NMFS - SWFSC - 549.

Craig, M.P. and T.J. Ragen. 1999. Body size, survival, and decline of juvenile Hawaiian monk seals, *Monachus schauinslandi*. Mar. Mamm. Sci. 15(3): 786–809.

Dailey, M.D. 2001. Parasitic diseases. pp. 357–379. *In*: L.A. Dierauf and F.M.D. Gulland [eds.]. CRC Handbook of Marine Mammal Medicine 2nd Edition. CRC Press, Boca Raton, FL, USA.

de Swart, R., P. Ross, L. Vedder, H. Timmerman, S. Heisterkamp, H. Van Loveren et al. 1994. Impairment of immune function in harbor seals (*Phoca vitulina*) feeding on fish from polluted waters. Ambio 23(2): 155–159.

DeLong, R.L., W.G. Gilmartin and J.G. Simpson. 1973. Premature births in California sea lions: Association with high organochlorine pollutant residue levels. Science 181(4105): 1168–1170.

Friedlander, A.M. and E.E. DeMartini. 2002. Contrasts in density, size, and biomass of reef fishes between the northwestern and the main Hawaiian Islands: the effects of fishing down apex predators. Mar. Ecol. Prog. Ser. 230: 253–264.

Gerrodette, T. and W.G. Gilmartin. 1990. Demographic consequences of changed pupping and hauling sites of the Hawaiian monk seal. Conserv. Biol. 4(4): 423–430.

Gilmartin, W.G., R.L. DeLong, A.W. Smith, L.A. Griner and M.D. Dailey. 1980. An investigation into unusual mortality in the Hawaiian monk seal, *Monachus schauinslandi*. pp. 32–41. *In*: R.W. Grigg and R.T. Pfund [eds.]. Proceedings on Status of Resource Investigation in the Northwestern Hawaiian Islands, University of Hawaii, Honolulu, UNIHI-SEAGRANT-MR-80-04.

Gobush, K.S., J.D. Baker and F.M.D. Gulland. 2011. Effectiveness of an antihelminthic treatment in improving the body condition and survival of Hawaiian monk seals. Endang. Species Res. 15(1): 29–37.

Goldstein, T., F.M.D. Gulland, R.C. Braun, G.A. Antonelis, L. Kashinsky, T.K. Rowles et al. 2006. Molecular identification of a novel gamma herpesvirus in the endangered Hawaiian monk seal (*Monachus schauinslandi*). Mar. Mamm. Sci. 22(2): 465.

Goodman-Lowe, G.D. 1998. Diet of the Hawaiian monk seal (*Monachus schauinslandi*) from the Northwestern Hawaiian Islands during 1991 to 1994. Mar. Biol. 132(3): 535–546.

Harting, A.L., J.D. Baker and T.C. Johanos. 2007. Reproductive patterns of the Hawaiian monk seal. Mar. Mamm. Sci. 23(3): 553–573.

Harting, A.L., T.C. Johanos and C.L. Littnan. 2014. Benefits derived from opportunistic survival-enhancing interventions for the Hawaiian monk seal: the silver BB paradigm. Endang Species Res. 25: 89–96.

Honnold, S.P., R. Braun, D.P. Scott, C. Sreekumar and J.P. Dubey. 2005. Toxoplasmosis in a Hawaiian monk seal (*Monachus schauinslandi*). J. Parasitol. 91(3): 695–697.

Iverson, S., J. Piché and W. Blanchard. 2011. Hawaiian monk seals and their prey: Assessing characteristics of prey species fatty acid signatures and consequences for estimating monk seal diets using Quantitative Fatty Acid Signature Analysis. Technical Memorandum NOAA-TM-NMFS-PIFSC-23. NMFS, Honolulu, HI.

Job, D.A., D.J. Boness and J.M. Francis. 1995. Individual variation in nursing vocalizations of Hawaiian monk seal pups, *Monachus schauinslandi* (Phocidae, Pinnipedia), and lack of maternal recognition. Can. J. Zool. 73(5): 975–983.

Johanos, T.C., B.L. Becker and T.J. Ragen. 1994. Annual reproductive cycle of the female Hawaiian monk seal (*Monachus schauinslandi*). Mar. Mamm. Sci. 10(1): 13–30.

Johanos, T.C., A.L. Harting, T.A. Wurth and J.D. Baker. 2014. Range-wide movement patterns of Hawaiian monk seals. Mar. Mamm. Sci. 30(3): 1165–1174.

Johnson, B.W. and P.A. Johnson. 1978. The Hawaiian Monk Seal on Laysan Island, 1977. National Technical Information Service, US Department of Commerce.

Johnson, B.W. and P.A. Johnson. 1984. Observations of the Hawaiian monk seal on Laysan Island from 1977 through 1980. US Department of Commerce, National Oceanic and Atmospheric Administration, National Marine Fisheries Service, Southwest Fisheries Center.

Karamanlidis, A.A., P. Dendrinos, P.F. de Larrinoa, A.C. Gücü, W.M. Johnson, C.O. Kiraç et al. 2016. The Mediterranean monk seal *Monachus monachus*: Status, biology, threats, and conservation priorities. Mamm. Rev. 46: 92–105.

Katz, A.R., V.E. Ansdell, P.V. Effler, C.R. Middleton and D.M. Sasaki. 2002. Leptospirosis in Hawaii, 1974–1998: Epidemiologic analysis of 353 laboratory-confirmed cases. Am. J. Trop. Med. Hyg. 66(1): 61–70.

Kenyon, K.W. and D.W. Rice. 1959. Life history of the Hawaiian monk seal. Pac. Sci. 13(3): 215–252.

Kenyon, K.W. 1972. Man versus the monk seal. J. Mammal. 53(4): 687–696.

Kenyon, K.W. 1977. Caribbean monk seal extinct. J. Mammal. 58(1): 97–98.

Kirch, P.V. 2011. When did the Polynesians settle Hawai'i? A review of 150 years of scholarly inquiry and a tentative answer. Hawaiian Archaeology 12(2011): 3–26.

Kovacs, K.M. and D.M. Lavigne. 1986. Maternal investment and neonatal growth in phocid seals. J. Anim. Ecol. 55(3): 1035–1051.

Kretzmann, M.B., W.G. Gilmartin, A. Meyer, G.P. Zegers, S.R. Fain, B.F. Taylor et al. 1997. Low genetic variability in the Hawaiian monk seal. Conservation Biology 11: 482–490. doi:10.1046/j.1523-1739.1997.96031.x.

Kretzmann, M.B., N.J. Gemmell and A. Meyer. 2001. Microsatellite analysis of population structure in the endangered Hawaiian monk seal. Conservation Biology 15: 457–466. doi:10.1046/j.1523-1739.2001.015002457.x.

Laws, R.M. 1956. Growth and sexual maturity in aquatic mammals. Nature 178: 193–194.

Littnan, C.L., B.S. Stewart, P.K. Yochem and R. Braun. 2007. Survey for selected pathogens and evaluation of disease risk factors for endangered Hawaiian monk seals in the main Hawaiian Islands. EcoHealth 3(4): 232–244.

Lopez, J., D. Boyd, G.M. Ylitalo, C. Littnan and R. Pearce. 2012. Persistent organic pollutants in the endangered Hawaiian monk seal (*Monachus schauinslandi*) from the main Hawaiian Islands. Mar. Pollut. Bull. 64(11): 2588–2598.

McClenachan, L. and A.B. Cooper. 2008. Extinction rate, historical population structure and ecological role of the Caribbean monk seal. Proc. R. Soc. B 275: 1351–1358.

National Marine Fisheries Service. 2007. Recovery Plan for the Hawaiian Monk Seal (*Monachus schauinslandi*), Second Revision. National Marine Fisheries Service, Silver Spring, MD, USA.

Parrish, F.A., M.P. Craig, T.J. Ragen, G.J. Marshall and B.M. Buhleier. 2000. Identifying diurnal foraging habitat of endangered Hawaiian monk seals using a seal-mounted video camera. Mar. Mamm. Sci. 16(2): 392–412.

Parrish, F.A., K. Abernathy, G.J. Marshall and B.M. Buhleier. 2002. Hawaiian monk seals (*Monachus schauinslandi*) foraging in deep-water coral beds. Mar. Mamm. Sci. 18(1): 244–258.

Parrish, F.A., G.J. Marshall, C.L. Littnan, M. Heithaus, S. Canja, B. Becker et al. 2005. Foraging of juvenile monk seals at French Frigate Shoals, Hawaii. Mar. Mamm. Sci. 21(1): 93–107.

Parrish, F.A., G.J. Marshall, B. Buhleier and G.A. Antonelis. 2008. Foraging interaction between monk seals and large predatory fish in the Northwestern Hawaiian Islands. Endang. Species Res. 4(3): 299–308.

Piché, J., S.J. Iverson, F.A. Parrish and R. Dollar. 2010. Characterization of forage fish and invertebrates in the Northwestern Hawaiian Islands using fatty acid signatures: Species and ecological groups. Mar. Ecol. Prog. Ser. 418: 1–15.

Polovina, J.J., G.T. Mithcum, N.E. Graham, M.P. Craig, E.E. DeMartini and E.N. Flint. 1994. Physical and biological consequences of a climate event in the central North Pacific. Fish. Oceanogr. 3(1): 15–21.

Polovina, J.J., G.T. Mitchum and G.T. Evans. 1995. Decadal and basin-scale variation in mixed layer depth and the impact on biological production in the Central and North Pacific, 1960–88. Deep Sea Res. Part 1 Oceanogr. Res. Pap. 42(10): 1701–1716.

Ragen, T.J. 1999. Human activities affecting the population trends of the Hawaiian monk seal. *In*: J.A. Musick [ed.]. Life in the Slow Lane: Ecology and Conservation of Long-Lived Marine Animals. Am. Fish. Soc. Symp (American Fisheries Society Symposium) 23: 183–194.

Reif, J.S., A. Bachand, A.A. Aguirre, D.L. Borjesson, L. Kashinsky, R.C. Braun et al. 2004. Morphometry, hematology, and serum chemistry in the Hawaiian monk seal (*Monachus schauinslandi*). Mar. Mamm. Sci. 20(4): 851–860.

Reif, J.S., M.M. Kliks, A.A. Aguirre, D.L. Borjesson, L. Kashinsky, R.C. Braun et al. 2006. Gastrointestinal helminths in the Hawaiian monk seal (*Monachus schauinslandi*): Associations with body size, hematology, and serum chemistry. Aquat. Mamm. 32(2): 157–167.

Reiter, J., N.L. Stinson and B.J. Boeuf. 1978. Northern elephant seal development: The transition from weaning to nutritional independence. Behav. Ecol. Sociobiol. 3(4): 337–367.

Repenning, C.A. and C.E. Ray. 1977. The origin of the Hawaiian monk seal. Proc. Biol. Soc. Wash. 89(58): 667–688.

Slater, G., D. Scheel, S. Kolokotronis, C. Potter, D. Rotstein, K. Tsangaras et al. 2014. Biogeography and taxonomy of extinct and endangered monk seals illuminated by ancient DNA and skull morphology. ZooKeys 409: 1–33.

Schultz, J.K., J.D. Baker, R.J. Toonen and B.W. Bowen. 2009. Extremely low genetic diversity in the endangered Hawaiian monk seal (*Monachus schauinslandi*). J. Hered. 100(1): 25–33.

Schultz, J.K., A.J. Marshall and M. Pfunder. 2010. Genome-wide loss of diversity in the critically endangered Hawaiian monk seal. Diversity 2(6): 863–880.

Schultz, J.K., B.L. Becker, T.C. Johanos, J.U. Lopez and L. Kashinsky. 2011. Dizygotic twinning in the Hawaiian monk seal. J. Mammal. 92(2): 336–341.

Stewart, B.S., G.A. Antonelis, J.D. Baker and P.K. Yochem. 2006. Foraging biogeography of Hawaiian monk seals in the northwestern Hawaiian Islands. Atoll Res. Bull. 543(13): 1–145.

Voigt, W., J. Perner, A.J. Davis, T. Eggers, J. Schumacher, R. Bährmann et al. 2003. Trophic levels are differentially sensitive to climate. Ecology 84(9): 2444–2453.

Watson, T.K., J.N. Kittinger, J.S. Walters and T.D. Schofield. 2011. Culture, conservation, and conflict: Assessing the human dimensions of Hawaiian monk seal recovery. Aquat. Mamm. 37(3): 386–396.

Westlake, R.L. and W.G. Gilmartin. 1990. Hawaiian monk seal pupping locations in the Northwestern Hawaiian Islands. Pac. Sci. 44(4): 366–383.

Wilcox, M.K., L.A. Woodward, G.M. Ylitalo, J. Buzitis, S. Atkinson and Q.X. Li. 2004. Organochlorines in the free-ranging Hawaiian monk seal (*Monachus schauinslandi*) from French Frigate Shoals, North Pacific Ocean. Sci. Total. Environ. 322: 81–93.

Wirtz, W.O. 1968. Reproduction, growth and development, and juvenile mortality in the Hawaiian monk seal. J. Mammal. 49(2): 229–238.

Worthy, G.A. and D.M. Lavigne. 1983. Energetics of fasting and subsequent growth in weaned harp seal pups, *Phoca groenlandica*. Can. J. Zool. 61(2): 447–456.

Ylitalo, G.M., J.E. Stein, T. Hom, L.L. Johnson, K.L. Tilbury, A.J. Hall et al. 2005. The role of organochlorines in cancer-associated mortality in California sea lions (*Zalophus californianus*). Mar. Pollut. Bull. 50(1): 30–39.

Ylitalo, G.M., M. Myers, B.S. Stewart, P.K. Yochem, R. Braun, L. Kashinksy et al. 2008. Organochlorine contaminants in endangered Hawaiian monk seals from four subpopulations in the Northwestern Hawaiian Islands. Mar. Pollut. Bull. 56: 231–244.

5

Hawaiian Monk Seal Conservation
Past, Present and Future

Charles Littnan,[1,*] *Michelle Barbieri,*[1,a] *Jessica Lopez Bohlander,*[1,b] *Tenaya Norris*[2] *and Stacie Robinson*[1,c]

Introduction

The Hawaiian monk seal is one of the world's rarest marine mammal species and is listed as depleted and endangered based on national and international criteria. As with many wildlife species, the natural and anthropogenic threats that monk seals face affect the population heterogeneously based on demographic group and geographic location. The following sections discuss the legal mandates driving monk seal conservation, some of the main sources of mortality and morbidity for seals across the Hawaiian Archipelago, and the conservation strategies, past and present, that managers and scientists use to mitigate these threats and ensure the species future.

---

[1] NOAA Fisheries Service, Hawaiian Monk Seal Research Program, 1845 WASP Blvd., Building 176 Honolulu, HI 96818.
[a] Email: michelle.barbieri@noaa.gov
[b] Email: jessica.bohlander@noaa.gov
[c] Email: stacie.robinson@noaa.gov
[2] The Marine Mammal Center, 2000 Bunker Road, Sausalito, CA 94965. Email: norrist@tmmc.org
* Corresponding author: charles.littnan@noaa.gov

Laws and Conservation Mandates

There are a number of laws at both U.S. Federal and State levels that work to protect monk seals and their habitats. However, the two primary legislative mandates to protect and conserve Hawaiian monk seals are the Endangered Species Act (ESA) and Marine Mammal Protection Act (MMPA).

The ESA was established to provide the means to conserve endangered and threatened species' ecosystems and provide programs for the conservation of such species. It is the policy of the ESA that all Federal agencies must seek to conserve threatened and endangered species and use their authorities to further the purposes of the ESA. It also requires Federal agencies, the National Marine Fisheries Service (NMFS) in the case of Hawaiian monk seals, to develop and implement a recovery plan for the conservation of these species. The ESA requires Federal agencies to consult on any action that it authorizes, funds, or undertakes to ensure such activities do not jeopardize the continued existence of any threatened or endangered species or result in destruction or adverse modification of critical habitat. As defined by the ESA, critical habitat is the geographic area that contains features essential to the conservation of a threatened or endangered species and that may require special considerations or protection.

The MMPA was established to ensure that marine mammal populations were maintained at levels that allowed them to continue as functioning parts of their respective ecosystems. The law made it U.S. policy to prohibit marine mammal "takes", which are activities that include the hunting, harassing, capturing, or killing of any marine mammal in the U.S. and territorial waters. Few exceptions exist to this take prohibition. Permits for *bona fide* scientific research on marine mammals and permits to enhance the survival or recovery of a species are required to authorize takes. There are further exceptions for hunting of marine mammals for some indigenous communities.

Historical Conservation and Challenges

As is the case of many species' recovery programs, the early days of Hawaiian monk seal conservation were hindered by a lack of good science, funding, and awareness of the species status. Despite the species' decline and precarious state being recognized in the 1950s, it was not until the 1970s that legislative action was taken to protect monk seals (Lowry et al. 2011). Based on estimates of a population numbering 1000 ± 500 seals and an ongoing decline, the U.S. Government listed Hawaiian monk seals as depleted under the MMPA and endangered under the ESA in 1976.

Following their listings under the MMPA and ESA, more concerted efforts to understand the species' population trends, biology, and threats were initiated and in 1980 a NMFS monk seal program was formally in place. Lowry et al. (2011) describe the initial research and conservation efforts as being hampered by inconsistent commitments of funding, the logistical difficulty and costs of working in the remote Northwestern Hawaiian Islands (NWHI), and difficulty in convincing managers to adopt protective measures due to inherent uncertainty of scientific information. Despite these challenges, which would recur over the years, there were a few key early successes that laid a foundation for the robust and proactive conservation program that currently exists. These include the drafting of the first recovery plan (Gilmartin 1983) and a 1988 designation of critical habitat in the NWHI that expanded marine protected areas and halted interactions with long-line fisheries. Both of these activities were required by the ESA. Fundamental research began with marking, identifying and counting seals that formed the foundation of studies to track trends in abundance, survival, reproduction, and age-sex composition at individual colonies. A full history of these early research and recovery efforts is detailed in Lowry et al. (2011).

Modern Conservation Program

In 2007, a revised recovery plan for the Hawaiian monk seal (NMFS 2007) was published serving as a transition point from what was largely a population monitoring program to a proactive research-based conservation program. Multi-faceted research provides the foundational data essential to support management and conservation decisions. Annual population monitoring, through the use of camps in the remote NWHI and citizen scientists in the populous MHI, helps track seal survival and abundance and collect data to understand the factors that impact the population (Fig. 1). The Hawaiian monk seal population monitoring program in the NWHI is one of the longest running, most intensive, and most comprehensive datasets for an endangered species. The program has been enumerating the population across the species range and tracking individual life histories since 1982 (NMFS, unpublished data).

In addition to monitoring population status, the Hawaiian monk seal conservation program collects biological and ecological data, identifies threats to monk seal survival, and develops mitigation strategies. This information has linked population trends to environmental phenomena such as climate and oceanographic cycles and has provided data on site-specific and age class/sex-specific survival rates and associated conservation challenges. This extensive dataset also helps to parameterize population dynamics models that can be used to assess extinction

Figure 1. Remote field camps deployed in the Northwestern Hawaiian Islands serve as the foundation for Hawaiian monk seal population assessment. During the peak breeding season field researchers monitor population trends and causes of mortality and intervene to help increase survival.

vulnerability and evaluate management options. For example, long term data and population dynamic models suggest that interventions to save female seals can have the greatest population-level impact due to contributions through breeding, as shown by the female survival estimates at French Frigate Shoals (FFS) from 1984 to 2015 (Fig. 2). Additional forms of data from telemetry instruments, animal-mounted video cameras, and ecological surveys, allow researchers to understand the species needs and use of habitat and nutritional resources. These data improve the understanding of the large-scale processes that impact the ecosystem, and the small-scale, localized interactions between monk seals and their environment, both of which are important to monk seal recovery efforts. Further, the long-term data that allow researchers to link population trends with ecosystem-level processes also helps assess future vulnerabilities to system perturbations (e.g., climate change, human disturbance, fishing) and forecast necessary adaptations in recovery actions. Monitoring individual seals and the events that threaten their welfare (e.g., entanglement, injury) identifies the factors that have the greatest survival impact on certain age groups and at certain sites. By identifying threats at the system and population levels, as well as having extensive data on individuals, it is possible to tailor conservation strategies to have the greatest impact on population recovery.

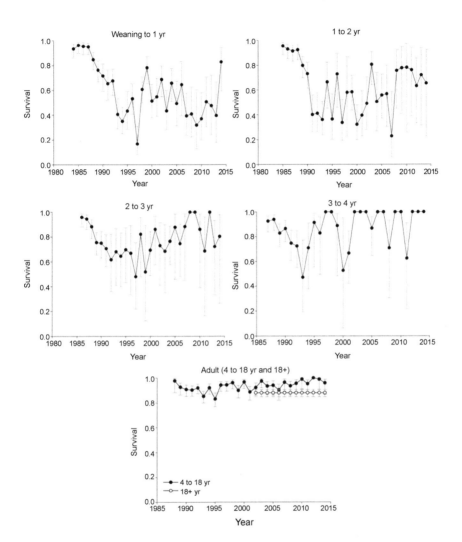

Figure 2. Robust scientific datasets are the foundation of the Hawaiian monk seal recovery effort. The population has been monitored intensively for more than three decades, resulting in one of the world's most detailed population datasets for a species. This allows scientists to determine trends in populations, how threats vary across the monk seal's range, and how to implement different conservation strategies. This figure shows the different age-sex structures and age-related survival rates for a number of monk seal populations.

Another crucial part of applied research in conservation is evaluating program success, and, in the case of an endangered species, progress towards species' recovery. In addition to using data to refine and improve protocols and mitigation strategies, continual data collection allows researchers to track successes and modify actions as necessary. The long-term data collected by researchers has demonstrated that interventions have benefited approximately one third of the existing population (Harting et al. 2014). These data supply important justification for the costs and efforts associated with a multi-faceted conservation and research program.

Threats and Mitigation: Conservation in Action

The threats to monk seal survival are well documented and, as mentioned previously, are highly variable across time and the species' range. The following sections describe these threats in greater detail and highlight some of the mitigation strategies that have previously been used or tested and those that are currently being applied or considered for future implementation.

Food limitation

Poor juvenile survivorship in the NWHI linked to food limitation is one of the primary drivers of the continued Hawaiian monk seal population decrease (Craig and Ragen 1999, Baker and Thompson 2007, Baker 2008). Monk seals less than two years old are particularly vulnerable to food limitation because, although young seals need substantial nutritional resources for growth, they lack the experience and body size to forage as effectively as adult seals and other predators (Fig. 3).

Figure 3. Emaciated juvenile monk seals are a common site in the Northwestern Hawaiian Islands. A variety of factors contribute to young seals' struggles to attain sufficient food resources necessary for survival. Translocations to areas with greater survival rates and rehabilitation in captive settings are two strategies to combat starvation.

Across the Hawaiian Archipelago, spatiotemporal variability in ocean productivity that impact the availability of monk seal prey are largely driven by Pacific Ocean basin-wide features, such as the position of the Transition Zone Chlorophyll Front (TZCF) and the modality of the El Niño-Southern Oscillation (ENSO) and Pacific Decadal Oscillation (PDO). Southward migration of the TZCF brings more productive waters farther into the northern portion of the NWHI, which has been correlated with greater survival of juvenile monk seals (Fig. 4; Baker et al. 2007). In addition, young monk seals at some locations are in better body condition and/or have greater survivorship during El Niño events and the positive phases of the PDO, both of which are characterized by increased productivity in waters around the Hawaiian Archipelago (Antonelis et al. 2003, Baker et al. 2012).

In addition to system-level drivers, the local assemblage of other apex predators (e.g., sharks, jacks, and large-bodied snappers) influences prey availability for monk seals through both direct and indirect competition. One study found apex predators comprised greater than 54% of the total fish biomass in the NWHI, compared with large predatory fishes making up only 3% of total fish biomass in the main Hawaiian Islands (MHI) (Friedlander and DeMartini 2002). This difference in apex predator biomass is likely the result of heavy fishing activity around the MHI whereas fishing for large predatory fishes has been historically less intensive, and recently prohibited, in the NWHI. Using seal-mounted underwater video cameras (National Geographic Crittercams©), monk seals in the NWHI have been observed foraging with large predatory fishes nearby and actively competing with, and often taking prey items from the seals (Parrish et al. 2008). These top-down controls work synergistically with bottom-up forcing to limit food availability in the NWHI, which primarily affects the survival of juvenile monk seals.

Since it is not be feasible to adjust large-scale ecosystem drivers or ecological communities, mitigation of this threat requires hands-on interventions that fall into two categories. The first category is intervening when young seal are already exhibiting signs of prey limitation as evident by emaciated or malnourished body condition. In these cases, rehabilitation is the preferred intervention strategy. The other category involves intervention before the effects of food limitation are obvious, and translocations are the most appropriate conservation strategy in these cases.

Rehabilitation has been used to promote weight gain and improve juvenile survivorship in an effort to restore the depleted NWHI monk seal population for several decades (Ragen and Lavigne 1999, Gilmartin et al. 2011, Norris et al. 2011, Schofield et al. 2011). Hawaiian monk seal rehabilitation facilities have ranged from shore pens at remote NWHI sites to a state-of-the-art hospital with veterinary staff in the MHI. From 1984–1995, 98 young underweight female monk seals were rehabilitated.

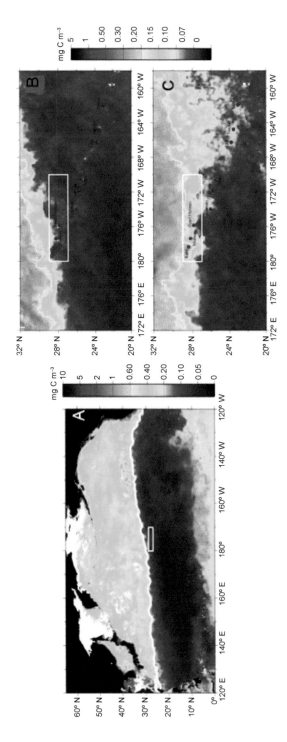

Figure 4. Large scale oceanographic factors, such as the Transition Zone Chlorophyll front (TZCF; A), influence the survival of young monk seals by temporarily increasing regional or local productivity at many Hawaiian monk seal subpopulations (B and C).

Survival for the rehabilitated seals was reduced compared with seals that were not rehabilitated ("controls") in the first year post-release, but survival in subsequent years was similar for rehabilitated and control seals. Monk seal rehabilitation efforts ceased temporarily in 1995 after 10 young seals developed an eye condition of unknown etiology, which led to partial blindness, and had to be placed in permanent captivity (NMFS 2007). Nine more monk seals were rehabilitated from 2003–2008, primarily using beach pens, but none of these seals were alive in the wild one year after release. To provide the best quality of care for monk seals needing rehabilitation, a dedicated monk seal hospital was built by The Marine Mammal Center in Kona, Hawaii in 2014. As of May 2016, 15 emaciated and malnourished juvenile monk seals from the NWHI have been rehabilitated at this facility and released back to the NWHI in excellent body condition.

A relatively new method to address the problem of prey limitation in the NWHI is through translocation, by moving young seals from sites with poor survivorship to sites with a greater probability of survival (Baker et al. 2011, NMFS, unpublished data). Translocation for this purpose is a viable mitigation strategy only when there is a strong survival differential between the donor and release sites. Because monk seals born at French Frigate Shoals (FFS) typically have the greatest mortality rates and are in poorer body condition than pups born elsewhere in the NWHI, weaned pups from this site have been the focus of this intervention strategy. Five weaned pups from FFS were translocated to Kure (1990), 12 were translocated to Nihoa Island (2008–2009), and 15 were translocated to Laysan (2012–2015). In 2014, two weaned pups each from Midway and Kure were also translocated to Lisianski. In all cases, control seals (non-translocated seals at the release site) had the greatest survival, but translocated seals had greater survival than those left at the donor site. The use of a two-stage translocation strategy is also being explored for the future, in which juvenile seals would temporarily be moved to a new location with better foraging conditions and then back to their original location after they have reached an appropriate age (e.g., > 2 years old) (Baker et al. 2013).

Shark predation

Predation is part of the natural landscape for pinnipeds, and the primary predator for monk seals are tiger sharks (*Galeocerdo cuvier*). However, a unique and extreme situation involving shark predation is occurring at FFS, once home to the largest monk seal subpopulation. In 1994, nursing and recently weaned pups started to disappear or were found with severe shark bite wounds (Fig. 5A). Experts found that a small cohort of Galapagos sharks (*Carcharhinus galapagensis*) had developed a new behavior of attacking pups by beaching themselves or attacking pups as they rested

Figure 5. In the mid-1990s, Galapagos sharks (*Carcharhinus galapagensis*) began preying on nursing and recently weaned monk seals pups. These images show the catastrophic injuries caused by predation attempts (A), and the near-shore predatory behavior of the sharks (B).

or played in shallow waters (Fig. 5B). Many pups are immediately killed, but others are permanently maimed by severe bites and subsequently die. While predation peaked in 1997–1999 with over 20 pups depredated per year, it still accounts for the death of approximately 5–10 pups annually (15–25% of the pup cohort) and is the primary source of pup mortality at FFS. Between 1997 and 2014, shark predation affected approximately 250 pups out of about 1000 born at FFS. Since 1997, NMFS has engaged in a variety of actions to address this threat with varying degrees of success.

Because Galapagos shark predation of young monk seal pups does not occur at all of the islets at FFS, one strategy that has proven successful at minimizing pup losses is to translocate young seals from islands where shark predation occurs to those where it does not. Research teams will take recently weaned pups, or sometimes forcibly wean pups if weaning is imminent, to islets that have never or rarely experienced shark predation. A translocation as short as 2 km can reduce the likelihood of Galapagos shark attacks to almost zero (Fig. 6).

Researchers have also tested the use of sound, electromagnetic devices, human presence, and other instruments to deter Galapagos sharks from patrolling the near-shore waters when pups are present. Deterrent studies have tested the application of visual implements and acoustic playbacks that mimic human activity to repel sharks from the immediate area. Results demonstrated that shark presence did not differ significantly among treatments, and that any benefit was likely due to the increased human presence needed to maintain the deterrents (Gobush and Farry 2012). The use of deterrents has been discontinued at this time.

Removing the few sharks exhibiting this behavior from the environment is considered the most effective means of preventing continued predation. Successful removal of these individuals could have a profound effect on the monk seal population at FFS while having negligible impact on the

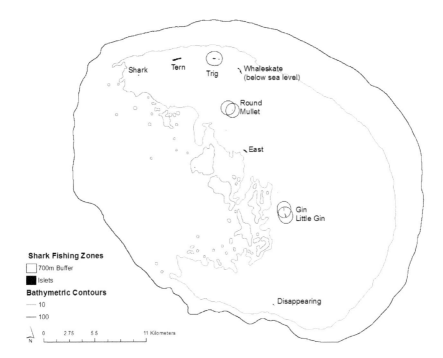

Figure 6. The predatory behavior by Galapagos sharks (*Carcharhinus galapagensis*) is unique to the French Frigate Shoals atoll and only observed at a few the islets within the lagoon. This map shows the atoll in its entirety and the sites where pup predation has been observed. It also shows Tern Island, a safe haven for young pups to be translocated to in order to avoid predation.

Galapagos shark population. The issuance of permits required to remove sharks from FFS waters is a sensitive issue for managers, particularly since it is part of the Papahānaumokuākea Marine National Monument, and the activity is reviewed each year before it occurs.

Intraspecific male aggression

The primary cause of adult female mortality affecting recovery potential in the monk seal population during the 1980s and early 1990s was injury, and often death, caused by intraspecific adult male aggression (Banish and Gilmartin 1992; Fig. 7). These attacks have been less frequent over the past 20 years, but occur when a number of males gather and repeatedly attempt to mount and mate with a single female. Multiple-male aggression has been observed at low levels throughout the population but is thought to be exacerbated when an imbalance in adult sex ratios occurs, with males outnumbering females, or by certain environmental factors such as small

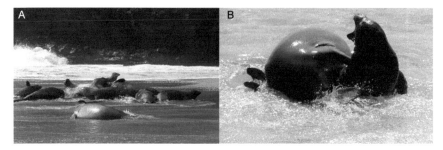

Figure 7. Images of the two types of male aggression observed in monk seals. Multiple males compete for access to a reproductive female (A). These interactions often end in severe injury or death of the female. An individual sub-dominant male attacks a juvenile monk seal (B).

breeding beaches. An example of this occurred at Laysan Island where the sex ratio was skewed to males by approximately 2:1 in 1983–1989 (Hiruki et al. 1993). Injury and death to female seals by aggressive males during this period was three to four times greater than the frequency observed elsewhere. In response to this unsustainable impact to the reproductive potential at Laysan, scientists and managers decided to take steps to equalize population sex ratios. Over a 10-year period, 37 sub-dominant adult males were translocated from Laysan Island to other locations in the species' range (Johanos et al. 2010). The sex ratio at Laysan was returned to approximately 1:1, and mortality of adult females was dramatically reduced and similar to levels observed elsewhere. The translocated males in their new region did not continue the aggressive behavior.

Attacks by single adult males have also resulted in several young monk seal mortalities. This form of single-male aggression occurs at most locations and involves behaviors that range from normal pinniped male harassment of younger animals to an aberrant level of focused aggression, especially directed toward weaned pups. This was most notable at FFS in 1997, where at least 8 pups died as a result of adult male aggression (Carretta et al. 2015), and more recently at Kure Atoll where one sub-adult male repeatedly attacked weaned pups. Many more pups have likely been killed in the same way, but the cause of their deaths could not be confirmed.

To mitigate this threat, the offending male may be removed from the population by relocation elsewhere in the species range or brought into permanent captivity, as was done in the Kure incident. There may be cases where lethal removal of an animal exhibiting aberrant behavior would be considered in order to protect multiple young seals. Chemical sterilization treatment also is an area where further research may yield additional tools to manage such a situation.

Infectious diseases

The greatest disease threats to Hawaiian monk seals are infectious pathogens that have the potential to cause mortality, spread rapidly, and for which population immunity is lacking. The lack of genetic diversity within the population further limits its flexibility in responding to a newly introduced disease. Increasing human populations in the MHI also favor growth of domestic and feral terrestrial animals that can shed pathogens into the marine ecosystem, potentially placing seals that use the MHI at greater risk than in the past.

Morbillivirus

Despite decades of surveillance, no exposure to morbilliviruses, including canine and phocine distemper and cetacean morbillivirus, has been documented in Hawaiian monk seals (Littnan et al. 2007, Aguirre et al. 2007, NMFS, unpublished data). Morbilliviruses are spread through respiratory secretions, requiring close contact, and have caused thousands of mortalities in other marine mammal species (Duignan et al. 2014, Van Bressem et al. 2014). Monk seal behavior is sufficient for virus spread between individuals. The likelihood of recovery for the relatively small Hawaiian monk seal population would be greatly diminished or eliminated in the wake of a morbillivirus outbreak of this magnitude.

Sources of morbillivirus exposure for monk seals include ocean- and land-dwelling mammals. Canine distemper is uncommon in Hawaii, but an infected dog has the potential to transmit the infection to a seal. Interactions between seals and dogs are reported several times annually and have resulted in sufficient trauma to cause mortality of one seal pup. While unlikely, the risk of exposure to morbillivirus from other marine mammal species cannot be dismissed. Asymptomatically-infected marine mammals could travel from other parts of the globe to Hawaii, including those that migrate through the region or those that occasionally swim off-course from their natural habitats.

West Nile virus

Monk seals are susceptible to West Nile virus (WNV) as demonstrated by the death of captive monk seals on the continental U.S. from the disease. Though it is not yet present in Hawaii, WNV is expected to make it to the state in the foreseeable future (Kilpatrick et al. 2004), and State programs conduct limited surveillance of imported animals. A mosquito vector is responsible for viral spread rather than seal-seal transmission. Mosquitoes are universally present in the MHI; hence spread is highly likely if the virus is introduced.

Influenza

While not detected in Hawaiian monk seals to date, influenza viruses are known to infect phocids (Osterhaus et al. 2000, Aguirre et al. 2007, Anthony et al. 2012, Greig et al. 2014, Krog et al. 2015). Several influenza virus subtypes have been hypothesized to play a role in mortality events in phocids. Infections may range from asymptomatic to chronically debilitating and diagnosis is sometimes complicated by secondary infections (Anthony et al. 2012, Goldstein et al. 2013). Given the propensity for influenza viruses to mutate and the possibility for cross-species transmission, the threats and impacts on monk seals, as well as other wildlife populations and humans, remain a topic of great scientific investigation.

Mitigation: biomedical sampling, disease surveillance, and proactive prevention

Early detection is the primary strategy for addressing emerging threats to Hawaiian monk seal health and their sequelae for humans and ecosystems. This strategy includes routine examination of all carcasses by gross necropsy and, when carcass condition permits, further examination through histopathology, immunohistochemistry, and molecular techniques. In live animals, the opportunistic collection of samples during research activities provides a mechanism for longitudinal evaluation of baseline health data as well as exposure to pathogens, biotoxins, and contaminants. Targeted surveillance for high-risk disease threats such as morbillivirus, WNV, and influenza is a key tenant of these opportunistic sampling protocols. Equally important is the archiving of samples for retrospective health studies and to investigate emerging threats to health through new assays and technologies as they are developed.

Disease outbreak preparedness and vaccination efforts for Hawaiian monk seals currently are focused on prevention of morbillivirus, though these plans can be modified to meet other vaccine-preventable disease threats, such as influenza. An epidemiological model of a morbillivirus outbreak in monk seals is underway, taking into consideration host- and virus-specific variability. The ongoing epidemiological modeling effort shows that as a sole strategy, vaccination in response to a morbillivirus outbreak is insufficient to contain an outbreak in monk seals. To be effective, vaccination efforts will require a three-pronged approach: (1) vaccination of seals handled during regular research activities, (2) preparedness for large-scale vaccination efforts in the face of an outbreak and (3) a robust quarantine program.

Entanglement

Marine debris, such as derelict fishing gear and nets, pose a serious threat to Hawaiian monk seals in the NWHI. Seals may become entangled in nets, line, and other debris, sometimes restricting mobility, foraging ability, or even leading to injury or death (Fig. 8). Hawaiian monk seals have the one of highest entanglement rates reported for pinnipeds (Henderson 2001, Boland and Donohue 2003; Fig. 9).

Data from 1982 through 1999 indicated that approximately 0.7% of monk seals in the population were entangled in marine debris each year (Henderson 2001). Updated information indicates that entanglement rates have increased. From 2000–2014, approximately 1% of the seal population became entangled in marine debris annually (Henderson in prep). This increase is despite regulations to reduce debris from the marine fishing industry (MARPOL), the establishment of the NWHI as a marine protected area, and extensive debris removal efforts. Pups (including weaned and nursing) experienced the greatest rate of entanglement (3.47%), followed by juveniles (1.46%), sub-adults (1.35%) and adults (0.74%).

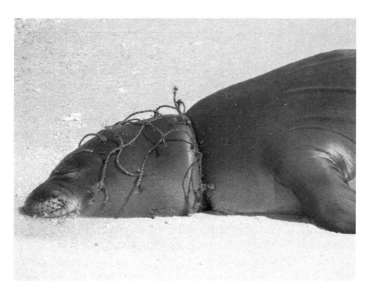

Figure 8. A young Hawaiian monk seal entangled in marine debris in the Northwestern Hawaiian Islands. Hawaiian monk seals have the highest entanglement rate for any pinniped species.

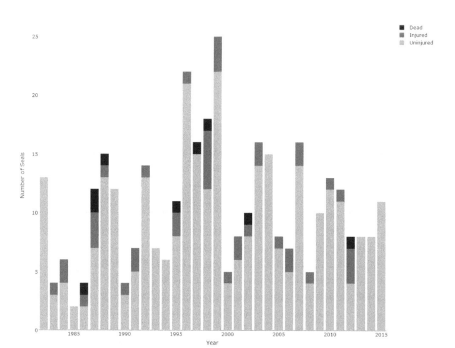

Figure 9. The record of entanglements observed in monk seals since 1982 including those animals that were injured or killed. The number of entanglements has remained relatively level despite a decline in the species population size.

The types of debris observed in monk seal entanglements provide insight into the most damaging sources of marine debris (from the seal perspective). Derelict fishing gear (nets and lines) were the most common entangling debris. Plastic cones used specifically for hagfish trapping also were common, and were usually stuck on the snouts of pups. Packing straps were another common entanglement. Despite some decrease of debris reported in Alaskan waters following regulatory actions such as MARPOL (Johnson 1994), no such improvement has been noted in the NWHI. It is unknown whether this is because dumping of fishing gear and nets persists despite regulation, debris accumulates as a result of accidental loss, or old debris simply remains in the system through continual circulation of debris in ocean currents.

Oceanographic data has demonstrated the importance of stable ocean current patterns in the persistence and concentration of marine debris (Howell et al. 2012). In the North Pacific, prevailing winds form two large oceanic gyres, each bounded by major current systems. Between the two major gyres is the North Pacific Transition Zone, formed by frontal systems to the north and south that are characterized by sharp gradients in water temperature and salinity. The North Pacific Subtropical Convergence Zone

(STCZ) is an area of debris accumulation where wind and current conditions lead to the aggregation of buoyant material (both organic and inorganic), and increase its retention at the surface.

The importance of climate-driven oceanographic patterns in the cycle of debris accumulation in the marine environment foreshadows potential impacts to entanglement threats under future climate change scenarios. The STCZ, which is regularly located between 30° N and 42° N latitude, increases spatially and temporally during El Niño years, extending as far south as 28° N, bringing the accumulation of debris closer to the NWHI. By examining 23 years of monk seal debris entanglement data in the context of sea surface temperature and wind vector data, Donohue and Foley (2007) were able to demonstrate increases in entanglement rates during periods characterized by El Niño events. This provides an example of the impact of physical oceanographic processes on entanglement risks. As climate conditions warm, more frequent El Niño conditions are expected, creating concern for increasing future entanglement rates.

Fortunately, few entanglements are fatal. However, this depends heavily on human intervention to remove entangling debris. Of the 347 observed entanglements (1982–2014), debris was removed from the seal in 277 (79%) instances, with 93 seals escaping unaided, 9 deaths, and 8 of unknown fate. It is also important to note that these figures represent only the observed entanglements, but it can be assumed that unobserved entanglements occur outside the field season. If death rates are similar, we might assume that marine debris accounts for some portion of the unobserved mortalities of monk seals. In addition to aiding entangled individuals, extensive multi-agency efforts are dedicated to eliminating debris from the environment. Approximately 511 metric tons of marine debris were removed from the NWHI between 1996 and 2006.

Direct fisheries interactions

Direct interactions between monk seals and long-line, bottom-fish, near-shore recreational and lobster fisheries have been documented. After designation of the Papahānaumokuākea Marine National Monument, all commercial fishing within the monument was phased out and permanently closed as of June 2011. Consequently, any direct interactions with commercial fisheries within the monument are now limited or nonexistent. Seals in the MHI, however, have regular interactions with recreational fishermen. Fisheries interactions occur most-often with shoreline fishermen targeting ulua (giant trevally, *Caranx ignobilis*). There were 132 incidents of monk seal hooking, and seven entanglements in active gillnets documented between 1976 and 2014 (NMFS, unpublished data). Most of the monk seals involved were 2 years old or younger and a quarter of them were hooked or entangled more than once. Hookings typically involved large circle hooks accompanied by slide-bait rigging (Fig. 10). Although direct fisheries interaction was implicated in 10 monk seal deaths and was slightly more frequent than other mortality factors in the MHI, overall, survival of monk seals with a documented fisheries interaction was similar to matched controls. Once a hooking is reported, interventions can range from manual restraint and removal of the hook in the field to bringing the seal into a facility for surgical hook removal, followed by monitoring and/or rehabilitation before the seal is returned to the wild. Additional mitigation methods also have been implemented to discourage seals from becoming accustomed to getting food from fishing vessels, docks, or off of fishing lines.

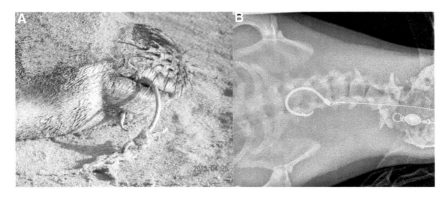

Figure 10. Interactions between monk seals and fisheries are becoming more common as the population of seals recovers in the main Hawaiian Islands. One of the most common interactions is between monk seals and the shore cast fishery that uses large circle hooks. Here are two examples of these interactions: a monk seal with a superficial hook inserted in its mouth (A) and a more serious interaction where a seal has swallowed a hook (B). These latter interactions usually require invasive surgery to remove the hook and save the seal.

Conservation in a Changing World

The current Hawaiian monk seal conservation program faces several new challenges in addition to the array of long-standing threats. Many of the new threats are a function of monk seals reestablishing themselves in the heavily human-populated MHI. In the same time frame that the seal population has begun to rebound in the MHI, the human population in the state of Hawaii has increased by nearly 40%. This puts an increased importance on coexistence in an environment of finite resources. Competition (real or perceived) between seals and human ocean users can lead to disdain of conservation regulation and conflict between people and wildlife. In the most extreme cases, negative feelings have led people to intentionally kill seals (7 documented to date). However, even well-intentioned interactions between humans and seals can lead to problems; some seals inhabiting popular swimming, snorkeling, or fishing spots have become habituated to people through socialization or food handouts. When seal-human interactions have escalated to jeopardize human safety, seals have been necessarily translocated, usually to remote locations with lower survival potential than their MHI habitat. Thus, outreach and education become crucial components of conservation management to dispel misperceptions and encourage positive coexistence.

Increasing disease risk is another unintended consequence of a growing MHI monk seal population. As seals share beaches with humans and domestic animals, they have greater chances of encountering pathogens from other species; toxoplasmosis from cats, distemper-type virus from dogs, leptospirosis from pigs and rodents are all health threats in a multi-species environment. The monk seal conservation program works to integrate public awareness along with increased capacity to care for sick animals or fight disease outbreaks.

Finally, large scale environmental changes present an ever shifting set of challenges for Hawaiian monk seal recovery. Halting oceanic and climate changes is unrealistic. Thus, it is important to assess vulnerabilities to future situations and maximize the ability to adapt conservation strategies and actions. We know that climate drivers, such as the PDO, play an important role in shaping ocean currents and forming the nutrient base upon which the monk seals' ecosystem depends. Positive PDO cycles and El Niño events have been associated with improved nutritional status and survival of young seals, but also with a greater influx of marine debris and entanglement rates in monk seals. It is uncertain how these factors will balance under future climate regimes in which more frequent El Niño conditions are expected. One of the more certain impacts of climate change will be sea level rise that, given the low profile of most monk seal haul-out beaches in the NWHI, may have a substantial impact on resting and pupping habitat available. Model outputs predict that up to 75% of terrestrial habitat could be lost at some sites by 2100 (Baker et al. 2006). Ongoing and upcoming research will further assess vulnerabilities to climate change and assess ways in which recovery actions may adapt to protect this species in a changing world.

Conclusion

The Hawaiian monk seal is still in decline with approximately 1,300 individuals left in the world. But despite this precarious state, there is reason to hope for the species' continued persistence. A proactive recovery program undertaking a broad suite of activities to help increase survival has had meaningful impact on the monk seal population. The species' rate of decline has been cut in half in recent years and up to 30% of the monk seals alive today owe their existence to human intervention (Carretta et al. 2015, Harting et al. 2014). These conservation activities coupled with the growing population of monk seal in the main Hawaiian Islands may provide the resilience necessary for the species to deal with future environmental uncertainties.

Keywords: Hawaiian monk seal, anthropogenic threats, conservation, infectious disease, Morbillivirus, West Nile virus, influenza, shark predation, fisheries interactions, Endangered Species Act, and Marine Mammal Protection Act

References

Aguirre, A.A., T.J. Keefe, J.S. Reif, L. Kashinsky, P.K. Yochem, J.T. Saliki et al. 2007. Infectious disease monitoring of the endangered Hawaiian monk seal. J. Wildl. Dis. 43(2): 229–241.

Anthony, S.J., J.A. St. Leger, K. Pugliares, H.S. Ip, J.M. Chan, Z.W. Carpenter et al. 2012. Emergence of fatal avian influenza in New England harbor seals. MBio. 3(4): e00166–12.

Antonelis, G.A., J.D. Baker and J.J. Polovina. 2003. Improved body condition of weaned Hawaiian monk seal pups associated with El Niño events: Potential benefits to an endangered species. Mar. Mamm. Sci. 19(3): 590–598.

Baker, J.D., C.L. Littnan and D.W. Johnston. 2006. Potential effects of sea level rise on terrestrial habitats of endangered and endemic epifauna of the Northwestern Hawaiian Islands. Endangered Species Research 4: 1–10.

Baker, J.D. and P.M. Thompson. 2007. Temporal and spatial variation in age-specific survival rates of a long-lived mammal, the Hawaiian monk seal. Proc. R. Soc. B 274(1608): 407–415.

Baker, J.D., J.J. Polovina and E.A. Howell. 2007. Effect of variable oceanic productivity on the survival of an upper trophic predator, the Hawaiian monk seal *Monachus schauinslandi*. Mar. Ecol. Prog. Ser. 346: 277–283.

Baker, J.D. 2008. Variation in the relationship between offspring size and survival provides insight into causes of mortality in Hawaiian monk seals. Endang. Species Res. 5: 55–64.

Baker, J.D., B.L. Becker, T.A. Wurth, T.C. Johanos, C.L. Littnan and J.R. Henderson. 2011. Translocation as a tool for conservation of the Hawaiian monk seal. Biological Conservation 144(11): 2692–2701. DOI: 10.1016/j.biocon.2011.07.030.

Baker, J.D., E.A. Howell and J.J. Polovina. 2012. Relative influence of climate variability and direct anthropogenic impact on a sub-tropical Pacific top predator, the Hawaiian monk seal. Mar. Ecol. Prog. Ser. 469: 175–189.

Baker, J.D., A.L. Harting and C.L. Littnan. 2013. A two-stage translocation strategy for improving juvenile survival of Hawaiian monk seals. Endang. Species Res. 21: 33–44.

Banish, L.D. and W.G. Gilmartin. 1992. Pathological findings in the Hawaiian monk seal. J. Wildl. Dis. 28(3): 428–434.

Boland, R.C. and M.J. Donohue. 2003. Marine debris accumulation in the nearshore marine habitat of the endangered Hawaiian monk seal, *Monachus schauinslandi* 1999–2001. Mar. Pollut. Bull. 46(11): 1385–1394.

Carretta, J.V., E.M. Oleson, D.W. Weller, A.R. Lang, K.A. Forney, J. Baker et al. 2015. U.S. Pacific Marine Mammal Stock Assessments: 2014. U.S. Department of Commerce, NOAA Technical Memorandum, NOAA - TM - NMFS - SWFSC - 549.

Craig, M.P. and T.J. Ragen. 1999. Body size, survival, and decline of juvenile Hawaiian monk seals, *Monachus schauinslandi*. Mar. Mamm. Sci. 15(3): 786–809.

Donohue, M. and D. Foley. 2007. Remote sensing reveals links among the endangered Hawaiian monk seal, marine debris, and El Nino. Mar. Mamm. Sci. 23(2): 468–473.

Duignan, P.J., M.F. Van Bressem, J.D. Baker, M. Barbieri, K.M. Colegrove, S. De Guise et al. 2014. Phocine distemper virus: Current knowledge and future directions. Viruses 6(12): 5093–5134.

Friedlander, A.M. and E.E. DeMartini. 2002. Contrasts in density, size, and biomass of reef fishes between the northwestern and the main Hawaiian Islands: The effects of fishing down apex predators. Mar. Ecol. Prog. Ser. 230: 253–264.

Gilmartin, W.G. 1983. Recovery plan for the Hawaiian monk seal, *Monachus schauinslandi*. US Department of Commerce, National Oceanic and Atmospheric Administration, National Marine Fisheries Service, Southwest Region.

Gilmartin, W.G., A.C. Sloan, A.L. Harting, T.C. Johanos, J.D. Baker, M. Breese et al. 2011. Rehabilitation and relocation of young Hawaiian monk seals (*Monachus schauinslandi*). Aquat. Mamm. 37(3): 332–341.

Gobush, K.S. and S.C. Farry. 2012. Non-lethal efforts to deter shark predation of Hawaiian monk seal pups. Aquat. Conserv. Mar. Freshw. Ecosys. 22(6): 751–761.

Goldstein, T., I. Mena, S.J. Anthony, R. Medina, P.W. Robinson, D.J. Greig et al. 2013. Pandemic H1N1 influenza isolated from free-ranging Northern Elephant Seals in 2010 off the central California coast. PLoS One 8(5): e62259.

Greig, D.J., F.M. Gulland, W.A. Smith, P.A. Conrad, C.L. Field, M. Fleetwood et al. 2014. Surveillance for zoonotic and selected pathogens in harbor seals *Phoca vitulina* from central California. Dis. Aquat. Organ. 111(2): 93–106.

Harting, A.L., T.C. Johanos and C.L. Littnan. 2014. Benefits derived from opportunistic survival-enhancing interventions for the Hawaiian monk seal: The silver BB paradigm. Endanger. Species Res. 25: 89–96.

Henderson, J.R. 2001. A pre-and post-MARPOL Annex V summary of Hawaiian monk seal entanglements and marine debris accumulation in the Northwestern Hawaiian Islands, 1982–1998. Mar. Pollut. Bull. 42(7): 584–589.

Hiruki, L.M., I. Stirling, W.G. Gilmartin, T.C. Johanos and B.L. Becker. 1993. Significance of wounding to female reproductive success in Hawaiian monk seals (*Monachus schauinslandi*) at Laysan Island. Can. J. Zool. 71(3): 469–474.

Howell, E.A., S.J. Bograd, C. Morishige, M.P. Seki and J.J. Polovina. 2012. On North Pacific circulation and associated marine debris concentration. Mar. Pollut. Bull. 65(1): 16–22.

Johnson, S.W. 1994. Deposition of trawl web on an Alaska beach after implementation of MARPOL Annex V Legislation. Mar. Pollut. Bull. 28(8): 477–481.

Johanos, T.C., B.L. Becker, J.D. Baker, T.J. Ragen, W.G. Gilmartin and T. Gerrodette. 2010. Impacts of sex ratio reduction on male aggression in the Critically Endangered Hawaiian monk seal *Monachus schauinslandi*. Endangered Species Research 11: 123–132.

Kilpatrick, A.M., Y. Gluzberg, J. Burgett and P. Daszak. 2004. Quantitative risk assessment of the pathways by which West Nile virus could reach Hawaii. EcoHealth 1(2): 205–209.

Krog, J.S., M.S. Hansen, E. Holm, C.K. Hjulsager, M. Chriél, K. Pedersen et al. 2015. Influenza A (H10N7) virus in dead harbor seals, Denmark. Emerg. Infect. Dis. 21(4): 684–687.

Littnan, C.L., B.S. Stewart, P.K. Yochem and R. Braun. 2007. Survey for selected pathogens and evaluation of disease risk factors for endangered Hawaiian monk seals in the main Hawaiian Islands. EcoHealth 3(4): 232–244.

Lowry, L.F., D.W. Laist, W.G. Gilmartin and G.A. Antonelis. 2011. Recovery of the Hawaiian monk seal (*Monachus schauinslandi*): A review of conservation efforts, 1972 to 2010, and thoughts for the future. Aquat. Mamm. 37(3): 397–419.

National Marine Fisheries Service. 2007. Recovery Plan for the Hawaiian Monk Seal (*Monachus schauinslandi*), Second Revision. National Marine Fisheries Service, Silver Spring, MD, 165 pp.

Norris, T.A., C.L. Littnan and F.M. Gulland. 2011. Evaluation of the captive care and post-release behavior and survival of seven juvenile female Hawaiian monk seals (*Monachus schauinslandi*). Aquat. Mamm. 37(3): 342–353.

Osterhaus, A.D.M.E., G.F. Rimmelzwaan, B.E.E. Martina, T.M. Bestebroer and R.A.M. Fouchier. 2000. Influenza B virus in seals. Science 288(5468): 1051–1053.
Parrish, F.A., G.J. Marshall, B. Buhleier and G.A. Antonelis. 2008. Foraging interaction between monk seals and large predatory fish in the Northwestern Hawaiian Islands. Endang. Species Res. 4(3): 299–308.
Ragen, T.J. and D.M. Lavigne. 1999. The Hawaiian monk seal: Biology of an endangered species. pp. 224–245. *In*: J.R. Twiss, Jr. and R.R. Reeves [eds.]. Conservation and Management of Marine Mammals. Smithsonian Institution Press, Washington, DC.
Schofield, T.D., G. Levine, F.M. Gulland, C.L. Littnan and C.M. Colitz. 2011. The first successful hand-rearing of a neonate Hawaiian monk seal (*Monachus schauinslandi*) and post-release management challenges. Aquat. Mamm. 37(3): 354–359.
Van Bressem, M.F., P.J. Duignan, A. Banyard, M. Barbieri, K.M. Colegrove, S. De Guise et al. 2014. Cetacean morbillivirus: Current knowledge and future directions. Viruses 6(12): 5145–5181.

6

Guadalupe Fur Seal Population Expansion and its Post-breeding Male Migration to the Gulf of Ulloa, México

David Aurioles-Gamboa, Nereyda Pablo-Rodríguez,[a]
M. Patricia Rosas-Hernández[b] and
Claudia J. Hernández-Camacho[c]*

Introduction

The Guadalupe fur seal *Arctocephalus townsendi* (GFS) is known in Latin America as the lobo fino de Guadalupe, oso marino de Guadalupe, or lobo de dos pelos de Guadalupe. Guadalupe Island (GI), off the coast of Baja California, México, is the typical locality where this species occurs; the taxon was first described by Merriam (1897) based on a specimen collected on the island's west coast. Presently, the GFS distribution is centered on

---

Laboratorio de Ecología de Pinnípedos "Burney J. Le Boeuf", Centro Interdisciplinario de Ciencias Marinas, Instituto Politécnico Nacional. Av. IPN s/n, Col. Playa Palo de Santa Rita, C.P. 23096, La Paz, Baja California Sur, México.
[a] Email: pr_syren@yahoo.com.mx
[b] Email: m.patricia.rosas@gmail.com
[c] Email: jcamacho@ipn.mx
* Corresponding author: dgamboa@ipn.mx

GI (Fleischer 1987, Gallo-Reynoso 1994, Belcher and Lee 2002, Aurioles-Gamboa et al. 2010); however, another colony has been growing rapidly on the San Benito Archipelago (SBA) since 1997 (Maravilla-Chavez and Lowry 1999, Aurioles-Gamboa et al. 2010, García-Capitanachi 2011). The SBA is a group of three islands (East, Center, and West) and several islets and was home to a large GFS rookery prior to the population's decimation by the sealing industry. From the late 1700s to 1848, thousands of animals were killed on several islands on the Pacific coasts of México and the United States (Hubbs 1979). A few remaining individuals were harvested in Mexican waters in the late 1800s (Townsend 1931), earning the species near extinct status (Hubbs 1956).

The past geographic range of the GFS extended from GI and the SBA to the islands of southern California. The major concentration of abundance was on San Miguel Island, where the GFS was also the pinniped species most frequently encountered in archaeological deposits (Walker and Craig 1979). Deposits dating from 3,500 BCE located between 32°–50° N latitude suggest that the highest density of GFSs occurred on the Channel Islands and southern parts of the mainland (34°–36° N), decreasing north of Point Conception (Rick et al. 2009).

The size of the GFS population prior to the seal exploitation period is not well known, but has been estimated at ca. 200,000 (Hubbs 1979). As a result of the intense exploitation, the species was reduced to 200–500 individuals by the 1950s (Hubbs 1956); however, by 2010 the population had recovered to ca. 20,000 (Aurioles-Gamboa et al. 2010, García-Capitanachi 2011). For most of the 20th century, the species only inhabited GI, with an annual population growth rate of 13.7% from 1955 to 1993 (Gallo-Reynoso 1994). In the 1980s, some individuals were observed frequently on San Miguel Island and other California islands, including the Farallon Islands off the coast of northern California (Bartholomew 1950, Stewart 1981, Stewart et al. 1987, 1992, Hanni et al. 1997).

In 1997, a small group of GFSs was discovered at the former rookery on the SBA, southwest of GI, near the Baja California coast. By 2010-2011, the population on the SBA had increased to ca. 3,000 individuals. However, few pups are born there, suggesting that this colony is mostly the result of immigration from GI (Aurioles-Gamboa et al. 2010). The species' distribution at sea is poorly known, but records from a few satellite-tracked adult females from GI suggest they may travel several hundred kilometers during feeding trips (Gallo-Reynoso 1994). During last two decades, GFS vagrants have been sighted and recorded stranded in California, Oregon, and Washington, USA (Hanni et al. 1997, Etnier 2002), and the Gulf of California, México (Aurioles-Gamboa et al. 1999), particularly during El Niño years. Based on a systematic effort to register all stranded marine mammals initiated in 2003, an increasing number of stranded GFSs have been recorded along a 45 km stretch of coast on the northwest margin of Magdalena Island (MI).

In this study, we analyzed data on the 2003–2015 GFS strandings along the west coast of MI to interpret the ecological significance of this phenomenon. We assessed GFS population numbers on the SBA, the sea surface temperature (SST) and chlorophyll "a" (Chl-a) surface pigment concentration anomalies around the SBA and in the center of the Gulf of Ulloa (GU) from 2003 to 2015, and the annual landings of the Magdalena Bay squid fishery from 2003 to 2014 as variables that might play a role in the increase of GFSs strandings on MI. Various squid species constitute ca. 90% of the GFS diet on the SBA (Aurioles-Gamboa and Camacho-Ríos 2007, Pablo-Rodríguez et al. 2015), including Humboldt squid (*Dosidicus gigas*) for which there is a fishery in Baja California.

We also analyzed GFS feeding habits around the SBA during 2000-2001 and 2007-2008, using scat analysis to identify possible diet shifts during the process of re-colonization. To explore GFS feeding habits further, we conducted carbon and nitrogen stable isotope analyses on vibrissae from dead animals on MI. We compared the stable isotope data with those from the sympatric California sea lion (CSL), a coastal predator. We hypothesized that the increase in the number of GFSs strandings on MI was related to the increasing GFS population on the SBA; and that as a consequence, the species has been increasingly exploiting the rich area of the GU, reflecting a shift from oceanic to coastal feeding habits. We used a generalized linear model to determine the effect of the oceanographic and biological variables mentioned on the recent pattern of GFSs stranded on MI.

Materials and Methods

Study area

Along with Santa Margarita Island, MI forms the west margin of Magdalena Bay, a large and productive fishery area on the Pacific coast of Baja California, México (Fig. 1). Santa Margarita Island harbors the only CSL rookery within 500 km, with a population size of ca. 3,000 individuals; the colony is the most likely source of the CSLs strandings on MI (Ascencio-Estrada 2010). At the southern limit of the stranding area is Cabo San Lazaro, a rocky point that is a CSL hauling out area occupied by 300–500 subadult and adult males (Fig. 1). The GU is a feeding area for CSLs and many other large vertebrates, including sharks, tunas, sea turtles, birds, dolphins, and whales (Blackburn 1969, Etnoyer et al. 2004, Wingfield et al. 2011).

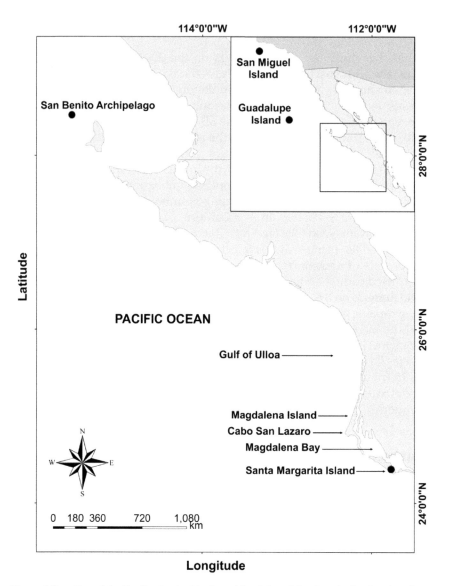

Figure 1. Location of the San Benito Archipelago, Magdalena Island, and other key locations related to the Guadalupe fur seal distribution off the coast of the Baja California Peninsula.

SST and Chl-a Data

The SST (in °C) and Chl-a (in mg*m^3) were inferred from Level 3 (L3) satellite imagery captured by the MODIS Aqua sensor (http://oceancolor.gsfc.nasa.gov/cgi/level3.pl), with a resolution of 4 x 4 km in a 100 km diameter around the SBA and the center of the GU off the Pacific coast of the Baja California Peninsula. Matlab (2013b) was used to obtain the values of the environmental parameters from hierarchical data format (HDF) files (http://hdfgroup.org).

We assessed inter-annual anomalies in SST and Chl-a from 2002 to 2014 in the two areas of interest to identify significant changes that might affect the GFS abundance in both areas. The anomalies for each environmental parameter were determined based on the difference between the value assigned to a given satellite image and the mean value of all images. Increased SST and low Chl-a are usually associated and may force animals to move far from their breeding grounds, reducing their total population number as observed onland surveys during warm years (Trillmich and Ono 1991, Aurioles and Le Boeuf 1991, Gallo-Reynoso 1994). Weak and emaciated pinnipeds more often strand during warming events (Greig et al. 2005). The mean SST and Chl-a anomaly values from 2002 to 2015 were obtained for the SBA and GU areas as variables associated with the numbers of live GFSs on SBA and stranded GFSs on MI.

GFS Strandings on MI

To the northwest, MI includes a sandy beach ca. 53 km in length of which 45 km were surveyed to register all marine mammals stranded from July 2003 to July 2015. The monthly and annual observation effort was quantified in days with two complete surveys along the beach each day (92 km). In order to apply a generalized linear model, we divided the total number of GFSs stranded during the study period by the observation effort (days per year) and then multiplied by 365 days to obtain an estimate of the number of GFSs stranded that year. All marine mammals were photographed, sexed, and measured for total length when possible. Stranded GFSs were identified in the field from body observation; when the animal was too decomposed for field determination, taxonomic status was confirmed via skull examination in the laboratory. We recorded the date and general condition of each GFS and collected fur, vibrissa, and skull samples in most cases. A total of 70 GFSs were recorded from July 2003 to July 2015, from which we recovered 59 skulls to date that allowed a more comprehensive definition of the sex and age structure of the GFSs stranded on MI.

Squid Fishery Data

Various squid species, including Humboldt squid (*Dosidicus gigas*), constitute ca. 90% of the GFS diet on the SBA (Aurioles-Gamboa and Camacho-Ríos

2007, Pablo-Rodríguez et al. 2016). A Humboldt squid fishery operates off the west coast of Baja California.

Since squid of different species are the main prey of GFSs off Baja California (Aurioles-Gamboa and Camacho-Ríos 2007, Pablo-Rodríguez et al. 2016), squid abundance may also influence the presence and abundance of GFSs in a given area. Humboldt squid are landed monthly by the Magdalena Bay fishery. We analyzed the *D. gigas* landings trend as a proxy of prey abundance in the GU and determined its relationship with GFS strandings on MI. Data on *D. gigas* landings were available from the office of the Comisión Nacional de Acuacultura y Pesca (SAGARPA, National Commission on Aquaculture and Fishing) at Puerto San Carlos, where all squid caught in Magdalena Bay are reported (http://www.conapesca.gob.mx/wb/cona/estadisticas_de_produccion_pesquerasitio). The squid landings include data on the monthly and annual catches of all squid species; the vast majority of squid are *D. gigas*.

GFS Population on the SBA

GFS population abundance data are available only for some years (1997, 2000, 2006–2008, 2010, and 2013-2014) (Maravilla-Chavez and Lowry 1999, Aurioles-Gamboa et al. 2010, García-Capitanachi 2011, Sierra-Rodríguez 2015). To make this database comparable to the annual GFS stranding data, we estimated the missing abundance data by fitting an exponential model ($r^2 = 0.9$) to the available data in order to describe population growth on the SBA. Currently, this colony is increasing at a rate of 15% (intrinsic rate of natural increase); slower than the rate of $r = 18\%$ reported in a previous study (Aurioles-Gamboa et al. 2010).

GFS Feeding Habits on the SBA

The GFS rookery closest to MI is located on the SBA (Fig. 1), where two feeding studies were conducted in 2001-2002 and 2007. We assessed and compared GFS feeding habits between the two periods to identify any possible shift in diet. The GFS diet was determined from scat samples collected over 15 d during the summer and winter of each year. Scat samples were selected from areas of the rookery where adult females were most abundant. By using fresh scat from the same time period, we eliminate the possible confounding effects of changes in this species' feeding habits throughout the year. The scat samples were stored in labeled plastic bags in the field and transferred to plastic containers with water and detergent for a period of 12 to 48 h upon returning to the laboratory to facilitate sieving. Subsequently, the samples were sorted using a sediment sieve set with varying mesh sizes (2.0, 1.19, and 0.71 µm) to separate fish otoliths, cephalopod beaks, and crustacean remains. A stereo microscope was used to identify fish otoliths through comparison with the Centro Interdisciplinario de Ciencias Marinas del Instituto Politécnico Nacional (CICIMAR-IPN; Interdisciplinary Center for Marine Science of the National Polytechnic Institute) Pinniped Ecology Laboratory photographic database, whereas identification keys were used to identify cephalopod beaks (Iverson and Pinkas 1971, Wolf 1982, 1984).

Data matrices were developed for each sample in order to estimate the frequency and relative abundance of each prey species using the Weighted Importance Index (IIMPi) (García-Rodríguez and Aurioles-Gamboa 2004). This measure is based on the occurrence of different prey species in each sampling unit (scat) and estimates the probability of finding a given prey species in a scat sample. The IIMPi values were analyzed using cluster analysis and principal component analysis (PCA) to identify different diets in the feeding spectrum of each species during both seasons.

The minimum sample size required to ensure the representativeness of the potential spectrum of prey consumed each season by GFSs was calculated following the technique used in previous studies (Ferry and Cailliet 1996, Ferry et al. 1997, Porras-Peters et al. 2008, Páez-Rosas and Aurioles-Gamboa 2010). The method calculates the accumulated average and standard deviation of a group of diversity curves generated from prey abundance data using the Shannon-Wiener Index (H') (Krebs 1999). These diversity curves are derived from a function in the Matlab routine that computes 500 permutations at random with all the original data and a 0.05 confidence interval (Matlab 2013).

To determine whether the species can be considered a specialist predator during either season, we calculated the "Bj", or Levin's index (Krebs 1999), using Levin's standardized formula. When values are < 0.6, a species is considered a specialist predator, while values close to 1 (> 0.6) indicate a generalist diet (Krebs 1999).

The species' trophic level (TL) is defined as the number of times energy or matter is transformed through a consumer's diet into other biomass as it moves along the predator's food chain (Williams and Martínez 2004). We used Christensen and Pauly's (1992) algorithm to estimate the TL based on the results of the scat analysis for each season.

Vibrissae Stable Isotope Analysis

One complete vibrissa was plucked from each of four GFSs and four CSLs; vibrissae were stored in paper envelopes prior to processing for carbon and nitrogen stable isotope analysis. The GFS vibrissae were collected from stranded animals in 2012, while the CSL vibrissae were sampled from adult females that were captured and released on Santa Margarita Island (Fig. 1) also in 2012. The vibrissae were rinsed once with a 2:1 chloroform:methanol solution to remove surface contaminants. Cleaned vibrissae were cut into segments to a weight of ca. 1.0 ± 0.2 mg using nail clippers to later be sealed into tin boats for isotope analysis. Carbon (δ^{13}C) and nitrogen (δ^{15}N) isotope values were determined using a PDZ Europa ANCA-GSL elemental analyzer interfaced to a PDZ Europa 20-20 isotope ratio mass spectrometer (Sercon Ltd., Cheshire, UK) at the Stable Isotope Facility at the University of California, Davis. The analytical error was 0.2‰ for ^{13}C and 0.3‰ for ^{15}N. Isotopic composition is expressed in the δ notation, as the deviation from standards in parts per thousand (‰) based on the following equation: δ^{15}N or δ^{13}C = [($R_{sample}/R_{standard}$) – 1] × 1000, where R is the ratio of ^{15}N/^{14}N or ^{13}C/^{12}C for the sample and standard, respectively. The standards were V-PDB (Vienna Pee Dee Belemnite) for carbon and air for nitrogen.

The GFL and CSL isotopic niches were estimated by calculating the standard ellipse area (SEAc) using the Bayesian approach Stable Isotope Bayesian Ellipses in R (SIBER; Jackson et al. 2011). Unlike others methods for estimating niche width (e.g., convex hulls; see Layman et al. 2007), SEAc calculations are less susceptible to outlying data points. In our analysis, we calculated isotopic niches for each species using isotope data for the most recently grown 20 vibrissae segments from each individual. A total of 155 segments were analyzed, 76 from GFSs and 79 from CSLs.

Data Analysis

We used a Kruskal-Wallis test to identify inter-annual differences in the Chl-a and SST anomaly values for the GU and around the SBA. Observation effort (days) and number of GFS stranded per semester and years were checked for departures from normality using the Shapiro-Wilks Normality test, while an F test was used to assess equality of variances. We then compared the observation effort and the number of GFSs stranded on MI during each half of the year, and between years from July to June 2003–2009 and 2010–2015. When normality and equality of variance assumptions were not met, we used a non-parametric test (Wilcoxon test).

We used a generalized linear model to evaluate how the following variables influenced the number of GFSs stranded each year: (1) survey year, (2) GFS abundance on SBA, (3) SST anomalies, (4) Chl-a anomalies around the SBA and the central area of the GU, and (5) annual landings of the Magdalena Bay squid fishery. All databases include data from 2003 (when we started recording strandings) to 2014; except annual landings of the squid fishery, which only includes data since 2005.

Before constructing the global model and the different candidate models, we tested for collinearity using Spearman's rank correlation coefficient. Since our response variable involves count data, we used the Poisson distribution (log link function). We used the explained residual deviance as a goodness-of-fit measure for the global models; models whose explanatory variables explained at least 70% of the variation in the number of stranded GFSs were considered a good fit (Zuur et al. 2009). Moreover, we estimated the over-dispersion ($\hat{c}$) for the global model from the goodness-of-fit chi-square statistic (χ^2) of the global model and its degrees of freedom (Burnham and Anderson 2002).

We then selected the most parsimonious models using the small-sample Akaike information criterion (AICc) (Burnham and Anderson 2002). Differences in AICc (Δ AICc) and Akaike weights (ω; relative likelihood of the model, given the data) were used to determine the level of support for each model. We considered models with Δ AICc < 2 and the highest ω to have substantial support. Models with Δ AICc = 4–7 were considered to have less support but still explain some variation in the data.

RESULTS

Chl-a Anomaly Trends

Figure 2 shows the Chl-a anomaly trends around the SBA (A) and in the GU (B), respectively. A Kruskal-Wallis test showed significant inter-annual differences in Chl-a and SST anomalies in the GU and around the SBA (H [12, N = 556] = 68.05, p < 0.000; H [12, N = 556] = 70.02, p < 0.000). For the GU and the SBA, the year 2014 was significantly different from 2001, 2003, 2004, 2006, 2007, 2008, and 2010.

SST Anomaly Trends

Figure 3 shows the SST anomaly trends around the SBA (C) and in the GU (D). Significant differences were found in SST anomalies in the GU between 2002 and 2011 (H (12, N = 568) = 35.238, p < 0.000), and around the SBA between 2002 and 2011, as well as between 2014 and the years 2004, 2005, 2007, 2009, and 2010 (H [12, N = 45.581] = 70.02, p < 0.000).

Monthly GFS strandings and squid landing trends

Trends in monthly observation effort (days) and the number of GFSs stranded on MI are shown in Fig. 4A. There is an apparent increase in the number of stranded GFSs from January to August, decreasing slightly toward December. The proportion of GFSs strandings was significantly

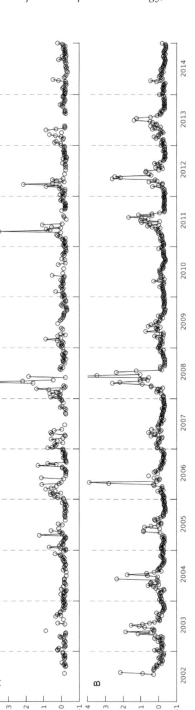

Figure 2. Inter-annual trends (2002–2015) in chlorophyll "a" concentration anomalies around the San Benito Archipelago (A) and in the center of the Gulf of Ulloa (B).

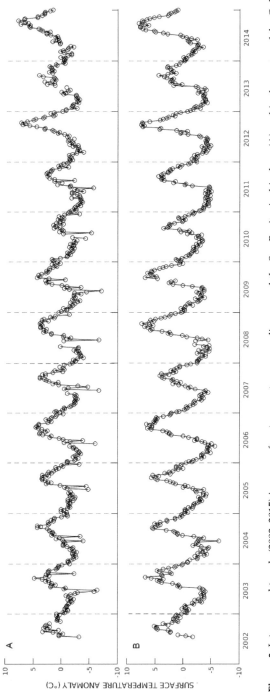

Figure 3. Inter-annual trends (2002–2015) in sea surface temperature anomalies around the San Benito Archipelago (A) and in the center of the Gulf of Ulloa (B).

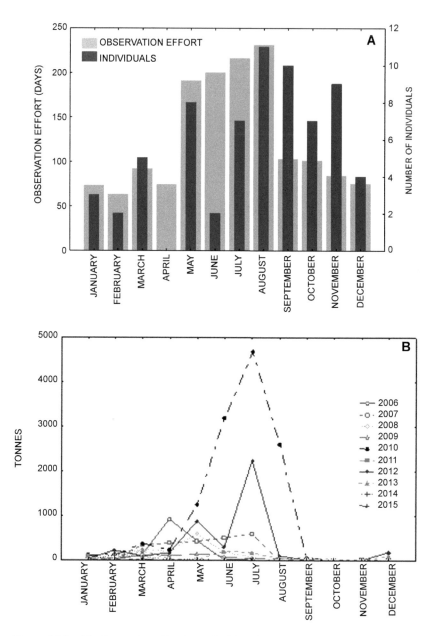

Figure 4. Monthly observation effort (gray shade) and number of Guadalupe fur seals (black shade) during 2003-2015 (A). Monthly fishery landings of Humboldt squid in Magdalena Bay (B).

higher in the second half of the year (July–December) ($t_{0.05}$ = –3.03, gl = 10, p = 0.0128), regardless of the reduced observation effort from September through December (W = 31, p = 0.2403).

Figure 4B shows the monthly trend in squid catches off MI. Squid catches are relatively low from January to April and then increase from May to September with the two largest records being reported during the month of July. Catches decrease markedly from October to December each year. The presence of stranded GFSs on MI was higher from August to September (Fig. 4B) when squid were less abundance around Magdalena Bay, as reflected by the fishery landings.

Annual GFS strandings and squid landing trends

The first cases of a GFS stranded on MI occurred in 2004 and then again in 2006 after which stranded GFSs have been recorded every year (Fig. 5A). The number of stranded GFSs increased steadily from 2008 to 2011, decreased for the period 2012-2013, and increased again for 2014 and 2015. The number of stranded GFSs increased significantly from 2010–2015 compared to 2003–2009 (W = 24, p = 0.0140). In contrast, the difference in observation effort was highly significant from 2003 (includes only six months) to 2010 (W = 63, p = 0.0012) and then decreased markedly in subsequent years except for 2015. A contrast was observed in 2014, when a high number of GFSs stranded but observation effort was low (Fig. 5A).

For comparison, the population trend of GFSs on the SBA is presented in Fig. 5B. Although the data are limited, the trend of population increase is evident, reaching 2,500 individuals by 2010 and a peak of 4,700 individuals in 2012. The annual trend of squid landings is also presented in Fig. 5B, showing considerable variability with the maximum squid catch occurring in 2010. The considerable increase in GFSs stranded in 2014 coincided with a low annual catch of squid that year.

Sex and age composition of stranded GFSs

A total of 59 skulls were recovered from GFSs stranded on MI from 2003 to 2015 (Fig. 6); of these, 46 were males (three juveniles, eight sub-adults, and 35 adults) and 13 were females (four juveniles and nine adults). The overall male:female ratio of the GFSs stranded on MI was 3.5:1, whereas the adult male:female ratio (operational sex ratio) was 3.8:1. No GFS pups have been

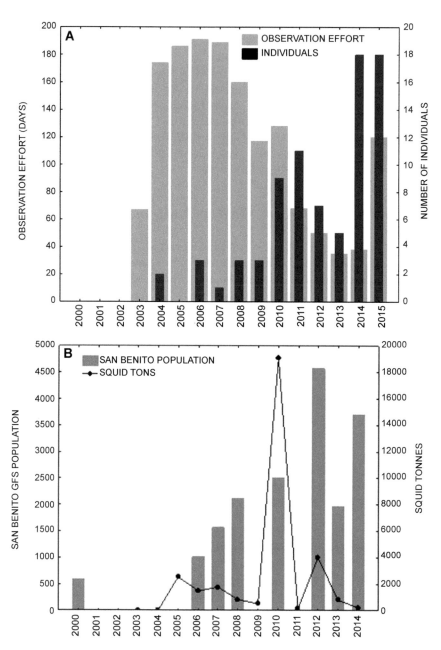

Figure 5. Annual trend in observation (gray shade) and number of Guadalupe fur seal sstranidngs (black shade) on Magdalena Island during 2003-2015 (A). Population size on the San Benito Archipelago and the annual trend in Humboldt squid landings in Magdalena Bay (B).

recorded on MI. Figure 6A shows the female (juvenile and adults) temporal stranding trend on MI which has not an apparent pattern; in contrast, Fig. 6B shows a subadult and adult male stranding trend for the second half of the year, with the highest number just after the breeding season and a steady decrease throughout January.

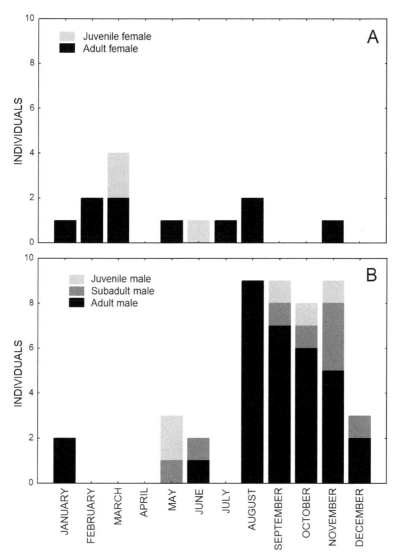

Figure 6. Sex-and-age categories of Guadalupe Fur seals stranded on Magdalena Island along the year based on recovered skulls (combined data of 12 years).

GFS diet on the SBA

The GFS diet on the SBA in 2001-2002 was composed principally of cephalopods dominated by *L. opalescens* followed by *D. gigas* and *Gonatus* sp. In Fig. 7, the primary GFS prey species are shown for the two periods of study. The importance of *D. gigas* increased markedly during summer 2002, when the species was almost as important as *L. opalescens*. The GFS diet diversity during 2001 and 2002 was relatively low (Aurioles-Gamboa and Camacho-Ríos 2007).

In the summer of 2007, the GFS diet included 19 species belonging to 14 families, of which 10 species were cephalopods and nine fish. In winter 2008, GFSs consumed 13 species belonging to 12 families, including seven cephalopod species and six fish species (Pablo-Rodríguez et al. 2015). In order to describe the GFS diet during the two seasons, those authors considered only the prey species that exceeded 3% of the IIMPi; based on this criterion, the GFS summer diet consisted mainly of cephalopod species, with the most important being *Loligo opalescens* (48.6%), *Dosidicus gigas* (16.1%), *Gonatus* spp. (9.0%), *Octopus* spp. (2.3%), and *Abraliopsis felis* (0.7%). The fish *Argentina sialis* was also an important prey (16.5%) during summer 2007.

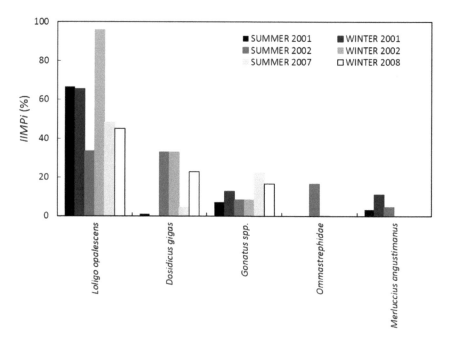

Figure 7. Principal prey identified in Guadalupe fur seal scat samples collected on the SBA in 2001-2002 and 2007-2008.

During winter, the diet consisted primarily of squid, including *L. opalescens* (53.6%), *D. gigas* (23.4%), and *Gonatus* spp. (18.3%), as well as the fish *Serranus aequidens* (1%).

During both periods of study (2001-2002 and 2007-2008), plant material identified as the seagrass *Phyllospadix* sp. was found in 40% and 83.3% of the scat samples, respectively (Aurioles-Gamboa and Camacho-Ríos 2007, Pablo-Rodríguez et al. 2015).

Based on the total prey found in the diet in 2001-2002, the GFS TL was 4.2 (Aurioles-Gamboa and Camacho-Ríos 2007), whereas the TL estimated for 2007 was 4.1 for both seasons (Pablo-Rodríguez et al. 2015). The trophic breadth, as calculated using Levin's index, indicated a specialized diet for both 2001-2002 and 2007-2008. In 2001-2002 and 2007-2008, GFS feeding habits involved consuming a similar prey composition during both summer and winter. The most important prey species during these studies were *L. opalescens* and *D. gigas* (Aurioles-Gamboa and Camacho-Ríos 2007, Pablo-Rodríguez et al. 2015).

Carbon and nitrogen isotope trends along GFS and CSL vibrissae

Figure 8A shows the $\delta^{13}C$ trends along the vibrissae of four GFSs stranded on MI. The $\delta^{13}C$ trends were highly variable in three individuals (GFS-2, GFS-3, and GFS-4), with $\delta^{13}C$ ranges of 0.98‰, 1.11‰, and 1.46‰, while individual GFS-1 showed a range of 0.59‰. Most of the $\delta^{13}C$ data from the four vibrissae fell between −16.0‰ and −15.0‰.

For comparison, the $\delta^{13}C$ trends for the four CSLs from Santa Margarita Island are shown in Fig. 8B. The ranges in $\delta^{13}C$ values were less variable (0.28‰, 0.59‰, 0.82‰, and 0.93‰) than the GFS $\delta^{13}C$ values. As in the case of GFS $\delta^{13}C$ values, most of the data fell between −16.0‰ and −15.0‰ except for some segments of the CSL-3 vibrissa, which showed a series of values indicative of a more pelagic habitat.

The $\delta^{15}N$ trends in GFS vibrissae are shown in Fig. 9A. Again, the variation in $\delta^{15}N$ values along the GFS vibrissae was high, with ranges of 1.79‰, 2.02‰, 2.12‰, and 3.48‰, but most of the $\delta^{15}N$ values fell between 17.0 and 20.0, which is representative of a single trophic level.

The $\delta^{15}N$ trends along vibrissae were similarly less variable for CSLs than GFSs (Fig. 9B) with $\delta^{15}N$ ranges of 0.75‰, 0.95‰, 1.28‰, and 1.35‰. GFSs $\delta^{15}N$ values fell within the limits of a single trophic level, but were more concentrated near the upper limit.

An overall representation of the isotope space occupied by each species is shown as the standard ellipse areas in Fig. 10. The isotope niche space is wider for GFSs than CSLs, mainly due to the $\delta^{15}N$ gradient, and the

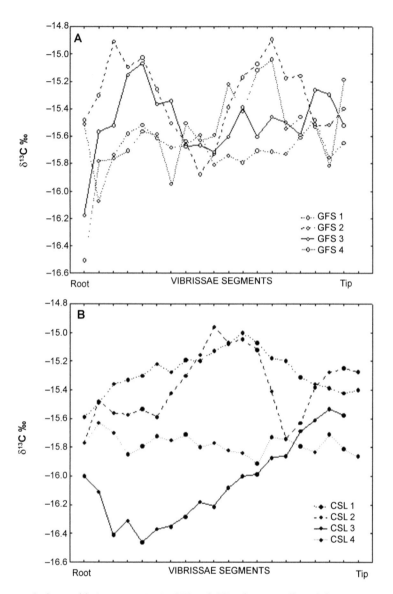

Figure 8. Carbon stable isotope ratios in GFS and CSL vibrissae collected during 2012.

difference in the isotope width between the two species was significant (W = 5762.5, p < 2.2e^{-16}). The δ^{13}C isotope space for both species was similar, suggesting the shared use of the habitat off the Baja California coast.

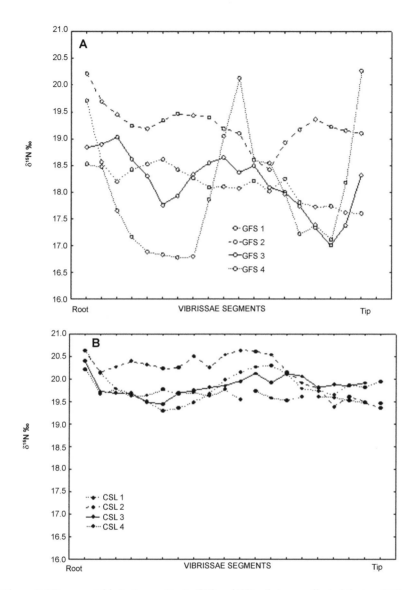

Figure 9. Nitrogen stable isotope ratios in GFS and CSL vibrissae collected during 2012.

Generalized linear model

Based on the results of the collinearity test, only two pairs of variables were correlated: "year" and "GFS abundance on the SBA," and "SST anomalies in the central area of the GU" and "SST anomalies around the

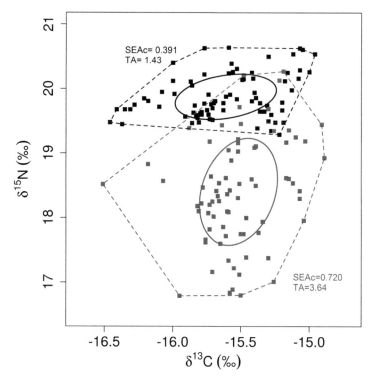

Figure 10. Convexhull and standard ellipse areas of stable isotope values from vibrissae of GFSs (SEAc=0.720, TA=3.64 and CSLs (SEAc = 0.391, TA=1.43) from the Gulf of Ulloa.

SBA" (Appendix 1). Although we predicted that GFS strandings would vary annually, we eliminated "year" from the posterior analysis because it was correlated with "GFS abundance on the SBA" and thus its interpretation would be ambiguous as unknown factors (e.g., environmental variables) are implicit on it.

We constructed two global models; the first global model included all variables except "SST anomalies in the central area of the GU" and the second global model included all variables except "SST anomalies around the SBA". Only the second global model fit the data (90% of the variation on the response variable was explained by this model) and there was no evidence of over-dispersion ($ĉ = 1.1$); as such, we did not adjust the AIC_c values. The first model was omitted from further consideration. Thus, we obtained a set of 31 candidate models with various combinations of variables; we had no reason to believe there were interactions between the variables.

The global model was the most parsimonious and accounted for 99% of the model weight (Table 1); the rest of the models were essentially unsupported. The relative importance of each variable was similar (ca. 1) and all variables were significant (Table 2). All variables positively influenced the annual number of GFS strandings except "surface pigment Chl-a concentration anomalies" around the SBA.

DISCUSSION

Chl-a and SST Anomaly Trends

Drastic declines in marine productivity are usually associated with increases in SST; events like the El Niño Southern Oscillation may trigger mass movements of marine mammals to cooler and more productive areas (Barber and Chavez 1983, Trillmich and Ono 1991). During the period of this study, shifts in marine productivity and increases in SST were relatively minor, except for 2015 which coincides with an anomalous dispersal of GFSs to northern locations, including California, Oregon, Washington (USA), and British Columbia (Canada) (http://www.nmfs.noaa.gov/pr/health/mmume/guadalupefurseals2015.html).

Most of the strandings reported in California during 2015 include GFS pups and juveniles. In contrast, of 18 individuals stranded on MI in 2015, only three were juveniles and no stranded pups have been recorded to date.

Monthly GFS strandings and squid landing trends

GFS strandings increase from January to August-September independent of the observation effort. After this peak, strandings decrease on MI but still contribute to there being a significantly higher occurrence of strandings during the second half of the year (Fig. 5A). The higher number of GFS strandings from July to December has a weak relationship with the abundance of squid in the GU, although this variable may have some effect when combined with other parameters. The monthly stranding trend of GFS on MI is more closely related to the species' breeding cycle. The GFS breeds from June to August after which time males abandon the breeding ground (Pierson 1987, Gallo-Reynoso 1994), dispersing or migrating in as yet unknown directions. The increase in GFS stranding son MI immediately after the breeding season corresponds well with when males are moving out of the rookery.

Table 1. Summary of the small-samples Akaike information criterion (AIC$_c$) for the 31 candidate models for Guadalupe fur seals stranded on Magdalena Island, Baja California Sur, México.

Model	AIC	K	AICC	Δ AICc	ω
N_{SBA} + Squid + Chl-a$_{GU}$ + Chl-a$_{SBA}$ + SST$_{GU}$	106.86	7	134.86	0.000	0.99
Squid + Chl-a$_{GU}$ + Chl-a$_{SBA}$ + SST$_{GU}$	147.82	6	164.62	29.76	3.45E-07
N_{SBA} + Squid + Chl-a$_{SBA}$ + SST$_{GU}$	180.81	6	197.61	62.75	6.86E-08
N_{SBA} + Squid + Chl-a$_{GU}$ + SST$_{GU}$	191.68	6	208.48	73.62	2.99E-10
N_{SBA} + Chl-a$_{GU}$ + Chl-a$_{SBA}$ + SST$_{GU}$	192.54	6	209.34	74.48	1.95E-10
N_{SBA} + Chl-a$_{SBA}$ + SST$_{GU}$	208.57	5	218.57	83.71	1.93E-12
Squid + Chl-a$_{SBA}$ + SST$_{GU}$	219.44	5	229.44	94.58	8.40E-15
N_{SBA} + Squid + SST$_{GU}$	236.76	5	246.76	111.90	1.46E-18
N_{SBA} + Chl-a$_{GU}$ + SST$_{GU}$	244.19	5	254.19	119.33	3.55E-20
N_{SBA} + SST$_{GU}$	258.44	4	264.15	129.29	2.43E-22
N_{SBA} + Squid + Chl-a$_{SBA}$	261.77	5	271.77	136.91	5.40E-24
N_{SBA} + Squid + Chl-a$_{GU}$ + Chl-a$_{SBA}$	256.18	6	272.98	138.12	2.95E-24
N_{SBA} + Chl-a$_{GU}$ + Chl-a$_{SBA}$	281.74	5	291.74	156.88	2.49E-28
Chl-a$_{GU}$ + Chl-a$_{SBA}$ + SST$_{GU}$	284.23	5	294.23	159.37	7.17E-29
Chl-a$_{SBA}$ + SST$_{GU}$	290.45	4	296.16	161.30	2.73E-29
N_{SBA} + Chl-a$_{SBA}$	290.74	4	296.45	161.59	2.36E-29
N_{SBA} + Squid + Chl-a$_{GU}$	307.96	5	317.96	183.10	5.04E-34
N_{SBA} + Squid	316.03	4	321.74	186.88	7.60E-35
Squid + Chl-a$_{GU}$ + SST$_{GU}$	343.12	5	353.12	218.26	1.17E-41
N_{SBA} + Chl-a$_{GU}$	359.48	4	365.19	230.33	2.79E-44
Squid + Chl-a$_{GU}$ + Chl-a$_{SBA}$	360.68	5	370.68	235.82	1.79E-45
N_{SBA}	370.02	3	373.02	238.16	5.58E-46
Squid + SST$_{GU}$	372.78	4	378.49	243.63	3.61E-47
Squid + Chl-a$_{SBA}$	378.04	4	383.75	248.89	2.60E-48
Chl-a$_{GU}$ + Chl-a$_{SBA}$	427.12	4	432.83	297.97	5.73E-59
SST$_{GU}$	442.13	3	445.13	310.27	1.22E-61
Chl-a$_{GU}$ + SST$_{GU}$	443.95	4	449.66	314.80	1.27E-62
Chl-a$_{SBA}$	473.52	3	476.52	341.66	1.87E-68
Squid + Chl-a$_{GU}$	537.78	4	543.49	408.63	5.35E-83
Squid	566.4	3	569.40	434.54	1.27E-88
Chl-a$_{GU}$	643.12	3	646.12	511.26	2.78E-105

Table 2. Summary of parameters and relative variable importance of the best model for the Guadalupe fur seals stranded on Magdalena Island.

Variable	Relative variable importance	Parameter estimate	S.E.	p-value
Intercept		–1.579	4.795e–01	0.000
N_{SBA}	0.999	4.224e–04	6.463e–05	0.000
Squid	1.000	2.413e–04	3.170e–05	0.000
$Chl\text{-}a_{GU}$	1.000	4.005	5.478e–01	0.000
$Chl\text{-}a_{SBA}$	1.000	–1.140e+01	1.383	0.000
SST_{GU}	1.000	9.841e–01	1.007e–01	0.000

N_{SBA} = GFS abundance on the SBA; Squid = tons/year; $Chl\text{-}a_{GU}$ = chlorophyll concentration anomalies in the GU; $Chl\text{-}a_{SBA}$ = chlorophyll concentration anomalies around the SBA; SST_{GU} = sea surface temperature anomalies in the GU; SST_{SBA} = sea surface temperature anomalies around the SBA.

Annual GFS stranding and squid landing trends

The inter-annual trend in GFS strandings on MI is also increasing, with a temporal decrease during 2012 and 2013. These two years are part of a period with higher SST in the region around SBA and the GU that also extended into 2014 and 2015. However more GFSs stranded on MI during this warming period. The observation effort did not markedly affect the stranding trend since 2013 and 2014 had very low effort but a considerable number of strandings. This inconsistency between effort and strandings suggests that more GFSs are stranding on MI but have not been recorded.

There is no clear relationship between the inter-annual abundance of squid and the presence of GFSs in the area or the number of GFSs strandings on MI. This lack of relationship may be explained by the synchrony between the abundance of squid in the GU and the breeding period, which both occur between June and August (Fig. 5B). Although squid were quite abundant in the GU during 2010 concomitant with a high number of GFSs strandings on MI, squid abundance greatly decreased after 2010 when the number of GFSs stranded on MI increased markedly. The maximum number of stranded GFSs in 2014 and 2015 occurred when the squid abundance was lowest.

We observed a relationship between the number of stranded GFSs and the size of the GFS population on the SBA. Although the effort to census the SBA during the breeding season is very limited; an increase in the population is evident from 2006 to 2014. Similarly, the number of GFS stranded on MI also increases, particularly after 2009. The relationship between both variables is not entirely consistent; in 2012, the maximum number of GFSs on the SBA does not coincide with the maximum number of stranded GFSs. The accuracy of the number of animals counted in 2012 is uncertain as most of the animals (78.35%) (García-Capitanachi 2011) were categorized as unidentified; however, the GFS population on the SBA seemed abundant that year.

Sex and age composition of stranded GFS

Most of the GFSs stranded on MI between 2003 and 2015 were males (78%) mostly adult males (60%) with adult females making up only 22% of the GFSs stranded on MI. This biased sex-ratio in the GFS stranding, along with its temporal concentration for the second half of the year on MI, reflects a post-breeding male migration into the GU.

Among otariids, the migratory pattern of subadult and adult males differs from that of adult females, which tend to stay in the breeding area or move in different directions than from males (Fry 1939, Mate 1975, Aurioles-Gamboa et al. 1983, Burton and Koch 1999, Giardino et al. 2016).

GFS diet on the SBA

In 2001-2002 and 2007, the GFS diet on the SBA showed no change in prey composition between different years or seasons. The most important prey during both periods was *L. opalescens*, followed by *D. gigas*. Esperón-Rodríguez and Gallo-Reynoso (2013) reported these species as the principal prey for juveniles and subadult males during summer. On GI, Gallo et al. (2008) reported different species of squid but included *L. opalescens* and *D. gigas* in the GFS diet. *L. opalescens* and *D. gigas* are part of the squid fauna of the California Current System (CCS) including the SBA and the continental shelf of Baja California and the oceanic habitat of GI (Young 1972, Okutani 1980, Allen et al. 1995, Roper et al. 1995).

The TL values for GFSs remained similar during both periods and their prey composition reflect a specialist diet focused on cephalopods, particularly squid with GFSs showing a pelagic feeding strategy (Gallo-Reynoso 1994, Aurioles-Gamboa and Camacho-Ríos 2007, Gallo-Reynoso et al. 2008).

Based on the scat analysis, the feeding habits of GFSs around the SBA indicate little change between 2001-2002 and 2007-2008; however, these results should be interpreted as representing animals that feed around the islands and return from local trips to defecate. Another consideration is that the scat samples were collected on the SBA prior to the marked increase in GFS strandings on MI after 2010.

Carbon and nitrogen isotope trends along GFS and CSL vibrissae

Carbon isotope information from GFSs stranded in 2012 show that these animals (males) are feeding in the same habitat as CSL females. The $\delta^{13}C$ values indicate high variation, but within 1‰. Some individuals showed a $\delta^{13}C$ suggesting a cycle that could be interpreted as diet change or habitat shift. In contrast, the CSL $\delta^{13}C$ values tend to be more stable along the entire vibrissa. CSL females tend to remain around the breeding area when they are nursing (Aurioles-Gamboa and Zavala-González 1994; Hernández-Camacho et al. 2008), while subadult and adult males migrate away after the breeding season (Odell 1975; Aurioles-Gamboa et al. 1983). GFS territorial males and subadult males also depart from the rookeries immediately after the breeding season (Pierson 1987, Gallo-Reynoso 1994), possibly changing feeding habitats during the annual cycle, which would explain the $\delta^{13}C$ pattern in their vibrissae.

The $\delta^{15}N$ patterns in GFS vibrissae were more variable than those of CSL females. Three GFSs showed an apparent cycle of nitrogen isotope values, while another had a slightly decreasing trend. However, most of the $\delta^{15}N$ data from these individuals fell within 3‰, which is indicative of feeding on prey within a single trophic level (Fig. 6). In contrast, the CSL female $\delta^{15}N$ values showed less variation along each vibrissa, suggesting consumption of prey from similar trophic levels. In general, CSL isotope values indicate the species is a coastal feeder year round, whereas GFS males use the coastal habitat during just part of the year. This interpretation is consistent with the stranding trend in which the period from July to December suggests a higher use of the coastal habitat.

Generalized linear model

Our analysis indicated that the model that best explained the variability in annual GFS strandings was the global model (model selection uncertainty was minimal). It is not advisable to base predictions on a single model unless $\omega i > 0.90$ (Burnham and Anderson 2002); in this case, $\omega 1 = 0.99$. Moreover, the evidence ratio ($\omega 1 / \omega 2 = 2.8e + 08$) of the best vs. second best model supports the interpretation that the global model is the best model.

Contrary to our predictions, no single variable was more important than any other. All variables influenced annual GFS strandings on MI. Although our results were conclusive, they should be interpreted with caution considering that our sample only included 12 years of data and high uncertainty was associated with the "GFS abundance on the SBA" data. We attribute the considerable variation in the GFS_{SBA} abundance data to two factors: (1) counts were made by different observers, and (2) the total number of individuals may have been overestimated in some years, as suggested by the large number of unidentified animals in those years relative to the number of animals in any other age- or sex-class. When we record such large numbers of unidentified individuals, it typically means that the animals were rushing into the water while the survey was being conducted; as a result, observers count more individuals than are actually present.

In every modeling situation, the usefulness of a model depends on the quality of the data used to generate it; thus, we strongly recommend developing future analyses of this phenomenon when long-term data are available.

Post-breeding GFS male migration

Our study shows that the GFSs strandings on MI are primarily (73%) subadult and adult males, with strandings occurring mostly during the second part of the year and particularly after the breeding season. The warming in 2015, during which many GFS pups and juveniles were stranded in California, did not have similar repercussions on the sex-and age-classes as the strandings on MI. GFS strandings on MI are apparently the result of a male post-breeding migration, in which the GU become a "new" feeding ground for subadult and adult males as the total GFS population, particularly from the SBA, has increased during the last decade. The use of the GU for feeding by the GFS males is confirmed by stable isotope values in vibrissae from GFSs beached in 2012 as contrasted with the typical coastal feeding area exploited by CSLs. Our findings highlight the urgent need to develop a conservation policy for the GU as an important feeding ground for a species recovering from extinction.

Appendix 1. Spearman's rank correlation matrix of different variables from the Gulf of Ulloa (GU) and the San Benito Archipelago (SBA). Values are the correlation coefficients; values in bold are significant (p < 0.05).

	Year	N_{SBA}	Squid	Chl-a_{GU}	Chl-a_{SBA}	SST_{GU}	SST_{SBA}
Year	1.000						
N_{SBA}	**0.818**	1.000					
Squid	−0.382	−0.139	1.000				
Chl-a_{GU}	−0.006	0.066	−0.261	1.000			
Chl-a_{SBA}	**−0.552**	−0.357	0.091	0.055	1.000		
SST_{GU}	−0.006	−0.224	−0.381	−0.030	0.055	1.000	
SST_{SBA}	0.321	0.006	−0.309	−0.066	−0.066	**0.757**	1.000

N_{SBA} = GFS abundance on the SBA
Squid = tons/year
Chl-a_{GU} = chlorophyll concentration anomalies in the GU
Chl-a_{SBA} = chlorophyll concentration anomalies around the SBA
SST_{GU} = sea surface temperature anomalies in the GU
SST_{SBA} = sea surface temperature anomalies around the SB

Keywords: Guadalupe fur seals, population, expansion, inter-annual variability, sea surface temperature, chlorophyll "a", carbon and nitrogen isotope ratios, Humboldt squid, Gulf of Ulloa

References

Allen G.R., M.L. Bauchot, D.R. Bellwood, G.Bianchi, W.A. Bussing, J.H. Caruso et al. 1995. *Peces óseos.* pp. 800–1200. *In*: W. Fischer, F. Krupp, W. Schneider, C. Sommer, K.E. Carpenter and N.H. Niem [eds.]. Guía FAO para la identificación de especies para los fines de la pesca. Pacífico centro-oriental.

Ascencio-Estrada, H. 2010. Evaluación de la Mortalidad del Lobo Marino de California *Zalophus californianus* en Isla Magdalena B.C.S. México. Tesis de Maestría en Ciencias. Centro Interdisciplinario de Ciencias Marinas, IPN. 98 p.

Aurioles-Gamboa, D., F. Sinsel, C. Fox, E. Alvarado and O. Maravilla. 1983. Winter migration of subadult male California sea lions (*Zalophus californianus*) in the southern part of Baja California. J. Mammal. 64(3): 513–518.

Aurioles-Gamboa, D. and B.J. Le Boeuf. 1991. Effects of the El Niño 1983 on the California sea lion population in México. pp. 112–118. *In*: F. Trillmich and F. Ono [eds.]. Pinnipeds and El Niño. Responses to Environmental Stress. Springer-Verlag, Berlin, Germany.

Aurioles-Gamboa, D. and A. Zavala-González. 1994. Ecological factors that determine distribution and abundance of the California sea lion *Zalophus californianus* in the Gulf of California. Cienc. Mar. 20: 535–553.

Aurioles-Gamboa, D., C.J. Hernández-Camacho and E. Rodriguez-Krebs. 1999. Notes on the southernmost records of the Guadalupe fur seal, *Arctocephalus townsendi*, in Mexico. Mar. Mammal. Sci. 15: 581–583.

Aurioles-Gamboa, D. and F.J. Camacho-Ríos. 2007. Diet and feeding overlap of two otariids, *Zalophus californianus* and *Arctocephalus townsendi*: Implications to survive environmental uncertainty. Aqua. Mamma. 33(3): 315–326.

Aurioles-Gamboa, D., F. Elorriaga-Verplancken and C.J. Hernández-Camacho. 2010. Guadalupe fur seal population status on the Islas San Benito, Mexico. Mar. Mammal. Sci. 26(2): 402–408.

Barber, R.T. and F.P. Chavez. 1983. Biological consequences of El Niño. Science 222: 1203–1210.

Bartholomew, G.A. 1950. A male Guadalupe fur seal on San Nicolas Island, California. J. Mammal. 31: 175–180.

Belcher, R.I. and T.E. Lee. 2002. *Arctocephalus townsendi*. Mamm. Spe. 700: 1–5.

Blackburn, M. 1969. Conditions related to upwelling which determine distribution of tropical tunas off western Baja California. Fish. Bull. 68(1): 147–176.

Burnham, K.P. and D.R. Anderson. 2002. Model Selection and Multimodel Inference. A Practical Information-Theoretic Approach. Springer, New York.

Burton, K.R. and P.L. Koch. 1999. Isotopic tracking of foraging and long-distance migration in northeastern Pacific pinnipeds. Oecologia 119: 578–585.

Christensen, V. and D. Pauly. 1992. ECOPATH II Software for balancing steady-state ecosystem models and calculating network characteristics. Ecol. Mod. 61: 169–185.

Etnoyer, P., D. Canny, B. Mate, L. Morgan and G. Ellen. 2004. Persistent pelagic habitats in the Baja California to Bering Sea (B2B) Ecoregion. Oceanography 17(1): 90–101.

Esperón-Rodríguez, M. and J.P. Gallo-Reynoso. 2013. Juvenile and subadult feeding preferences of the Guadalupe fur seal (*Arctocephalus townsendi*) at San Benito Archipelago, Mexico. Aqua. Mamm. 39: 125–131.

Etnier, M.A. 2002. Occurrence of Guadalupe fur seals (*Arctocephalus townsendi*) on the Washington coast over the past 500 years. Mar. Mammal. Sci. 18(2): 551–556.

Ferry, L.A. and G.M. Cailliet. 1996. Sample size and data analysis: Are we characterizing and comparing diet properly? pp. 71–80. *In*: D. Mackinlay and K. Shearen [eds.]. Feeding

Ecology and Nutrition in Fish. Symposium Proceeding. American Fisheries Society, San Francisco, USA.

Ferry, L.A., S.L. Clark and G.M. Cailliet. 1997. Food habits of spotted sand bass (*Paralabrax maculatofasciatus*, Serranidae) from Bahia de Los Angeles, Baja California. Bull. So. Calif. Acad. Sci. 96(1): 1–21.

Fleischer, L.A. 1987. Guadalupe fur seal, *Arctocephalus townsendi*. pp. 43–48. *In*: J.P. Croxall and R.L. Gentry [eds.]. Status, Biology, and Ecology of Fur Seals. NOAA Technical Report NMFS 51.

Fry, D.H. 1939. A winter influx of sea lions from lower California. California Fish and Game 25: 245–250.

Gallo-Reynoso, J.P. 1994. Factors affecting the population status of the Guadalupe fur seal, *Arctocephalus townsendi* (Merriam, 1897), at Isla Guadalupe, Baja California, México. PhD thesis. University of California, Santa Cruz, California, USA.

Gallo-Reynoso, J.P., A.L. Figueroa-Carranza and B.J. Le Boeuf. 2008. Foraging behavior of lactating Guadalupe fur seal females. pp. 505–614. *In*: C. Lorenzo, E. Espinoza and J. Ortega [eds.]. Avances en el estudio de los mamíferos de México. Publicaciones Especiales, Vol. II. Asoc Mex Mast, A.C., México, D.F.

García-Capitanachi, B. 2011. Estado de la población de lobo fino de Guadalupe (*Arctocephalus townsendi*) en Isla Guadalupe e Islas San Benito. Tesis de Maestría en Oceanografía Costera, Facultad de Ciencias, UABC. 104 p.

García-Rodríguez, F.J. and D. Aurioles-Gamboa. 2004. Spatial and temporal variation in the diet of the California sea lion (*Zalophus californianus*) in the Gulf of California, Mexico. Fish. Bull. 102(1): 47–62.

Giardino, V.G., M.A. Mandiola, J. Bastida, P.E. Denuncio, R.O. Bastida and D.H. Rodríguez. 2016. Travel for sex: Long-range breeding dispersal and winter haulout fidelity in southern sea lion males. Mamm. Biol. (Zeitschrift für Säugetierkunde) 81(1): 89–95.

Greig, D.J., F.M.D. Gulland and C. Kreuder. 2005. A decade of live California sea lion (*Zalophus californianus*) stranding along the central California coast: Causes and trends, 1991–2000. Aquat. Mamm. 31(1): 11–22.

Hanni, K.D., D.J. Long, R.E. Jones, P. Pyle and L.E. Morgan. 1997. Sightings and strandings of Guadalupe fur seals in central and northern California, 1988–1995. J. Mammal. 78: 684–690.

Hernández-Camacho, C.J., D. Aurioles-Gamboa and L.R. Gerber. 2008. Age specific birth rates of California sea lions (*Zalophus californianus*) in the Gulf of California, Mexico. Mar. Mammal. Sci. 24: 664–67.

Hubbs, C.L. 1956. Back from oblivion, Guadalupe fur seal still lives. Zoonoz, San Diego. Zool. Soc. 29(12): 6–9.

Hubbs, C.L. 1979. Guadalupe fur seal. pp. 24–27. *In*: Mammals in the Seas, Vol. II: Pinniped Species Summaries and Report on Sirenians, FAO Fisheries No. 5, Rome, Italy.

Iverson, I. and L. Pinkas. 1971. A pictorial guide to beaks of certain Eastern Pacific cephalopods. Fish. Bull. 152: 83–105.

Jackson, A.L., R. Inger, A.C. Parnell and S. Bearhop. 2011. Comparing isotopic niche widths among and within communities: SIBER–Stable Isotope Bayesian Ellipses in R. J. Anim. Ecol. 80: 595–602.

Krebs, C.J. 1999. Ecological Methodology. Addison Wesley, California. 620 pp.

Layman, C.A., D.A. Arrington, C.G. Montaña and D.M. Post. 2007. Can stable isotope ratios provide for community-wide measures of trophic structure? Ecology 88(1): 42–48.

Maravilla-Chavez, M.O. and M.S. Lowry. 1999. Incipient breeding colony of Guadalupe fur seals at Isla Benito del Este, Baja California, Mexico. Mar. Mammal. Sci. 15(1): 239–241.

Mate, B.R. 1975. Annual migrations of the sea lions *Eumetopias jubatus* and *Zalophus californianus* along the Oregon Coast. Rapp. P.V. Reun. Cons. Int. Explor. Mer. 169: 455–461.

Matlab. 2013. The MathWorks Inc., Natick, Massachusetts, USA.

Merriam, C.H. 1897. A new fur-seal or sea-bear (*Arctocephalus townsendi*) from Guadalupe island, off Lower California. Proc. Biol. Soc. Washington 11: 175–178.

Odell, D.K. 1975. Breeding biology of the California sea lion, *Zalophus californianus*. Rapp P.V. Reun. Cons. Int. Explor. Mer. 169: 374–378.
Okutani, T. 1980. Calamares de las aguas mexicanas. Secretaría de Pesca. México.
Pablo-Rodríguez, N., D. Aurioles-Gamboa and J. Montero-Muñóz. 2016. Niche overlap and habitat use at distinct temporal scales among the California sea lions (*Zalophus californianus*) and Guadalupe fur seals (*Arctocephalus philippii townsendi*). Mar. Mammal. 32(2): 466–489.
Páez-Rosas, D. and D. Aurioles-Gamboa. 2010. Alimentary niche partitioning in the Galapagos sea lion, *Zalophus wollebaeki*. Mar. Biol. 157: 2769–2781.
Pierson, M.O. 1987. Breeding behavior of the Guadalupe fur seal, *Arctocephalus townsendi*. pp. 83–94. *In*: J.P. Croxall and R.L. Gentry [eds.]. Status, Biology, and Ecology of Fur Seals. NOAA Technical Report NMFS 51.
Porras-Peters, H., D. Aurioles-Gamboa, V. Cruz and P. Koch. 2008. Trophic level and overlap of California sea lions, *Zalophus californianus* in the Gulf of California. Mar. Mammal. Sci. 24(3): 554–576.
Rick, T.C., R.L. Delong, J.M. Erlandson, T.J. Braje, T.L. Jones and D.J. Kennett. 2009. A trans-Holocene archaeological record of Guadalupe fur seals (*Arctocephalus townsendi*) on the California Coast. Mar. Mammal. Sci. 25: 487–502.
Roper, C.E.F., M.J. Sweeney and F.G. Hochberg. 1995. Cefalópodos. pp. 305–353. *In*: W. Fischer, F. Krupp, W. Schneider, C. Sommer, K.E. Carpenter and V.H. Niem [eds.]. Guía FAO para la identificación de especies para los fines de la pesca 1. Pacífico centro-oriental.
Sierra-Rodríguez, G. 2015. Recolonización y hábitos alimentarios maternos del lobo fino de Guadalupe (*Arctocephalus philippii townsendi*) del Archipiélago San Benito, B. C., México." Master Thesis. CICIMAR-IPN, Mexico.
Stewart, B.S. 1981. The Guadalupe fur seal (*Arctocephalus townsendi*) on San Nicolas Island, California. Bull. South Cal. Aca. Sci. 80: 134–136.
Stewart, B.S., P.K. Yochem, R.L. DeLong and G.A. Antonelis, Jr. 1987. Interactions between Guadalupe fur seals at San Nicolas and San Miguel Islands California. pp. 103–106. *In*: J.P. Croxall and R.L. Gentry [eds.]. Status, Biology, and Ecology of Fur Seals. NOAA Technical Report. NMFS 51.
Stewart, B.S., P.K. Yochem, R.L. DeLong and G.A. Antonelis, Jr. 1992. Trends in abundance and status of pinnipeds on the southern California Channel Islands. pp. 501–516. *In*: F.G. Hochberg [ed.]. Proceedings of the Third California Islands Symposium: Recent Advances in Research on the California Islands. Santa Barbara Museum of Natural History, Santa Barbara, CA, USA.
Townsend, C.H. 1931. The fur seal of the California islands with new descriptive and historical matter. Zoologica 9: 442–457.
Trillmich Fritz and K. Ono. 1991. Pinnipeds and El Niño. Responses to Environmental Stress. Ecological Studies 88, Springer Springer, Berlin, Germany. 293 p.
Walker, P.L. and S. Craig. 1979. Archaeological evidence concerning the prehistoric occurrence of sea mammals at Point Bennet, San Miguel Island. California Fish and Game 65: 50–54.
Williams, R.J. and N.D. Martinez. 2004. Limits to trophic levels and omnivory in complex food webs: Theory and data. The Amer. Naturalist 163(3): 458–468.
Wingfield, K.D., H. Peckham, D.G. Foley, D.M. Palacios, B.E. Lavaniegos, R. Durazo et al. 2011. The making of a productivity hotspot in the coastal ocean. PLoS One 6(11): e27874.
Wolf, G.A. 1982. A beak key for eight eastern tropical Pacific cephalopod species with relationships between their beak dimensions and size. Fish Bull. 80: 357–370.
Wolf, G.A. 1984. Identification and estimation of size from the beaks of eighteen species of cephalopods from the Pacific Ocean. NOOA Tech. Rep. NMFS. 17.
Young, R.E. 1972. The systematics and areal distribution of pelagic cephalopods from the seas of Southern California. Smithson. Contrib. Zool. 97: 1–159.
Zuur, A.F., E.N. Ieno, N.J. Walker, A.A. Saveliev and G.M. Smith. 2009. Mixed Effects Models and Extensions in Ecology with R. Springer, New York, NY, USA.

7

Population Status, Anthropogenic Stressors and Conservation of the Galapagos Fur Seal (*Arctocephalus galapagoensis*)
An Overview

Juan José Alava,[1,2,3,*] *Judith Denkinger,*[4]
Pedro J. Jiménez,[3] *Raúl Carvajal*[3,5] *and Sandie Salazar*[4]

Introduction

The Galapagos fur seal (*Arctocephalus galapagoensis*) is the smallest otariid of the world and a marine mammal species endemic to the Galapagos Islands Archipelago (Fig. 1). Galapagos fur seals inhabit most of the western islands (Isabela, Fernandina, Marchena, Genovesa, Santiago, Seymour), but are occasionally observed throughout the Galapagos Archipelago. They prefer to live and rest on rocky shores in close proximity to deep productive water, and use crevices, caves, large boulders and rock ledges for shade and shelter (Trillmich 1987, Godwin 1990, Merlen 1995, Salazar et al. 2011; Figs. 1A–1D). Vagrant individuals were recorded in Manabi, Guayas, Santa Elena and Esmeraldas Provinces along the Ecuadorian coast (Alava and Salazar

---
Affiliation at the end of the chapter.

Figure 1. Different age categories of endangered Galapagos fur seals (*Arctocephalus galapagoensis*) were observed dwelling their typical rocky shore habitat in the western part of Galapagos Islands during an expedition conducted on March 2008. A pup hiding among lava rocks (cave) at Marchena Island (A); a juvenile observed at Fernandina Island (B); a gang of juveniles and subadults/adults moving away through a rocky shore of lava at Merchena Island (C); and, adults (male and female) interacting on rocky shores close to ocean waters (Merchena Island) (D). Photo credits: J.J. Alava.

2006, Félix et al. 2007). The species was formerly referred to as *Arctocephalus australis galapagoensis* (Repenning et al. 1971). New genetic information suggests that this species is indeed very close to *A. australis* and may also be categorized as *A. australis galapagoensis* (Wolf et al. 2007). The population shows strong matrilineal structure where geographically close colonies display separate mitochondrial haplotypes, but male dispersal causes gene flow through all colonies (Lopes et al. 2015). Both at the national and global levels, Galapagos fur seals are listed as Endangered under criterion A2ab because of several environmental threats including natural and anthropogenic stressors (Salazar et al. 2011, Trillmich 2015). The population dynamics of the species are influenced by extreme oceanographic-climatic events such as those occurring during El Niño Southern Oscillations Events (ENSOs) which have resulted in high mortality of fur seals. While natural and density independent factors are fairly well understood (Trillmich and Limberger 1985, Trillmich and Dellinger 1991, Alava and Salazar 2006,

Salazar and Denkinger 2010, Trillmich 2015), assessment of anthropogenic stressors affecting the species population is a conservation issue deserving more attention. In this overview, we provide a brief update and insights on the population status and anthropogenic stressors affecting the conservation of Galapagos fur seals with implications for the management of the species.

Population Status

Galapagos fur seal populations were drastically reduced by intense sealing during the 19th century, but recovered during the 20th century. The first census in 1977-1978 gave an estimate of about 30,000 fur seals (Trillmich 1987). Since then, fur seals were fully protected in the Galapagos but the population fluctuated greatly because of the impacts of the ENSOs of 1982-1983 and 1997-1998, when the population showed a dramatic decline (Alava and Salazar 2006, Trillmich 2015).

Currently the colonies of Fernandina and Isabela Islands, close to cold and nutrient rich waters, are the largest and healthiest (Salazar et al. 2011). Data on population size are ambiguous, ranging from 2,500 (Godwin 1990) to 40,000 individuals (Trillmich 1979). According to a census conducted in November 2001, a total of 2,733 animals were counted at haul-outs and rookeries of Fernandina and Isabela Islands, giving a population estimate of 6,000–8,000 individuals (Salazar 2002, Alava and Salazar 2006). According to this estimate, the Galápagos fur seal population declined by 80% from 1977/78 (i.e., 40,000 individuals) to 2001. Further population monitoring conducted over the 2002–2005 time period suggests that the population has reached 10,000 mature individuals or a total population estimate of 15,000 animals (Salazar and Michuy 2008, Salazar and Denkinger 2010, Trillmich 2015), which means a reduction of more than 60% in the last 37 years relative to counts from 1977/1978, but still accounting for more than 50% population loss over the last three generations (Trillmich 2015). The highest population numbers are reported from the western and northern regions of the Archipelago (Alava and Salazar 2006). However, several central islands are used as resting sites, including Baltra and North Seymour islands (Salazar et al. 2011). The lack of quantitative population data for the past 12 years, information gaps and associated uncertainty preclude a complete assessment of the population trend of this species.

Emigration from the Galapagos Islands is unlikely. However during El Niño events, vagrant individuals were found along Ecuador's mainland coast, including more recent sightings in the Gulf of Guayaquil (e.g., Puna Island, Playas Villamil, Anconcito), Libertador Bolívar, Bahía de Caraquéz (Manabí Province) and Sua (Esmeralda Province) (Alava and Salazar 2006,

Felix et al. 2007, J. Denkinger, pers. observ., P. Jimenez, pers. observ.) and possibly southern Mexico (Aurioles-Gamboa et al. 2004), and Isla Foca, Peru (Collyns 2010, Wade 2010, C. Yaipen-Llanos/Organisation for Research and Conservation of Aquatic Animals (ORCA), pers. comm.). These individuals included two pregnant females that gave birth at the coast of Esmeraldas in 2004 and another at the coast of Posorja in 2005 (Felix et al. 2007). However, most of these isolated findings have yet to be confirmed according to Trillmich (2015).

Anthropogenic Impacts and Conservation Threats

Historically, hunting of fur seals was the major human-caused stressor and caused severe population declines of fur seals in the Galapagos Archipelago (Table 1). Currently, ocean warming combined with ENSO events are considered the major stressors for Galapagos fur seals (Salazar and Denkinger 2010). Population recovery from these events becomes even more complicated considering other indirect threats such as introduction of alien species, feral animals, spillovers of pathogens, marine pollution and bycatch mainly from artisanal fisheries (Alava and Salazar 2006, Merlen and Salazar 2007, Salazar et al. 2011, Alava et al. 2014), as shown in Table 1. Fortunately, most of the fur seal colonies are situated far from human settlements, which decreases the risk caused by introduced species and pathogen transfer from feral animals. Although information on pathogens in pinnipeds is scarce, the presence of common diseases in dogs and cats has been reported at Isabela Island (Levy et al. 2008).

Table 1. Identification of main environmental threats, both natural and anthropogenic, affecting Galapagos fur seal populations. Based on Alava and Salazar (2006), Salazar et al. (2011) and Alava et al. (2014).

Major threats	Past	Present	Future (potential impacts)
Hunting (sealing for pelts) and trade	√		
Fisheries interactions	√	√	√
Introduced species and feral animals	√	√	√
Biological pollution: Emerging Infectious Diseases (EID)		√	√
Chemical pollution: oil spills and Persistent Organic Pollutants (POPs)	√	√	√
El Niño event and climate change	√	√	√
Habitat loss and disturbance		√	√

Here, we focus on biological (i.e., pathogen spillovers) and chemical pollution driven by human activities.

Biological Pollution

In the Galapagos, the presence of rats, domestic dogs and feral cats is a continuing threat for most fur seal colonies as these animals can act as reservoir hosts or infected carriers of emerging infectious diseases (EID) (e.g., toxoplasmosis, leptospirosis, distemper virus) that could be transmitted to native species of marine mammals (Alava et al. 2014, Denkinger et al. 2015). Even though fur seals live out of the range of most domestic species, feral cats are found on most of the Islands. The first cases of zoonotic transfer of leptospirosis and morbillivirus has been reported in Galapagos sea lions of San Cristobal Island (J. Denkinger et al. in rev.). Galapagos sea lions share the same range and haul-out sites with fur seals, and thus can function as a vector to fur seals. For instance, a serological survey and DNA screening assessment for infectious disease pathogens conducted in Isabela Island (where sea lion colonies are found just off Puerto Villamil) revealed that domestic dogs and cats are exposed to many pathogens, including parvovirus, parainfluenza virus, adenovirus, distemper virus, *Dirofilaria immitis*, *Wolbachia pipiens*, *Bartonella* sp., *Ehrlichia/Anaplasma* spp., and *Mycoplasma haemocanis* in dogs and panleukopenia virus (67%), *Toxoplasma gondii* (63%), calicivirus (44%), and herpesvirus 1 in cats (Levy et al. 2008). A serological survey determined that five out of six (83%) domestic dogs were seropositive to *Leptospira* on Santa Cruz in 2001-2002 (Cruz et al. 2002). This implies that Galapagos pinnipeds may be at risk of infection by this bacterial pathogen. Subsequent health surveys showed that Galapagos sea lions were susceptible to the bacterium *Leptospira* (Guevara 2011) and Galapagos fur seals were susceptible to two strains of *Leptospira* (Alava et al. 2014). Using PCR analysis, the presence of *Leptospira* DNA was confirmed in 70% of tissue samples (i.e., kidney and placenta) collected from dead sea lions, including the placentas of three newborn pups, in San Cristobal (Guevara 2011). Despite events of canine distemper virus (CDV) in domestic dogs found in the Galapagos Islands, serological surveys of sea lions from different colonies of the Islands in 2001 revealed that no CDV antibodies were present in this species (Salazar et al. 2001, Alava and Salazar 2006). This indicates that they were not exposed to any recent infection by morbilliviruses at that time. However, a new study reports that DNA of CDV was found in carcasses of sea lion pups at different colonies at Wreck Bay (San Cristobal Island) using PCR analysis (J. Denkinger et al. in rev.).

Both CDV and leptospirosis could have played a role in unusual mortality events of Galapagos sea lion pups in 2009, 2011 and 2012, where about 50% of the pups died (Denkinger et al. 2015). These outbreaks may indicate that Galapagos fur seal may be vulnerable to potential infection due to the fact that Galapagos sea lions' regional movements are likely to contribute to the spread of infectious pathogens, eventually affecting neighboring colonies of fur seals with potential impact on the survival of this species.

Chemical Pollution

Marine pollution including chemical assaults by oil spills, hydrocarbons, metals and persistent organic pollutants (POPs) was identified and assessed as a major anthropogenic impact jeopardizing the health of Galapagos otariids and endemic marine fauna of the Galapagos (Alava and Salazar 2006, Alava et al. 2014). In particular, toxic and bioaccumulative POPs pose a threat to this species as polychlorinated biphenyls (PCBs), dichlorodiphenyltrichloroethanes (DDTs) and several other organochlorine pesticides were detected in a sympatric species, the Galapagos sea lions (*Zalophus wollebaeki*), in which biomagnification of POPs has recently been assessed (Alava and Gobas 2012, Alava et al. 2014). Both the past and modern use of DDT with the concomitant detection of this pollutant as the predominant POP in Galapagos sea lions in this species have raised potential health concerns for the Galapagos fur seal (Alava et al. 2011a, 2011b, 2014).

Indeed, contaminant data for POPs in three fur seals sampled in 2005 in the Galapagos revealed that this species was also exposed to DDTs, mirex, dieldrin, beta-hexachlorocyclohexane (β-HCH) and trans-nonachlor (J.J. Alava, unpublished data; Table 2). For instance, p,p-DDE, the major metabolite of DDT and a potent anti-androgenic chemical, was detected in all individuals, exhibiting an average ± SD concentration of 393 ± 130.5 µg/kg lipid, ranging from 270 to 530 µg/kg lipid. Similarly, p,p-DDD, mirex and trans-nonachlor were detected in all samples, while β-HCH was detected in two of the three samples (range: 56–540 µg/kg lipid) (Table 2). Total concentrations of DDT (ΣDDT) in the individuals sampled ranges from 277 to 1890 µg/kg lipid. The use of large volumes of DDT to control the malaria biological vector (*Anopheles* mosquito) and associated excessive use and illegal application of other pesticides in agriculture to control pests in Ecuador and other South American countries (e.g., Colombia and Peru) may have contributed to the emissions of this pesticide into the environment with subsequent volatilization and regional atmospheric transport (global distillation "grasshopper effect") to remote oceanic regions such as the Galapagos Islands (Alava et al. 2009, 2011a).

Table 2. Biological and field data and POP concentrations (organochlorine pesticides) detected in muscle-blubber biopsy samples of three Galapagos fur seals in March 2005.

Sample code	Fur Seal- IMAJ-02	Fur Seal- IMAJ-03	Fur Seal- HAP-07
Biological/Field Data			
Sampling site	Cabo Marshall, Isabela Island	Cabo Marshall, Isabela Island	Cabo Hammond, Fernandina Island
Length (cm)	93	91	94
Weight (kg)	18	14	15
Age (months)	12	24	12
Sex	M	M	F
Lipid% (sample)	63.1	78	76
POP Concentrations (µg/kg lipid)			
p,p-DDE	380	530	270
p,p-DDD	39.0	54.5	7.00
o,p-DDE	< MDL	< MDL	< MDL
o,p-DDD	< MDL	39.0	< MDL
o,p-DDT	< MDL	50.0	< MDL
p,p-DDT	1470	1297	< MDL
ΣDDT	1890	1970.5	277
Mirex	5.30	5.0	4.0
Dieldrin	< MDL	5.0	< MDL
β-HCH	< MDL	540	56
trans-nonachlor	52	45	32

MDL: Method of Detection Limit, which is calculated as the mean of procedural blanks run during the sample analysis plus three times the standard deviations of the procedural blanks. The MDL is then subtracted from the quantified contaminant data and only concentrations above the MDL were reported for this work.

Conservation and Management

Most of fur seal species in the world have suffered pressure and exploitation from human activities some centuries ago. Galapagos fur seals were no exception. In the nineteenth century, Galapagos fur seals were hunted indiscriminately for their fur. Whalers and sealers culled Galapagos fur seals extensively and few animals were left by 1900 (Trillmich 1987). Townsend (1934) documented that between 1816 and 1933 no less than 22,508 pelts were harvested by sealers, based on logbook data of ships destined for California; skins transported by British and Spanish ships were not documented (Trillmich 1987). The size of the population at that time is unknown, but it is presumed that their population levels were low (Salazar et al. 2011).

The Galapagos fur seal is protected by the Ecuadorian government since 1936 when hunting and extraction of native animals was banned from the Islands (Salazar et al. 2011) and with the subsequent creation of the Galapagos National Park in 1959. Several other legal and regulatory efforts have ratified the designation of protected species (i.e., Emergency Decree Law No. 17 of July 20, 1959; Official Gazette No. 166 of April 9, 1999; Official Register No. 5 of January 28, 2000 and Official Gazette No. 679 of October 8, 2002). The latter declaration indefinitely prohibits the capture, hunting, marketing and transport of live specimens, constituent elements and products throughout the national territory. The species is also protected by the Special Law of the National Park and Galapagos Marine Reserve. Since 1991, Ecuador has supported an Action Plan for the Conservation of Marine Mammals in the Southeast Pacific, which includes Galapagos fur seals and sea lions as research priority species (CPPS/PNUMA 1998). Internationally, the IUCN also considers Galapagos fur seals as endangered, i.e., facing a very high risk of extinction in the wild in the medium term (Trillmich 2015). Conversely, CITES (2008) included this species in Appendix II, under which a species can be traded as long as the authority of the country of origin certifies that this activity does not affect the survival and that the animals were legally obtained; however, this provision contradicts the Special Law for Galapagos, according to which no endemic species of the Galapagos Archipelago is subject to any trade.

In this context, the following conservation and management strategies for the species in order of priority are recommended (based on Alava and Salazar 2006, and Salazar et al. 2011):

1. Strengthen patrolling and monitoring of main breeding colonies in the western part of the Galapagos Islands.
2. Improve monitoring of the species by implementing and optimizing a control and surveillance system, including nocturnal surveys in feeding areas (i.e., radar), where the interaction of Galapagos fur seals with the squid fisheries in international waters close to the Galapagos Marine Reserve is basically unknown, as well as diurnal surveys in their breeding colonies (i.e., overflights, navigation).
3. Increase epidemiological surveillance efforts to control invasive species and pathogens (viruses, bacteria and parasites) to the Galapagos Islands.
4. Develop measures and preventive public health programs to mitigate the potential health risks and impacts by introduced and feral mammals (i.e., dogs, cats and rats).
5. Implement a marine pollution program to assess and track the impact and fate of anthropogenic chemicals in endemic marine wildlife and food webs of the Galapagos Marine Reserve.

6. Undertake global action to mitigate climate change, overfishing and the disturbance of terrestrial and marine habitats.
7. Implement specific regulations to address negative effects of tourism and misuse of visiting sites, such as seasonal and spatial closures of sites during the reproductive peaks.
8. Constant environmental education campaigns and social mobilization should be undertaken with local people, including the fisheries and tourism sectors, schools and colleges to reduce the impacts that threaten the species conservation.
9. Monitor compliance with the laws that protect the species and enforce sanctions for violations and perpetrators.
10. Create scientific research scholarships for university students and researchers to help improve the knowledge and science for this species.

Conclusions

Galapagos fur seals are endemic to the Galapagos Islands and endangered because of severe population decline during the past decades. Even though this species is protected within the Galapagos Marine Reserve and trade of any wildlife from the Galapagos Islands is illegal, they still face imminent anthropogenic threats that are scarcely studied, such as pathogen transfer and exposure to pollutants. These threats in combination with ocean warming due to climate change and severe El Niño events such as the 2015/2016 ENSO event can cause the populations to become unviable, reaching their tipping points and facing a high risk of extinction.

Acknowledgements

We are in debt with Dr. S. Villegas-Amtmann and Dr. D.P. Costa for collecting and shipping the biopsy samples for contaminant analysis in 2005. Special thanks to Dr. M. Ikonomou for the chemical analysis of samples to measure POPs in Galapagos fur seals at the Institute of Ocean Science (IOS, Fisheries and Ocean Canada). Funding for contaminant analysis was provided by Dr. F.A.P.C. Gobas (School of Resource and Environmental Management, Simon Fraser University). The first author (J.J. Alava) thanks the Institute for the Ocean and Fisheries (IOF, University of British Columbia) and the Coastal Ocean Research Institute (CORI), Vancouver Aquarium Marine Science Centre, for academic support and their continued efforts for ocean conservation. The authors acknowledge Fundación Ecuatoriana para el Estudio de Mamíferos Marinos (FEMM) for the continued commitment to conduct marine mammals' research during the last three decades in Ecuador.

We thank Dr. F. Trillmich for reading our chapter and Dr. V. Otton for proof reading the English text and providing comments.

Keywords: Galapagos fur seal, population, anthropogenic threats, biological pollution, viruses, canine distemper virus, *Leptospira*, persistent organic pollutants, PCBs, DDTs, conservation

References

Alava, J.J. and S. Salazar. 2006. Status and conservation of Otariids in Ecuador and the Galapagos Islands. pp. 495–519. *In*: A.W. Trites, S.K. Atkinson, D.P. De Master, L.W. Fritz, T.S. Gelatt, L.D. Rea and K.M. Wynne [eds.]. Sea Lions of the World. Alaska Sea Grant College Program, Fairbanks, Alaska, USA.

Alava, J.J., M.G. Ikonomou, P.S. Ross, D.P. Costa, S. Salazar, D. Aurioles-Gamboa et al. 2009. Polychlorinated biphenyls and polybrominated diphenyl ethers in Galapagos sea lions (*Zalophus wollebaeki*). Environ. Toxicol. Chem. 28: 2271–2282.

Alava, J.J., P.S. Ross, M.G. Ikonomou, M. Cruz, G. Jimenez-Uzcategui, S. Salazar et al. 2011a. DDT in endangered Galapagos sea lions (*Zalophus wollebaeki*). Mar. Pollut. Bull. 62: 660–671.

Alava, J.J., S. Salazar, M. Cruz, G. Jimenez-Uzcategui, S. Villegas-Amtmann, D. Paez-Rosas et al. 2011b. DDT strikes back: Galapagos sea lions face increasing health risks. Ambio 40: 425–430.

Alava, J.J. and F.A.P.C. Gobas. 2012. Assessing biomagnification and trophic transport of persistent organic pollutants in the food chain of the Galapagos sea lion (*Zalophus wollebaeki*): Conservation and management implications. pp. 77–108. *In*: A. Romero and E.O. Keith [eds.]. New Approaches to the Study of Marine Mammals. InTech, Rijeka, Croatia.

Alava, J.J., C. Palomera, L. Bendell and P.S. Ross. 2014. Pollution as a threat for the conservation of the Galapagos marine reserve: Environmental impacts and management perspectives. pp. 247–283. *In*: J. Denkinger and L. Vinueza [eds.]. The Galapagos Marine Reserve: A Dynamic Socio-ecological System. Springer Science and Business Media, New York.

Aurioles-Gamboa, D., Y. Schramm and S. Mesnick. 2004. Galapagos fur seals, *Arctocephalus galapagoensis*, in México. Lat. Am. J. Aquat. Mamm. (LAJAM) 9: 77–90.

CITES. 2008. Convention on International Trade in Endangered Species of Wild Fauna and Flora. Appendices I, II, III. www.cites.org/eng/app/appendices.shtml.

CPPS/PNUMA. 1998. Plan de acción para la conservación de los mamíferos marinos en el Pacífico sureste. Comisión Permanente del Pacífico Sur. Quito.

Collyns, D. 2010. Galapagos fur seals head for Perú waters. BBC News. 2010/02/08 http://news.bbc.co.uk/2/hi/americas/8503397.stm.

Cruz, D., M. Cruz and M. Velasquez. 2002. La epidemia de Distemper (moquillo) canino en las Islas Galápagos. *In*: Fundacion Natura and World Wildlife Fund (WWF), Informe Galapagos 2001-2002. Quito, Ecuador.

Denkinger, J., L. Gordillo, I. Montero-Serra, J.C. Murillo, N. Guevara, M. Hirschfeld et al. 2015. Urban life of Galápagos sea lions (*Zalophus wollebaeki*) on San Cristobal Island, Ecuador: Colony trends and threats. J. Sea Res. 105: 10–14.

Félix, F., P. Jiménez, J. Falconí and O. Echeverry. 2007. New cases and first births of the Galápagos fur seal, *Arctocephalus galapagoensis* (Heller, 1904) from the mainland coast of Ecuador. Rev. Biol. Mar. Oceanogr. 42(1): 77–82.

Godwin, S. 1990. Seals. A Friedman Group Book. Mallard Press, New York.

Guevara, N.C. 2011. Línea base del estado de salud y detección de *Leptospira* patógena por PCR en lobos marinos de Galápagos (*Zalophus wollebaeki*) de la Isla San Cristóbal. Bachellor Dissertation, Universidad San Francisco de Quito, Quito, Ecuador.

Levy, J.K., P.C. Crawford, M.R. Lappin, E.J. Dubovi, M.G. Levy, R. Alleman et al. 2008. Infectious diseases of dogs and cats on Isabela Island, Galapagos. J. Vet. Intern. Med. 22: 60–65.

Lopes, F., J.I. Hoffman, V.H. Valiati, S.L. Bonatto, J.B.W. Wolf, F. Trillmich et al. 2015. Fine-scale matrilineal population structure in the Galapagos fur seal and its implications for conservation management. Conserv. Genet. 16(5): 1099–1113.

Merlen, G. 1995. A Field Guide to the Marine Mammals of the Galápagos. Instituto Nacional de Pesca, Guayaquil, Ecuador.

Merlen, G. and S. Salazar. 2007. Estado y efectos antropogénicos en los mamíferos marinos de Galápagos. pp. 70–76. In: F. Felix [ed.]. Memorias del Taller de Trabajo sobre el Impacto de las Actividades Antropogénicas en Mamíferos Marinos en el Pacifico Sudeste, Bogotá, Colombia, 28-29 Noviembre 2006. CPPS/PNUMA, Guayaquil, Ecuador.

Repenning, C.A., R.S. Peterson and C.L. Hubbs. 1971. Contributions to the systematics of the southern fur seals, with particular reference to the Juan Fernández and Guadalupe species. pp. 1–34. In: W.H. Burt [ed.]. Antarctic Pinnipedia. Antarctic Research Series 18, American Geophysical Union, New York, USA.

Salazar, S., M. Cruz and D. Cruz. 2001. Canine distemper in the Galapagos Islands: Possible consequences and prevention plan. Charles Darwin Foundation–Galapagos National Park Service. Technical report February 2001, presented to Agriculture Ministry (CESA), Santa Cruz, Galapagos, Ecuador.

Salazar, S. 2002. Lobo marino y lobo peletero. pp. 267–290. In: E. Danulat and J. Edgar [eds.]. Reserva Marina de Galapagos. Linea Base de la Biodiversidad, Fundacion Charles Darwin/Servicio Parque Nacional Galapagos, Santa Cruz, Ecuador.

Salazar, S. and V. Michuy. 2008. Estado poblacional y conservación de los pinnípedos de las islas Galápagos. pp. 196. In: XIII Reunión de Trabajo de Especialistas en Mamíferos Acuáticos de América del Sur y 7° Congreso SOLAMAC. Montevideo, Uruguay.

Salazar, S. and J. Denkinger. 2010. Possible effects of climate change on the populations of Galapagos pinnipeds. Galapagos Res. 67: 45–49.

Salazar, S., J.J. Alava, V. Utreras and D.G. Tirira. 2011. Lobo Marino de Galapagos (*Zalophus wollebaeki*) [Galapagos sea lion (*Zalophus wollebaeki*)]. pp. 136–137. In: D.G. Tirira [ed.]. Libro Rojo de los mamíferos del Ecuador [The Red Book of Ecuadorian Mammals]. Fundación Mamíferos y Conservación, Pontificia Universidad Católica del Ecuador y Ministerio del Ambiente del Ecuador. Quito, Ecuador.

Townsend, C.H. 1934. The fur seal of the Galapagos Islands, *Arctocephalus galapagoensis* Heller. Zoologica (Bulletin of the New York Zoological Society) 18(2): 43–56.

Trillmich, F. 1979. Galápagos sea lions and fur seals. Noticias de Galapagos 29: 8–14.

Trillmich, F. and D. Limberger. 1985. Drastic effects of El Niño on Galápagos pinnipeds. Oecologia 67: 19–22.

Trillmich, F. 1987. Galapagos fur seal, *Arctocephalus galapagoensis*. pp. 23–27. In: J.P. Croxall and R.L. Gentry [eds.]. Status, Biology, and Ecology of Fur Seals. NOAA Technical Report.

Trillmich, F. and T. Dellinger. 1991. The effects of El Niño on Galápagos pinnipeds. pp. 66–74. In: F. Trillmich and K.A. Ono [eds.]. Pinnipeds and El Niño: Responses to Environmental Stress. Springer Verlag, Heidelberg, Germany.

Trillmich, F. 2015. *Arctocephalus galapagoensis*. The IUCN Red List of Threatened Species 2015. Version 2015.2. www.iucnredlist.org.

Wade, T. 2010. Galapagos fur seals gain foothold in warming Peru. Planet Ark. http://planetark.org/wen/56882.

Wolf, J.B.W., D. Tautz and F. Trillmich. 2007. Galapagos and Californian sea lions are separate species: Genetic analysis of the genus *Zalophus* and its implications for conservation management. Front. Zool. 4: 20 doi: 10.1186/1742-9994-4-20.

[1] Institute for the Oceans and Fisheries, University of British Columbia, 2202 Main Mall, Vancouver, BC V6T 1Z4, Canada.
[2] Ocean Pollution Research Program, Coastal and Ocean Research Institute, Vancouver Aquarium, Marine Science Centre, P.O. Box 3232, Vancouver, BC V6B 3X8, Canada.
[3] Fundacion Ecuatoriana para el Estudio de Mamíferos Marinos (FEMM), Ecuador.
[4] Universidad San Francisco de Quito, Galapagos Science Center, Campus Cumbaya, Quito, Ecuador.
[5] Conservación Internacional, Ecuador, Catalina Aldáz N34-181 y Portugal. Edif. Titanium II. Ofc. 402. Quito, Ecuador.
Email: jdenkinger@usfq.edu.ec, peterjoe01@yahoo.es, r.carvajal@conservation.org
* Corresponding author: j.alava@oceans.ubc.ca

8

Diving Physiology, Foraging and Reproductive Behavior of the Galapagos Sea Lion (*Zalophus wollebaeki*)

*Stella Villegas-Amtmann** and *Daniel P. Costa*

Introduction

As Darwin (Darwin 1859) stated: "It is not the strongest of the species that survive, nor the most intelligent, but the one most responsive to change", thus species that are more capable of adapting to changing environmental conditions will have a greater capability of responding to long term changes in climate. Seasonal changes in temperature and light level, affect primary production and the abundance, distribution and behavior of higher trophic levels. Many taxa have evolved behavioral and physiological responses to avoid unsuitable seasonal change by migrating (whales and birds) (Brodie 1975, Mate et al. 1999, Tremblay et al. 2006) or by means of hibernation and aestivation (bears, badgers and frogs) (Schooley et al. 1994, Tanaka 2006, Tracy et al. 2007). Life history patterns of other animals such as seals and sea

---

University of California Santa Cruz, Long Marine Lab, 115 McAllister Way, Santa Cruz, CA 95060, USA.
Email: costa@ucsc.edu
* Corresponding author: stella.villegas@gmail.com

lions; described as central place foragers (Orians and Pearson 1979, Costa 1991), respond to seasonal change by shifting their diet and/or foraging behavior. The magnitude of seasonal change differs with latitude; it is more pronounced at higher latitudes than in equatorial regions, where there are typically only two seasons: a wet and a dry one. In equatorial regions environmental temperature remains high year round, there is a constant 12 hr period of daylight throughout the year and oceanic primary productivity is lower (Longhurst et al. 1995) (http://www.marine.rutgers.edu). At higher latitudes, colder marine systems are typically more productive and thus have a greater abundance of prey (Longhurst et al. 1995).

These latitudinal differences influence animal populations in many ways, e.g., foraging efficiency of Australian forest mammals is lower in tropical than in temperate ecosystems (Johnson 1998), tropical island terns reduce clutch investment at lower latitudes (Hockey and Wilson 2003) and pinnipeds (seals, fur seals and sea lions) living in equatorial and temperate environments are more at risk of extinction than species living at higher latitudes (Ferguson and Higdon 2006, Davidson et al. 2012). These studies conclude that latitudinal differences in population size result from lower food availability and lower oceanic primary productivity, and prey depletion associated with resident-central place foraging behavior along with the reduced seasonality at lower latitudes. Among the pinnipeds, the phocid with the lowest latitudinal distribution, the Hawaiian monk seal (*Neomonachus schauinslandi*), living in a low productive environment, is endangered (Litnan et al. 2015). Of the 15 Otariidae (fur seals and sea lions) species, population size declines with decreasing latitude and only two species live in equatorial regions, the Galapagos sea lion (*Zalophus wollebaeki*) and Galapagos fur seal (*Arctocephalus galapagoensis*).

The Galapagos sea lion (GSL) is an endangered (Trillmich 2015b) and endemic species of the Galapagos Islands. The population is widely distributed among this archipelago, with the highest density of individuals at the central and southern islands (Salazar 2005). Current population estimates lie between 14,000–16,000 individuals (Alava and Salazar 2006). An estimate of 20,000 to 50,000 individuals in 1963 suggests a dramatic decline in the population over the last 20 yr (Heath 2002). The current population trend is decreasing (Trillmich 2015b), fluctuates in size and is negatively affected by El Niño events (Trillmich and Limberger 1985). The GSL is an important vector for the transport of marine nutrients to the terrestrial ecosystem (Fariña et al. 2003). Knowledge about this species is therefore, vital not only for its own conservation but also for the health of the ecosystem in which it lives.

Here we compile the work we have carried on the physiology and foraging ecology of Galapagos sea lions over the last decade.

Galapagos Islands Environmental Conditions

The Galapagos Archipelago is a dynamic ecosystem characterized by two contrasting seasons, a cool and dry season (May–December) and a warm and wet season (December–May). Sea surface and ambient temperature vary ~ 10–12°C between seasons (http://coastwatch.pfeg.noaa.gov/). Sea surface temperature (SST) around the central part of the archipelago ranges from 19–28°C and chlorophyll-a from 0.3–0.8 mg/m^3 during different seasons each year (http://coastwatch.pfeg.noaa.gov/). The warm season is characterized by lower upwelling, warmer water temperatures (~ 25°C), reduced productivity and lower prey abundance (Pak and Zanveld 1973, Feldman 1986). Seasonal prey abundance and distribution changes in the Galapagos archipelago may influence sea lions' diving behavior and foraging success, while water temperature changes may influence its physiological response. Sea lions living in ecological and thermally variable environments require behavioral and physiological adaptations to stay in energy balance.

Foraging Ecology and Diving Behavior

Non-migratory resident species, such as GSL, should be capable of modifying their foraging behavior to accommodate changes in prey abundance and availability associated with a changing environment. Populations that are better adapted to change will have higher foraging success and greater potential for survival in the face of climate change.

Three fundamental foraging patterns in air-breathing marine vertebrates have been described: epipelagic, mesopelagic and benthic. Many sea lion species with access to extensive continental shelves have been described as benthic foragers (Arnould and Costa 2006, Costa et al. 2006). Coincidently these species are considered threatened. The GSL, a top predator in the Galapagos Islands, is also considered threatened in this ecosystem. Sea lions at the central part of the archipelago have access to a vast continental shelf, whereas at their western distribution, where they coexist with Galapagos fur seals, the access to the continental shelf is significantly reduced.

Prey species found in the diet of *Zalophus wollebaeki*, from multiple colonies include epipelagic fish (Clupeidae) of surface and coastal upwelling waters, mesopelagic fish (Myctophidae) found in deep waters of the open ocean, and demersal, benthopelagic or pelagic fish (Chlorophtalmidae, Serranidae and Mugilidae) found over muddy and sandy bottoms of the continental shelf or in shallow and deep waters between rocks (Salazar 2005, Froese and Pauly 2006).

Effective protection and conservation of this species requires knowledge of their foraging patterns and habitat utilization. We investigated the diving behavior and spatial habitat utilization of lactating female GSL, of a centrally located colony situated inside the highest density area of the population (Caamaño Islet, during March 2005 and August 2006), and a west, marginally located population (Fernandina Island, during March and October 2009), using time-depth recorders and satellite telemetry/GPS. To determine their response to environmental change, we studied and compared their diving behavior, habitat utilization and diet (by measuring C and N isotope ratios) during a warm (March) and a cold (August) season, at the centrally located colony.

Different modes of marine predator foraging patterns or strategies (epipelagic, mesopelagic and benthic) (Costa et al. 2004) are typically associated with different geographic regions or habitats. Rarely are all three foraging patterns observed in one population, one location, or one sex and age group. Most species tend to exhibit one foraging pattern exclusively or tend to vary geographically.

GSL at their central distribution, exhibited three foraging patterns or strategies during the warm season, when resources are less abundant; and individuals utilizing each pattern foraged in different locations. The three foraging strategies persisted during the cold, productive season (Fig. 1).

Sea lions exhibited epipelagic, mesopelagic and benthic dives, but these dive types were not exclusively associated with a foraging pattern (Villegas-Amtmann et al. 2008, Villegas-Amtmann and Costa 2010). GSL foraging strategies during the warm season were classified as shallow, deep and bottom divers. Shallow divers mostly dived on the shallow benthos, and presented some epipelagic dives (mean dive depth: 45.5 ± 20.1 m and duration: 2.9 ± 1.2 min). They dived mostly at night, likely feeding on vertically migrating prey. The deep divers presented characteristically deep mesopelagic dives, exhibited the highest percentage of dives deeper than 200 m (15%), over 500 m waters depth (mean dive depth: 108.2 ± 89.4 m and duration: 3.8 ± 2.1 min). Lastly, the behavior of bottom divers was consistent with benthic foraging (mean dive depth: 100.9 ± 39.1 m and duration: 4.9 ± 1.5 min). Dives were similar to benthic dives observed in Australian (Costa and Gales 2003), New Zealand (*Phocarctos hookeri*) (Costa and Gales 2000) and Southern sea lions (*Otaria Flavescens*) (Thompson et al. 1998, Villegas-Amtmann et al. 2008).

Minor differences were observed in the three foraging strategies during the cold season. The shallow divers dived during the day and night (mean dive depth: 38.8 ± 27.7 m and duration: 2.8 ± 1.4). The deep divers group, referred to as 'deep bottom divers' during the cold season, dived deeper than during the warm season (cold: 126.6 ± 47.1 m, duration: 5.5 ± 2.1 min) and likely benthically. These animals foraged in the same

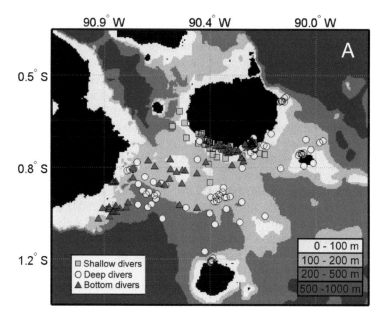

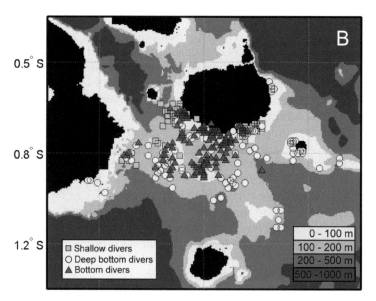

Figure 1. Satellite locations of female Galapagos sea lions from Caamaño Islet instrumented during A. a warm and B. a cold season (March 2005 and August-September 2006 respectively). N = 9 for each season. Different colors/shapes show the three dive type groups (Villegas-Amtmann et al. 2008, Villegas-Amtmann and Costa 2010).

areas but dived deeper, longer, exhibited longer bottom time duration and a higher percentage of similar depths on consecutive dives (indicative of benthic diving). The bottom divers exhibited similar characteristics than during the warm season (mean dive depth: 83.3 ± 38.1 m and duration: 4.1 ± 1.2 min) (Villegas-Amtmann and Costa 2010).

Dive types found in these groups are consistent with prey species found in the diet of the Galapagos sea lion. These include epipelagic Clupeids, mesopelagic Myctophids found in deep waters (200 to 1000 m), shallow and deep benthic Serranids, and shallow benthic Mugilids (0 to 120 m) among others (Salazar 2005). The persistence of the same foraging behavior in GSL between seasons, at their central distribution, is further supported by their same trophic level exhibited between seasons, in their $\delta^{15}N$ and $\delta^{13}C$ isotopic signatures, as they showed no change between seasons (Fig. 2) (Villegas-Amtmann et al. 2011).

However, their diet changes temporally as expected from changes in primary productivity, during the warm season they feed primarily on snake eels (*Ophichthus* sp.), herring (*Opisthonema* sp.) and fish from the family Sciaenidae, and during the cold season they feed on myctophids, sardines (*Sardinops* sp.) and fish from the family Ophididae (Salazar-Aldás 2007).

Although primary productivity changes seasonally in the Galapagos Archipelago, GSL individual specialization suggests the persistence of intraspecific competition throughout the year, reflected in the three foraging strategies found during both seasons, and the diversity of foraging areas utilized (Fig. 1) (Villegas-Amtmann et al. 2008, Villegas-Amtmann and Costa

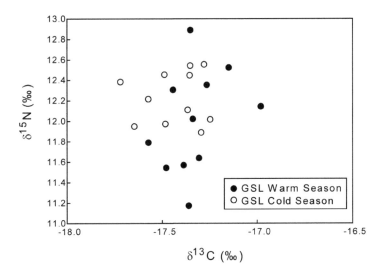

Figure 2. Galapagos sea lions $\delta^{13}C$ and $\delta^{15}N$ values for a warm (Mar. 2005, n = 11) and a cold (Aug-Sep. 2006, n = 11) season (Villegas-Amtmann et al. 2011).

2010). This competition indicates that resources might be limited year-round in this area of the Galapagos Islands, which has been described as a low productivity system (Martin et al. 1994, Sakamoto et al. 1998). These results are consistent with foraging observation on sea otters (*Enhydra lutris*) that suggests that intra-specific competition for limited resources is an ecological prerequisite for foraging specializations (Estes et al. 2003).

Similarly, the diving behavior of GSL from their western distribution, at Cabo Douglas, Fernandina island, also exhibited three foraging strategies, classified in a similar way as: shallow (mean dive depth: 19.7 ± 3.6 m, duration: 1.8 ± 0.2 min), intermediate (mean dive depth: 53.9 ± 13.9 m and duration: 2.5 ± 0.4 min) and deep divers (mean dive depth: 103.0 ± 18.3 m and duration: 3.6 ± 0.8 min). Sea lions from all groups exhibited day and night diving. The west coast of the archipelago is the most productive area of the Galapagos Islands (Martin et al. 1994, Sakamoto et al. 1998). The sea lion rookeries are located within the upwelling region of the cold Cromwell countercurrent. However, the consistent finding of three foraging strategies in the GSL western distribution, suggests that food resources might be limited or hardly accessible for sea lions at the west as well as for their central distribution (Villegas-Amtmann et al. 2008, Villegas-Amtmann and Costa 2010). This indicates that intra-specific competition may be a common feature in the foraging behavior of GSL (Villegas-Amtmann et al. 2013).

Foraging Trips

Lactating female GSL spent more time away from their pups (2.5 to 8 d) than the 21 h to 5 d reported for other sea lions species (Feldkamp et al. 1989, Merrick and Loughlin 1997, Thompson et al. 1998, Costa and Gales 2000, 2003, Kuhn 2006). As a consequence, their pups fasted longer than those of other species, indicating they may face higher nutritional stress. Longer trips are in part caused by the propensity of GSL females to haul out at locations other than their rookery. Between foraging trips females hauled out more frequently on other islands than they did on their breeding colony (Villegas-Amtmann et al. 2008). The necessity to haul out at other areas could be caused by shark predation risk and thermoregulatory energetic costs. It is energetically expensive to swim to shore but these factors might make it more expensive to remain at sea, as hypothesized by Thompson et al. (1998). Longer trip durations from the rookery indicate that females have to spend longer times at sea to locate food and meet energy requirements, causing them to haul out at areas away from their pups. This creates a conflict between continuing to forage and returning to the rookery to suckle their pup (Costa 2008). Longer fasting durations for pups influence their growth, survival rates, maternal dependence and, ultimately, population trends. GSL populations may increase slowly or decrease quickly, as seen during El Niño events (Trillmich and Limberger 1985), as a result of pups having to sustain long fasting periods.

Otariid trip durations from the rookery are expected to be shorter when pups are younger and not able to sustain long fasting periods (Merrick and Loughlin 1997, Mattlin et al. 1998, Beauplet et al. 2004). GSL females were nursing younger pups during the warm season, compared to the cold one. However, their trip durations from the rookery were not significantly different between seasons (warm: 4.18 ± 2 days, cold: 6.56 ± 4.7 days), and neither was the percent time they spent on land and at sea between the different seasons (Villegas-Amtmann and Costa 2010). This indicates that food resources are not sufficiently abundant during the warm season to sustain shorter trip durations from the rookery when pups are younger. When compared to California sea lions, its closest related species, GSL pup survival rates are lower (Hernandez-Camacho et al. 2008), with females successfully rearing pups only every other year and on average for the population at Caamaño Islet only one pup every three years (Müller, personal communication). Compared to California sea lions, GSL might be limited in their ability to invest resources in pup rearing (Mueller et al. 2011) because productivity is absolutely lower in the Galapagos, resulting in higher nutritional stress and lower pup survival in GSL pups.

Implications of Individual Specialization, Management and Conservation

GSL show individual specialization in the form of three distinct foraging strategies at their central distribution (where sea lions are the only pinniped present) (Villegas-Amtmann et al. 2008, Villegas-Amtmann and Costa 2010), and at their wester distribution (where they coexist with fur seals) (Villegas-Amtmann et al. 2013). Individual specialization has implications for demographic, ecological and evolutionary dynamics and is expected to be higher in environments with fewer competing species (Bolnick et al. 2003). Individual specialization can produce a delayed response to fluctuations in prey availability (Werner et al. 1981), which might explain why the GSL population, appears to be so sensitive to El Niño events, when prey availability is significantly diminished (Trillmich and Limberger 1985).

Sea lions living in equatorial regions face a more unpredictable oceanic system with lower productivity and lower range of variation than sea lions living at higher latitudes. Additionally, GSL encounter stronger El Niño events that lead to even lower productivity. These conditions faced by GSL make adjustments in their behavior challenging, favoring individual prey specialization and probably making them more vulnerable to climate change. GSL are already living on the edge of their habitat and any further shifts are likely to be difficult for this species to accommodate to.

Furthermore, individual specialization is diversifying (Roughgarden 1972), as populations with individual specialists may be open to future evolutionary diversification (Rosenzweig 1978, Lomnicki 1984, Doebeli 1996). Therefore, the high degree of individual specialization shown in GSL holds vast evolutionary potential to increase the great diversity that already characterizes this archipelago (Wolf et al. 2008, Jeglinski et al. 2015).

GSL are an endangered species with a fluctuating population (Trillmich 2015b) and not only their breeding, but also their foraging and other terrestrial habitats they use require protection for the species to survive. The foraging range of these individuals is wide and situated within a commercially exploited area, particularly for the centrally located ones. The fact that individuals adopt a range of strategies, hunting prey in epipelagic, mesopelagic and benthic environments, suggests that they do not have a strong, reliable prey resource, or that their foraging behavior is quite plastic and adaptable as a consequence of living in a low productive environment. Interactions with fisheries represent a great threat to the species (Kovacs et al. 2012), particularly because the number of registered fishers in Galapagos has shown a consistent increase over time (Baine et al. 2007) and some of the sea lion prey species are targeted by fisheries. The identification of three distinct foraging strategies with distinct geographic and depth zones found among this population has direct implications for management, particularly with regard to fisheries interactions. The information provided can be incorporated into new regulations to manage spatial and temporal aspects of the local fishing effort. Of particular interest are the areas around Santa Cruz Island from Isla Sin Nombre to Islas Plazas, around Santa Fe Island within the 100 m isobaths and waters between Santa Cruz, Floreana and Isabela Islands below 0° 40' S and within the 500 m isobath. In addition, north and west coastal areas around Fernandina Island, waters of the Bolivar Channel and southeast of Punta Mangle, south of Fernandina Island.

GSL exhibited a broad terrestrial habitat use as well. Centrally located females used several haul-out areas in addition to their breeding rookery-Caamaño islet, including Islas Plazas and Las Palmas in Santa Cruz Island, Los Cuatro Hermanos Islets, Tortuga Island, Santa Fe Island, Floreana Island and Punta Veintimilla at Isabela Island. Management plans should take into consideration the sea lions' wide foraging range, as well as their broad terrestrial habitat use.

Demography and diet studies should be continued to monitor population status and fisheries interactions, particularly during El Niño and La Niña years.

Diving Physiology and Behavior

Foraging success, influenced by ecological and physiological factors, is a key aspect in the persistence of many populations. GSL like all air-breathing marine vertebrates must have a finite amount of oxygen stores that constrains the duration of their dives (Costa 2007). Thus, the primary physiological factors that determine their diving capability, are its oxygen stores and the rate at which they are used (Kooyman 1989).

Given this constraint, it is not surprising that diving mammals have a higher oxygen storage capacity in their blood and muscle compared to terrestrial mammals of similar size (Kooyman 1989). Oxygen storage capacity also varies across this group of diving mammals according to their dive durations as observed in different pinniped species (seals, sea lions and fur seals) (Costa et al. 2004). The behavior and physiology of an animal are also affected by environmental conditions, such as ambient temperature and prey availability.

In general, seasonal differences in the behavior and physiology of marine tetrapods might be expected to be greater in species living at high latitude compared with those living closer to the equator. Seasonal variation in diving behavior, examined in penguins (Kirkwood and Robertson 1997), seals (Bennett et al. 2001, Burns and Kooyman 2001, Burns et al. 2004); and sea lions (Merrick and Loughlin 1997, Costa and Gales 2003), showed pronounced seasonal variation in their diving behavior. We investigated intra and inter-seasonal changes in female GSL oxygen stores (Hematocrit—red blood cell count, hemoglobin concentration, blood volume and myoglobin concentration) coupled with their diving behavior, during a warm (March 2005) and cold season (August 2006) at Caamaño Islet, Galapagos.

GSL exhibited plasticity in their diving physiology, by means of an intra and inter-seasonal variability in their oxygen stores. Within each season, physiological plasticity is attributed to dive behavior, given that animals with longer dive durations exhibited higher oxygen storage capacity. Oxygen stores of the three foraging strategies found in Galapagos sea lions (haematocrit (Hct), hemoglobin (Hb) and myoglobin (Mb) during the warm season and Hb, Mb and blood volume (Bv) during the cold season), increased according to dive durations. Furthermore, linear regression of ungrouped individuals' dive duration and Hb showed a significant relationship during both seasons (warm: $r^2 = 0.535$, $P = 0.02$; and cold $r^2 = 0.499$, $P = 0.05$) (Villegas-Amtmann and Costa 2010).

Oxygen stores and diving capability of marine vertebrates are known to increase during development and to vary between age and sex classes (Burns and Castellini 1996, Horning and Trillmich 1997a, Ponganis et al. 1999, Noren et al. 2001, Burns et al. 2005, Noren et al. 2005, Fowler et al.

2007, Weise and Costa 2007, Trillmich et al. 2008). GSL exhibited variability in oxygen stores and diving capability within the same sex and age class. Diverse foraging strategies with differing dive durations found in these sea lions, are probably conditioning the variability and enhancement of their oxygen stores (Villegas-Amtmann and Costa 2010), as it has been observed in muskrats (*Ondatra zibethicus*), harbour seals (*Phoca vitulina*) and tufted ducks (*Aythya fuligula*) (Kodama et al. 1977, Stephenson et al. 1989, MacArthur et al. 2003).

Physiological changes in oxygen stores between seasons are not attributed to diving behavior, given that animals exhibited similar dive behavior between them. These changes are mostly attributed to pregnancy as most females were pregnant during the cold season (see following section). GSL exhibited contrasting seasonal changes in oxygen stores: Hb and Bv were significantly higher while Mb was significantly lower during the warm season compared to the cold one. Pregnancy is associated with major physiological adjustments in mammals and could be contributing to the observed seasonal physiological changes in the GSL. Nine of 12 females during the cold season were in a late pregnancy status (~ seven months pregnant) (Villegas-Amtmann et al. 2009). These females face a higher oxygen demand to supply the fetus especially under hypoxic conditions when diving, as maternal perfusion of the placenta might continue during diving (Elsner et al. 1970, Liggings et al. 1980). The observed decrease in blood oxygen stores (Bv and Hb concentration) during the cold season was compensated for by an increase in muscle oxygen stores (Mb) (Fig. 3). This shift from blood to muscle oxygen stores may be related to the increased oxygen demands of a growing fetus. Furthermore, it may also be indicative of iron deficiency anemia, associated with iron limitation in their diet, and/or an increased demand for iron during pregnancy (Halvorsen and Halvorsen 1973, Hallberg 1988).

Total oxygen stores and the relative contribution of its different components also changed between seasons. Total body oxygen stores were significantly higher during the warm season (77.4 ± 6.2 ml/kg) compared to the cold season (61.5 ± 7.6 ml/kg) (t-test P = 0.01) (Fig. 3). During the warm season 8.2%, 66.9% and 24.9% of total oxygen was stored in lung, blood and muscle, while during the cold season it was 9.8%, 54.6% and 33.6% respectively (Villegas-Amtmann and Costa 2010).

The relative distribution of oxygen in the tissues of GSL is consistent with a deep-prolonged diver like a Weddell seal (*Leptonychotes weddellii*) (5%, 66%, 29%) or a sperm whale (*Physeter macrocephalus*) (10%, 58%, 34%) in contrast to that reported for a 'typical' otariid, the California sea lion (21%, 45%, 34%) (Kooyman and Ponganis 1998). Furthermore, GSL Hb and Mb values found are the highest reported for an otariid and amongst the highest reported for any diving vertebrate (Kooyman 1989, Costa et al. 1998,

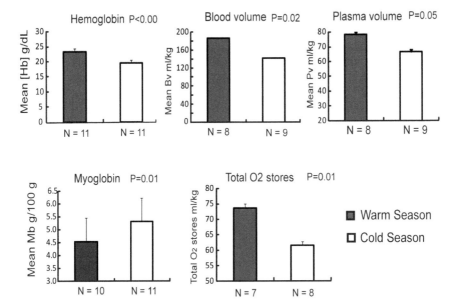

Figure 3. Galapagos sea lion females' hemoglobin, blood volume, plasma volume, myoglobin and total oxygen stores for a warm and cold season (March 2005 and August-September 2006) at Caamaño Islet. P-values showing significant differences are presented at the top right (Villegas-Amtmann and Costa 2010).

Kooyman and P.J. 1998, Noren and Williams 2000, Ponganis and Kooyman 2000, Fowler et al. 2007, Weise and Costa 2007).

Knowledge of an organism's physiological plasticity is important to assess its vulnerability to climate change. The documented physiological plasticity coupled with diving behavior in the same sex and age class of an air-breathing diving vertebrate, the GSL, shows that otariid species might be physiologically more resilient to environmental change than previously thought.

Reproductive Physiology

The reproductive biology of pinnipeds has several special features. In most species of seals and sea lions, a tight synchrony of births ensures that pups are born at the optimal time of year for weather, prey abundance, and general availability of resources necessary for maximal survival of offspring (Boyd 1991, Atkinson 1997). Lactation and mating strategies have evolved to ensure that adult females have the ability to reproduce annually.

The GSL, breeding biology differs from typical otariid reproductive biology. While most otariid species reproduce annually and have narrowly defined breeding seasons (~ two months) (Atkinson 1997), GSL exhibit a protracted breeding season lasting about four months and varying between two and 11 months among the different islands (Wolf and Trillmich 2007). Furthermore, one- and two-year-old pups often continue to suckle if their mothers fail to give birth that year and pups are weaned at a time when they are independently foraging—usually around two to three yr old (Trillmich 1986, Heath 2002, Jeglinski et al. 2012).

The GSL breeding season is much longer than for sympatric Galapagos fur seals (*Arctocephalus galapagoensis*) or closely related California sea lions (*Zalophus californianus*). Births occur from June to March, but the peak pupping period varies among rookeries and years. Their reproductive period shifts slightly from year to year and from island to island. Most pinnipeds of temperate zones aggregate on land mainly during short, highly synchronized reproductive periods. In contrast, otariids inhabiting the Galapagos are non-migratory and maintain large terrestrial colonies throughout the year (Trillmich 1979).

A high plane of nutrition will provide a female with greater energy stores, facilitating a pregnancy that result in the birth of a large, healthy pup (Atkinson 1997). For example, it has been suggested that the prolonged gestation periods of Australian sea lions may be due to the lack of available resources in a benthic forager (Gales et al. 1997, Costa and Gales 2003, Fowler et al. 2007). Similar factors could be influencing the protracted and low synchrony observed in the breeding season of GSL.

Progesterone levels have been used to differentiate pregnant from non-pregnant sea lions (Greig et al. 2007) because the corpus luteum of pinnipeds produces substantial amounts of progesterone throughout gestation (Craig 1964, Hobson and Boyd 1984) and declines rapidly in circulation just before parturition (Raeside and Ronald 1981, Boyd 1983, 1984, Reijnders 1990). We studied the reproductive physiology of wild female GSL in contrasting seasons, at Caamaño islet during March 2005 and August 2006 by examining progesterone and estrogen concentrations in serum and plasma.

The known peak pupping season for GSL at Caamaño islet occurs from late October to mid-December. Progesterone levels are high throughout gestation (Craig 1964, Hobson and Boyd 1984) and differences in its concentration between pregnant and non-pregnant pinnipeds may not be detectable until one-two months after implantation (Daniel 1981, Reijnders 1990, Gardiner et al. 1999, McKenzie et al. 2005a). New Zealand fur seals (*Arctocephalus forsteri*) with progesterone values of 6 ng/ml were unable to carry a pregnancy to term (McKenzie et al. 2005b) and California sea lions exhibited progesterone concentrations of 7.44 ± 2.77 ng/ml during mid-pregnancy (Greig et al. 2007). We found a greater proportion of females with high progesterone concentrations (> 8 ng/ml) during August, two–four months before the known pupping season (six-seven months after implantation), than during March, seven–nine months before the known pupping season (before or within one month of implantation), as expected. Field examinations confirmed the pregnancy status of these females (Fig. 4) (Villegas-Amtmann et al. 2009).

In March, within one month of implantation, and assuming GSL exhibit a seasonal and synchronous reproductive cycle as do other otariid species, we would not expect to find high progesterone concentrations. However, one-half of the females exhibited high progesterone concentrations (8.4–15.8 ng/ml), whereas the other half exhibited contrastingly low concentrations (0.2–0.7 ng/ml). Females with high progesterone concentrations were categorized as pregnant (Fig. 4) (Villegas-Amtmann et al. 2009).

Mean estrogen concentrations were neither significantly different between seasons (0.6 ± 0.2 and 0.7 ± 0.5 ng/ml), nor between females visually identified as pregnant versus non-pregnant (Villegas-Amtmann et al. 2009).

Similar and high progesterone concentrations as well as palpably pregnant female GSL during two seasons that are five months apart, provides strong evidence for remarkably low synchrony and minor seasonality in the breeding cycle of this species. This feature is unique among otariids, because it does not follow the typical pinniped pattern, or that of

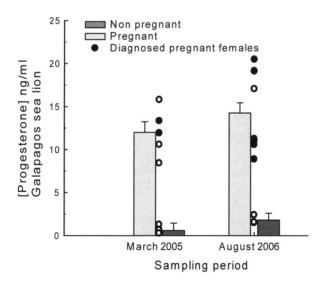

Figure 4. Mean and individual progesterone concentrations (ng/ml ± SE) from female Galapagos sea lions diagnosed as pregnant (P) and non-pregnant (NP) based on progesterone concentrations and field examination during March 2005 (P: n = 5, NP: n = 6) and August 2006 (P: n = 9, NP: n = 3) at Caamaño Islet. Darker dots indicate females diagnosed as pregnant during field examinations (Villegas-Amtmann et al. 2009).

its closest relative, the California sea lion (Odell 1975). In most pinniped species, implantation occurs when day length is declining; a photoperiod of about 12 h may provide the signal for implantation in some species (Boyd 1991). In the Galapagos Islands, situated at the equator, seasons are generally less pronounced and predictable than at higher latitudes, and the season of births can be protracted (Kenyon 1981, Trillmich and Majluf 1981). This phenomenon is observed in Australian sea lions, which show a 17.5-month breeding cycle (Ling and Walker 1978, Higgins 1993) and may have little dependence on photoperiod. In the Galapagos Islands, day length is 12 h year-round, thus pinnipeds living here lack a photoperiodic cue and must rely on different signals for implantation.

Additionally, the low synchrony found in the breeding cycle of GSL could be an adaptation to an environment with variable productivity and prey availability, as is the breeding asynchrony observed among colonies of Australian sea lions (Goldsworthy et al. 2004, McKenzie et al. 2005a). Nutritional intake and reproductive success are closely linked in mammalian reproduction (Widdowson 1981). Unpredictable prey resources in conjunction with seasons of low productivity in the Galapagos might make it difficult for the sea lion population to maintain a short and well-defined reproductive season. Adjusting reproductive behavior to allow for flexibility in their breeding season to match that of resource availability might be more energetically cost effective.

Measurements of progesterone concentrations between years, for example, El Niño Southern Oscillation versus non–El Niño Southern Oscillation years, could be used to infer whether pregnancy rates are affected by such environmental variations.

Foraging Niche Overlap

Survival of a population is achieved through foraging success, that will influence energy allocation to offspring and therefore, population growth and reproductive success. Being successful at acquiring prey is determined by prey abundance, accessibility, and species interactions, specifically competition for prey resources. Ecological niche separation can occur by organisms differing in their breeding chronology, foraging behavior, foraging time, prey type, trophic position, and life history strategies (Macarthur and Levins 1967, Hutchinson 1978, Trivelpiece et al. 1987, Miller et al. 2010, Young et al. 2010, Wilson 2010).

Inter-specific competition, defined as reciprocal negative effects of one species on another, is an important process determining the structure of natural communities (Cody and Diamond 1975, Diamond and Case 1986, Morin 1999, Chase et al. 2002). Considering intra-specific dynamics on a spatial and temporal scale, when studying species interactions, allows the detection of inter-specific interactions on a finer scale. Closely related species with similar life-history strategies often share similar niches such as fur seals and sea lions. In general, where fur seals and sea lions live in sympatry, the fur seal population is typically larger and they appear to outcompete sea lions (Costa et al. 2006).

In conditions of limited resources, competition between species implies a reduction in some population attributes, such as growth, survival or fecundity (Begon et al. 1996). The observation that most sea lion populations are smaller when sympatric with fur seals suggests that some interspecific competition may occur. A number of studies have examined potential competition between sympatric fur seal and sea lions with mixed results. Some have found ecological segregation with no trophic overlap (Antonelis et al. 1990, Dellinger and Trillmich 1999, Jeglinski et al. 2013), while others have found some dietary overlap (Szteren et al. 2004, Aurioles-Gamboa and Camacho-Rios 2007, Waite et al. 2012, Pablo-Rodriguez et al. 2015). However, most competition studies between species are conducted from a population-level approach and few have examined inter-specific competition in conjunction with intra-specific competition, with an individual-based approach.

The Galapagos fur seal and the GSL coexist on several islands and potentially compete. Both species are endangered, and although the sea lions are more abundant (14–16,000 indiv.) than fur seals (6,000–8,000 indiv.) (Alava and Salazar 2006, Trillmich 2015a, b), the local fur seal population is usually larger where they occur sympatrically. Such is the case at Cabo Douglas, Fernandina Island, where the fur seal population is larger than that of sea lions (fur seals: 215 ± 25 and sea lions: 42 ± 11 individuals (Jeglinski et al. 2013)). On average, fur seals are known to dive shallower than sea lions, which holds true for Galapagos otariids (Kooyman and Trillmich 1986a, Horning and Trillmich 1997b, Villegas-Amtmann et al. 2008). GSL and fur seals diving behavior and trophic position (stable isotopes), at the population level, shows little to no overlap (Paez-Rosas et al. 2012, Jeglinski et al. 2013). However, acknowledging individual-level variation can benefit ecological studies as it represents a more complete description of a biological system. Here we examined potential foraging niche overlap in sympatric Galapagos fur seals and sea lions at Cabo Douglas, Fernandina Island, from simultaneously collected dive and movement data to examine spatial and temporal inter-specific competition during March and October 2009.

Galapagos Sea Lions and Fur Seals Diving Behavior

At Cabo Douglas, Fernandina Island, Galapagos fur seals exhibited a uniform diving behavior with only one foraging strategy and diving predominantly at night, between 0–80 m depths. In contrast, GSL exhibited greater individual variability in their diving behavior, presenting 3 distinct foraging strategies (shallow, intermediate and deep divers), as previously mentioned. Sea lions dived mostly at night (63% of dives) between 0–40 m, within fur seals' diving depth range (Villegas-Amtmann et al. 2013). Sea lions dove deeper, longer and spent more time at the bottom of their dives than fur seals, as expected from previous work (Kooyman and Trillmich 1986a, b). Sea lions are significantly larger than fur seals and larger animals have proportionately greater oxygen stores and breath-hold capacity (Costa 1993, Costa et al. 2004, Weise and Costa 2007).

The west coast of the archipelago, where this study was carried out, is the most productive area of the Galapagos archipelago (Martin et al. 1994, Sakamoto et al. 1998). The fur seal and sea lion rookeries are located within the upwelling region of the cold Cromwell countercurrent. Although in a productive area, sea lions exhibited greater foraging effort as shown by the greater percentage of time spent diving compared to fur seals, possibly due to reduced prey availability or different prey distribution. This greater foraging effort and the existence of three foraging strategies in the diving behavior of GSL at their central and western distribution, suggests that greater intra-specific competition may be a common feature in the foraging behavior of GSL. Furthermore, food resources might be limited or less accessible for sea lions at their west as well as for their central distribution in the Galapagos Islands (Villegas-Amtmann and Costa 2010, Villegas-Amtmann et al. 2008, 2013).

Sea lions in the western part of the archipelago are known to feed on deep water pelagic and demersal fish such as sardines (*Sardinops sagax*), jack mackerel (*Trachurus symmetricus*) and *Chlorophtalmus* sp. (Dellinger and Trillmich 1999, Paez-Rosas et al. 2012). In contrast, fur seals feed on shallower waters, on prey from the deep scattering layer when they migrate to the surface during night, e.g., myctophids, bathylagids and cephalopods (Clarke and Trillmich 1980, Dellinger and Trillmich 1999, Paez-Rosas et al. 2012). Furthermore, sea lions are known to forage over the shelf (Villegas-Amtmann et al. 2008) and in the western region of the Galapagos archipelago the shelf habitat is very limited.

Costa and Gales (2003) postulated that increased foraging effort may explain why many pinnipeds and penguins that feed benthically have small stable or declining populations, while the many epipelagic divers have large stable and or increasing populations. This appears to hold true at Fernandina Island, where the fur seal population is significantly larger than the sea lion population. Additionally, there are fundamental differences in prey availability and habitat characteristics required for these species. The west Coast of Fernandina Island is an ideal habitat for fur seals, facilitating pelagic foraging on prey from the deep scattering layer, and is located within the region of highest offshore productivity. However, it is also a region with minimal continental shelf or benthic habitat, offering an environment with minimum optimal conditions for sea lions, representing marginal habitat for this species.

Foraging Niche Overlap

Foraging niche overlap between Galapagos fur seals and sea lions occurs at night when fur seals forage and most sea lion dives occur (63% of dives). Overlapping dive depths at night between these two species occurred in the first 30 m of the water column, suggesting that both species could be pursuing vertically migrating prey. Although overall sea lions dive deeper (day and night) than fur seals, most of their night dives were shallower and occurred within the foraging depth range of fur seals. Sea lions exhibited the greatest percentage of their night dives (22% of total dives) just before sunrise and after sunset (around 5 and 18 h); when fur seals dove the least. Fur seals and sea lions foraging behavior overlapped at 19 and 21 h. Fur seals exhibited the greatest percentage of night dives between 19–22 h (42%) and 18.5% of sea lion dives also occurred at 19 and 21 h (Fig. 5) (Villegas-Amtmann et al. 2013).

Considering sea lions' foraging strategies, sea lions from the deep diving strategy exhibited the greatest percentage of night dives (80.3%) and foraging overlap with fur seals, in time (19–21 h), depth during overlapping time (0–40 m) and foraging range (37.7%) (Figs. 5 and 6). Fur seal dives overlapped with sea lion shallow divers at 21 h and 22 h, between 0–30 m depths; and with sea lion intermediate divers at 19 h and 21 h, within 0–40 m (Fig. 5) (Villegas-Amtmann et al. 2013). It is possible that fur seals and sea lion deep divers are hunting for similar prey such as myctophids and cephalopods at these shallower depths. Myctophids and cephalopods are main prey items in the diet of Galapagos fur seals (Dellinger and Trillmich 1999, Paez-Rosas et al. 2012) and although they have not been identified in the diet of GSL at their western distribution, they are part of their diet in their central, southern and eastern distribution (Salazar 2005, Paez-Rosas and Aurioles-Gamboa 2010).

Overall, there is a spatial niche separation between species as fur seals foraging range is significantly larger than that of sea lions (Fig. 6). This is partly explained by differences in their provisioning strategies (Trillmich 1986), given that fur seals foraging trip durations are longer than those of sea lions (March: 35.3 ± 14.6 hr vs. 15.1 ± 5.3 hr, October: 18.5 ± 8.9 hr vs. 12.9 ± 4.3 hr, respectively). However, the fur seal area of highest diving density (north of the rookery–Cabo Douglas) is small and mostly located within the sea lion area of highest diving density. Therefore, the coastal area just north of Cabo Douglas is a foraging "hot spot" for both species where competition might occur (Fig. 6) (Villegas-Amtmann et al. 2013).

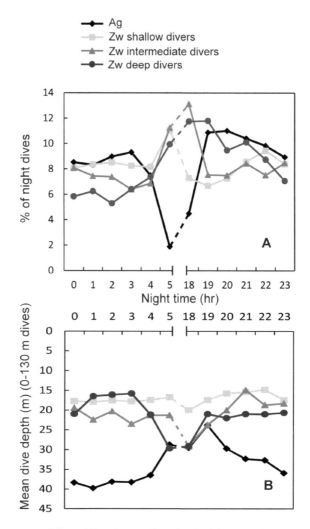

Figure 5. Percentage of dives (A) and mean dive depth of dives (B) for three Galapagos sea lion diving strategies and Galapagos fur seals, covering the range of fur seals dive depth (0–130 m) during night time (Mar. & Oct. 2009) at Cabo Douglas, Fernandina Island (Villegas-Amtmann et al. 2013).

By studying ecological interactions with an individual-based approach, we were able to detect foraging niche overlap on a finer scale that was previously overlooked. Individual specialization should be incorporated into models of food webs, competition, predator-prey and host-parasite interactions (Bolnick et al. 2003).

Furthermore, niche breadth is increased with increased environmental uncertainty and with decreased productivity (Macarthur and Levins 1967).

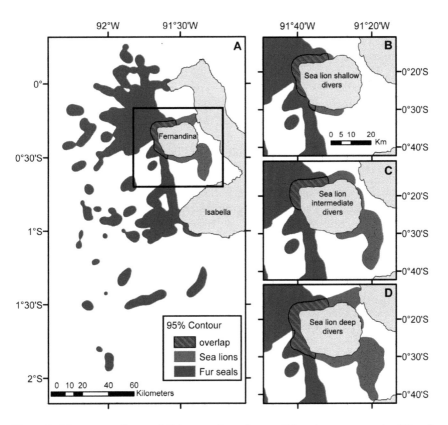

Figure 6. Galapagos sea lions and Galapagos fur seals overall foraging range overlap (A) and overlap between fur seals and each sea lion diving strategy (B–D) based on 95% contour kernel analysis (Mar. & Oct. 2009) at Cabo Douglas, Fernandina Island (Villegas-Amtmann et al. 2013).

Thus, the potential for foraging niche overlap between Galapagos fur seals and sea lions is possibly greater during warmer periods when prey availability is lower, such as it has been observed in California sea lions and Guadalupe fur seals (Aurioles-Gamboa and Camacho-Rios 2007, Pablo-Rodriguez et al. 2015). Therefore, if oceans continue on a warming trend, the continuation of conservation programs for these species becomes crucial.

Acknowledgements

We thank the TOPP (Tagging of Pacific Pelagics) Program supported by the Sloan, Packard and Moore Foundations, the ONR, UC MEXUS, CONACYT, NSF Office of Polar Programs, CRS (Center for Remote Sensing) at UCSC, the Biology Department at UCSC, the Society for Marine Mammalogy, the Alaska SeaLife Center Pinniped Research Program and the National Marine Fisheries Service for funding support. We thank The Charles Darwin Foundation and Parque Nacional Galapagos for their logistics and fieldwork support. We also thank D. Anchundia, S. Atkinson, D. Aurioles, D. Casper, E. Garcia Bartholomei, P. Howorth, L. Huckstadt, J. Jeglinski, C. Kuhn, C. Martínez, S. Maxwell, B. McDonald, V. Michuy, B. Müller, D. Páez, A. Parás, P. Robinson, M. Rutishauser, S. Salazar, S. Simmons, M. Szphegyi, E. Stetson, F. Trillmich, M. Zavanelli, and volunteers from the Darwin Station for their help in the field. Thanks to K. Mashburn and J. Ramsay at the Alaska SeaLife Center for their help in the laboratory.

Keywords: Conservation, diving behavior, diving physiology, foraging behavior, foraging ecology, foraging niche overlap, Galapagos sea lion, oxygen stores, physiological ecology, reproduction, *Zalophus wollebaeki*

References

Alava, J.J. and S. Salazar. 2006. Status and conservation of Otariids in Ecuador and the Galapagos Islands. pp. 495–520. *In*: A.W. Trites, S.K. Atkinson, D.P. Demaster, L.W. Fritz, T.S. Gelatt, L.D. Rea and K.M. Wynne [eds.]. Sea Lions of the World. Fairbanks: Alaska Sea Grant College Program, Alaska, USA.

Antonelis, G.A., B.S. Stewart and W.F. Perryman. 1990. Foraging characteristics of female Northern fur seals (*Callorhinus ursinus*) and California sea lions (*Zalophus californianus*). Can. J. Zool./Rev. Can. Zool. 68: 150–158.

Arnould, J.P.Y. and D.P. Costa. 2006. Sea lions in drag, fur seals incognito: Insights from the otariid deviants. pp. 309–323. *In*: A.W. Trites, S.K. Atkinson, D.P. Demaster, L.W. Fritz, T.S. Gelatt, L.D. Rea and K.M. Wynne [eds.]. Sea Lions of the World. Alaska Sea Grant College Program, University of Alaska Fairbanks.

Atkinson, S. 1997. Reproductive biology of seals. Rev. Reprod. 2: 175–194.

Aurioles-Gamboa, D. and F.J. Camacho-Rios. 2007. Diet and feeding overlap of two otariids, *Zalophus californianus* and *Arctocephalus townsendi*: Implications to survive environmental uncertainty. Aquat. Mamm. 33: 315–326.

Baine, M., M. Howard, S. Kerr, G. Edgar and V. Toral. 2007. Coastal and marine resource management in the Galapagos Islands and the Archipelago of San Andres: Issues, problems and opportunities. Ocean Coast. Manage. 50: 148–173.

Beauplet, G., L. Dubroca, C. Guinet, Y. Cherel, W. Dabin, C. Gagne et al. 2004. Foraging ecology of subantarctic fur seals *Arctocephalus tropicalis* breeding on Amsterdam Island: Seasonal changes in relation to maternal characteristics and pup growth. Mar. Ecol. Prog. Ser. 273: 211–225.

Begon, M., M. Mortimer and D. Thompson. 1996. Population Ecology. A Unified Study of Animals and Plants. Blackwell Science, Cambridge.

Bennett, K.A., B.J. Mcconnell and M.A. Fedak. 2001. Diurnal and seasonal variations in the duration and depth of the longest dives in southern elephant seals (*Mirounga leonina*): Possible physiological and behavioural constraints. J. Exp. Biol. 204: 649–662.

Bolnick, D.I., R. Svanbäck, J.A. Fordyce, L.H. Yang, J.M. Davis, C.D. Hulsey et al. 2003. The ecology of individuals: Incidence and implications of individual specialization. Am. Nat. 161: 1–28.

Boyd, I.L. 1983. Luteal regression, follicle growth and the concentration of some plasma steroids during lactation in grey seals (*Halichoerus-Grypus*). J. Reprod. Fertil. 69: 157–164.

Boyd, I.L. 1984. Development and regression of the corpus-luteum in grey seal (*Halichoerus-Grypus*) ovaries and its use in determining fertility rates. Can. J. Zool.-Rev. Can. Zool. 62: 1095–1100.
Boyd, I.L. 1991. Environmental and physiological factors controlling the reproductive-cycles of pinnipeds. Can. J. Zool.-Rev. Can. Zool. 69: 1135–1148.
Brodie, P.F. 1975. Cetacean energetics, an overview of intraspecific size variation. Ecology 56: 152–161.
Burns, J.M. and M.A. Castellini. 1996. Physiological and behavioral determinants of the aerobic dive limit in Weddell seal (*Leptonychotes weddellii*) Pups. J. Comp. Physiol. B 166: 473–483.
Burns, J.M. and G.L. Kooyman. 2001. Habitat use by Weddell seals and emperor penguins foraging in the Ross Sea, Antarctica. Am. Zool. 41: 90–98.
Burns, J.M., D.P. Costa, M.A. Fedak, M.A. Hindell, C.J.A. Bradshaw, N.J. Gales et al. 2004. Winter habitat use and foraging behavior of crabeater seals along the Western Antarctic Peninsula. Deep-Sea Res. Pt. II 51: 2279–2303.
Burns, J.M., D.P. Costa, K. Frost and J.T. Harvey. 2005. Development of body oxygen stores in Harbor seals: Effects of age, mass, and body composition. Physiol. Biochem. Zool. 78: 1057–1068.
Chase, J.M., P.A. Abrams, J.P. Grover, S. Diehl, P. Chesson, R.D. Holt et al. 2002. The interaction between predation and competition: A review and synthesis. Ecol. Lett. 5: 302–315.
Clarke, M.R. and F. Trillmich. 1980. Cephalopods in the diet of fur seals of the Galapagos Islands J. Zool. 190: 211–215.
Cody, M.L. and J.M. Diamond. 1975. Ecology and Evolution of Communities. Belknap Press of Harvard University Press, Cambridge, Massachusetts.
Costa, D.P. 1991. Reproductive and foraging energetics of high-latitude penguins, albatrosses and pinnipeds—implications for life-history patterns. Am. Zool. 31: 111–130.
Costa, D.P. 1993. The Relationship Between Reproductive and Foraging Energetics and the Evolution of the Pinnipedia. Oxford University Press, Symposium Zoological Society of London, London.
Costa, D.P., N.J. Gales and D.E. Crocker. 1998. Blood volume and diving ability of the New Zealand sea lion, *Phocarctos hookeri*. Physiol. Zool. 71: 208–213.
Costa, D.P. and N.J. Gales. 2000. Foraging energetics and diving behavior of lactating New Zealand sea lions, *Phocarctos hookeri*. J. Exp. Biol. 203: 3655–3665.
Costa, D.P. and N.J. Gales. 2003. Energetics of a benthic diver: Seasonal foraging ecology of the Australian sea lion, *Neophoca cinerea*. Ecol. Monogr. 73: 27–43.
Costa, D.P., C.E. Kuhn, M.J. Weise, S.A. Shaffer and J.P.Y. Arnould. 2004. When does physiology limit the foraging behavior of freely diving mammals? Int. Congr. Ser. 1275: 359–366.
Costa, D.P., M.J. Weise and J.P.Y. Arnould. 2006. Potential influences of whaling on the status and trends of pinniped populations. pp. 344–359. *In*: J.A. Estes, D.P. Demaster, D.F. Doak, T.M. Williams and R.L.J. Rownell [eds.]. Whales, Whaling, and Ocean Ecosystems. University of California Press, Berkeley, Los Angeles, London.
Costa, D.P. 2007. Diving physiology of marine vertebrates. pp. 1–7. *In*: Encyclopedia of Life Sciences. John Wiley & Sons, Ltd. www.els.net.
Costa, D.P. 2008. A conceptual model of the variation in parental attendance in response to environmental fluctuation: Foraging energetics of lactating sea lions and fur seals. Aquat. Conserv. 17: S44–S52.
Craig, A.M. 1964. Histology of reproduction and the estrus cycle in the female fur seal, *Callorhinus-Ursinus*. J. Fish. Res. Board Can. 21: 773–811.
Daniel, J.C. 1981. Delayed implantation in the northern fur seal (*Callorhinus ursinus*) and other pinnipeds. J. Reprod. Fertil. Suppl. 29: 35–50.
Darwin, C. 1859. On the Origin of Species by Means of Natural Selection or the Preservation of Favoured Races in the Struggle for Life. Murray J, ed. London.
Davidson, A.D., A.G. Boyer, H. Kim, S. Pompa-Mansilla, M.J. Hamilton, D.P. Costa et al. 2012. Drivers and hotspots of extinction risk in marine mammals. Proc. Natl. Acad. Sci. USA 109: 3395–3400.

Dellinger, T. and F. Trillmich. 1999. Fish prey of the sympatric Galapagos fur seals and sea lions: Seasonal variation and niche separation. Can. J. Zool. 77: 1204–1216.
Diamond, J. and T.J. Case. 1986. Community Ecology. Harper and Row, New York.
Doebeli, M. 1996. An explicit genetic model for ecological character displacement. Ecology 77: 510–520.
Elsner, R., D.D. Hammond and H.R. Parker. 1970. Circulatory responses to asphyxia in pregnant and fetal animals: A comparative study of Weddell seals and sheep. Yale J. Biol. Med. 42: 202–217.
Estes, J.A., M.L. Riedman, M.M. Staedler, M.T. Tinker and B.E. Lyon. 2003. Individual variation in prey selection by sea otters: Patterns, causes and implications. J. Anim. Ecol. 72: 144–155.
Fariña, J.M., S. Salazar, K.P. Wallem, J.D. Witman and J.C. Ellis. 2003. Nutrient exchanges between marine and terrestrial ecosystems: The case of the Galapagos sea lion *Zalophus wollebaeki*. J. Anim. Ecol. 72: 873–887.
Feldkamp, S.D., R.L. Delong and G.A. Antonelis. 1989. Diving patterns of California sea lions, *Zalophus californianus*. Can. J. Zool. 67: 872–883.
Feldman, G.C. 1986. Patterns of Phytoplankton Production around the Galapagos Islands. Springer-Verlag, Berlin, Germany.
Ferguson, S.H. and J.W. Higdon. 2006. How seals divide up the world: Environment, life history, and conservation. Oecologia 150: 318–329.
Fowler, S.L., D.P. Costa, J.P.Y. Arnould, N.J. Gales and J.M. Burns. 2007. Ontogeny of oxygen stores and physiological diving capability in Australian sea lions. Funct. Ecol. 21: 922–935.
Froese, R. and D. Pauly. 2006. FishBase. In: R. Froese and D. Pauly [eds.]. Accessed Oct. 2006. www.fishbase.org.
Gales, N.J., P. Williamson, L.V. Higgins, M.A. Blackberry and I. James. 1997. Evidence for a prolonged postimplantation period in the Australian sea lion (*Neophoca cinerea*). J. Reprod. Fertil. 111: 159–163.
Gardiner, K.J., I.L. Boyd, B.K. Follett, P.A. Racey and P.J.H. Reijnders. 1999. Changes in pituitary, ovarian, and testicular activity in harbour seals (*Phoca vitulina*) in relation to season and sexual maturity. Can. J. Zool. 77: 211–221.
Goldsworthy, S.D., P. Shaughnessy and R. Mcintosh. 2004. Plasticity in Gestation Length in Response to Environmental Conditions in Australian Sea Lions *Neophoca cinerea*: New Hypotheses to an Enigmatic Life History. Abstracts of the 22nd Wakefield Fisheries Symposium, Sea lions of the world: Conservation and Research in the 21st Century, Anchorage, Alaska.
Greig, D.J., K.L. Mashburn, M. Rutishauser, F.M.D. Gulland, T.M. Williams and S. Atkinson. 2007. Seasonal changes in circulating progesterone and estrogen concentrations in the California sea lion (*Zalophus californianus*). J. Mammal. 88: 67–72.
Hallberg, L. 1988. Iron balance in pregnancy. pp. 115–127. In: H. Berger [ed.]. Vitamins and Minerals in Pregnancy and Lactation. Raven Press, New York.
Halvorsen, K. and S. Halvorsen. 1973. Early anemia: Its relation to postnatal growth rate, milk feeding, and iron availability: Experimental study in rabbits Arch. Dis. Child. 48: 842–849.
Heath, C.B. 2002. California, Galapagos, and Japanese sea lions. *Zalophus californianus, Z. wollebaeki,* and *Z. japonicus*. pp. 180–181. In: W.F. Perrin, B. Wursig and J.G.M. Thewissen [eds.]. Encyclopedia of Marine Mammals. Academic Press, London.
Hernandez-Camacho, C.J., D. Aurioles-Gamboa, J. Laake and L.R. Gerber. 2008. Survival rates of the California sea lion, *Zalophus californianus*, in Mexico. J. Mammal. 89: 1059–1066.
Higgins, L.V. 1993. The nonannual, nonseasonal breeding cycle of the Australian sea lion, *Neophoca-cinerea*. J. Mammal. 74: 270–274.
Hobson, B.M. and I.L. Boyd. 1984. Gonadotropin and progesterone concentrations in placentae of grey seals (*Halichoerus-Grypus*). J. Reprod. Fertil. 72: 521–528.
Hockey, P.a.R. and W.A. Wilson. 2003. Reproductive traits of marine terns (Sternidae): Evidence for food limitation in the tropics? Ostrich 74: 110–116.
Horning, M. and F. Trillmich. 1997a. Development of hemoglobin, hematocrit, and erythrocyte values in Galapagos fur seals. Mar. Mamm. Sci. 13: 100–113.

Horning, M. and F. Trillmich. 1997b. Ontogeny of diving behaviour in the Galapagos fur seal. Behaviour 134: 1211–1257.
Hutchinson, G.E. 1978. An Introduction to Population Ecology. Yale University Press, New Haven/London.
Jeglinski, J.W.E., C. Werner, P.W. Robinson, D.P. Costa and F. Trillmich. 2012. Age, body mass and environmental variation shape the foraging ontogeny of Galapagos sea lions. Mar. Ecol. Prog. Ser. 453: 279–296.
Jeglinski, J.W.E., K.T. Goetz, C. Werner, D.P. Costa and F. Trillmich. 2013. Same size – same niche? Foraging niche separation between sympatric juvenile Galapagos sea lions and adult Galapagos fur seals. J. Anim. Ecol. 82: 694–706.
Jeglinski, J.W.E., J.B.W. Wolf, C. Werner, D.P. Costa and F. Trillmich. 2015. Differences in foraging ecology align with genetically divergent ecotypes of a highly mobile marine top predator. Oecologia 179: 1041–1052.
Johnson, C.N. 1998. Rarity in the tropics: Latitudinal gradients in distribution and abundance in Australian mammals. J. Anim. Ecol. 67: 689–698.
Kenyon, K.W. 1981. Monk Seals, Monachus Flemming, 1822. Academic Press, New York.
Kirkwood, R. and G. Robertson. 1997. Seasonal change in the foraging ecology of emperor penguins on the Mawson Coast, Antarctica. Mar. Ecol. Prog. Ser. 156: 205–223.
Kodama, A.M., R. Elsner and N. Pace. 1977. Effects of growth, diving history, and high-altitude on blood-oxygen capacity in harbor seals. J. Appl. Physiol. 42: 852–858.
Kooyman, C.A. and P.J. Ponganis. 1998. The Physiological Basis of Diving to Depth: Birds and Mammals. Annual Reviews Inc, Palo Alto, California, USA.
Kooyman, G.L. and F. Trillmich. 1986a. Diving behavior of Galapagos fur seals. pp. 186–195. In: R.L. Gentry and G.L. Kooyman [eds.]. Fur Seals Maternal Strategies on Land and at Sea. Princeton University Press, Princeton, NJ, USA.
Kooyman, G.L. and F. Trillmich. 1986b. Diving Behavior of Galapagos sea lions. pp. 209–219. In: R.L. Gentry and G.L. Kooyman [eds.]. Fur Seals Maternal Strategies on Land and at Sea. Princeton University Press, NJ, USA.
Kooyman, G.L. 1989. Diverse Divers: Physiology and Behavior. Springer-Verlag, Berlin Heidelberg.
Kovacs, K.M., A. Aguilar, D. Aurioles, V. Burkanov, C. Campagna, N. Gales et al. 2012. Global threats to pinnipeds. Mar. Mamm. Sci. 28: 414–436.
Kuhn, C.E. 2006. Measuring feeding rates to understand the foraging behavior of pinnipeds. PhD Thesis, University of California Santa Cruz, Santa Cruz, CA, USA.
Liggings, G.C., J. Qvist, P.W. Hochachka, B.J. Murphy, R.K. Creasy, R.C. Schneider et al. 1980. Fetal cardiovascular and metabolic responses to simulated diving in the Weddell seal. J. Appl. Physiol. 49: 424–430.
Ling, J.K. and G.E. Walker. 1978. 18-Month breeding cycle in Australian sea lion. Search 9: 464–465.
Litnan, C., A. Harting and J. Baker. 2015. *Neomonachus schauinslandi*. The IUCN Red List of Threatened Species 2015: e.T13654A45227978. http://dx.doi.org/10.2305/IUCN.UK.2015 2.RLTS.T13654A45227978.en. Downloaded on 06 November 2015.
Lomnicki, A. 1984. Resource partitioning within a single species population and population stability: A theoretical model. Theor. Popul. Biol. 24: 21–28.
Longhurst, A., S. Sathyendranath, T. Platt and C. Caverhill. 1995. An estimate of global primary production in the ocean from satellite radiometer data. J. Plankton Res. 17: 1245–1271.
Macarthur, J.W. and R. Levins. 1967. Limiting similarity convergence and divergence of coexisting Species. Am. Nat. 101: 377–385.
MacArthur, R.A., G.L. Weseen and K.L. Campbell. 2003. Diving experience and the aerobic dive capacity of muskrats: Does training produce a better diver? J. Exp. Biol. 206: 1153–1161.
Martin, J.H., K.H. Coale, K.S. Johnson, S.E. Fitzwater, R.M. Gordon, S.J. Tanner et al. 1994. Testing the iron hypothesis in ecosystems of the equatorial Pacific-Ocean. Nature 371: 123–129.

Mate, B.R., B.A. Lagerquist and J. Calambokidis. 1999. Movements of North Pacific blue whales during the feeding season off southern California and their southern fall migration. Mar. Mamm. Sci. 15: 1246–1257.

Mattlin, R.H., N.J. Gales and D.P. Costa. 1998. Seasonal dive behavior of lactating New Zealand fur seals (*Arctocephalus forsteri*). Can. J. Zool. 76: 350–360.

McKenzie, J., S.D. Goldsworthy, P.D. Shaughnessy and R. Mcintosh. 2005a. Understanding the impediments to the growth of Australian sea lion populations. South Australian Research and Development Institute (SARDI) Aquatic Science Publication No. RD04/0171. SARDI Res Rep Ser No. 74. South Australian Research and Development Institute (Aquatic Sciences), Adelaide.

McKenzie, J., L.J. Parry, B. Page and S.D. Goldsworthy. 2005b. Estimation of pregnancy rates and reproductive failure in New Zealand fur seals (*Arctocephalus forsteri*). J. Mammal. 86: 1237–1246.

Merrick, R.L. and T.R. Loughlin. 1997. Foraging behavior of adult female and young-of-the-year Steller sea lions in Alaska waters. Can. J. Zool. 75: 776–786.

Miller, A.K., M.A. Kappes, S.G. Trivelpiece and W.Z. Trivelpiece. 2010. Foraging-niche separation of breeding Gentoo and Chinstrap Penguins, South Shetland Islands, Antarctica. Condor 112: 683–695.

Morin, P.J. 1999. Community Ecology. Blackwell Science, Inc, Malden, MA.

Mueller, B., U. Porschmann, J.B.W. Wolf and F. Trillmich. 2011. Growth under uncertainty: The influence of marine variability on early development of Galapagos sea lions. Mar. Mamm. Sci. 27: 350–365.

Noren, S.R. and T.M. Williams. 2000. Body size and squeletal muscle myoglobin of cetaceans, adaptations for maximizing dive duration. Comp. Biochem. Physiol. A: Mol. Integr. Physiol. 126: 181–191.

Noren, S.R., T.M. Williams, D.A. Pabst, W.A. Mclellan and J.L. Dearolf. 2001. The development of diving in marine endotherms: Preparing the skeletal muscles of dolphins, penguins, and seals for activity during submergence. J. Comp. Physiol. B 171: 127–134.

Noren, S.R., S.J. Iverson and D.J. Boness. 2005. Development of the blood and muscle oxygen stores in gray seals (*Halichoerus grypus*): Implications for juvenile diving capacity and the necessity of a terrestrial postweaning fast. Physiol. Biochem. Zool. 78: 482–490.

Odell, D.K. 1975. Breeding biology of the California sea lion (*Zalophus californianus*). Rapports et Proces-Verbaux des Réunions Conseil International pour l'Exploration de la Mer 169: 374–378.

Orians, G.H. and N.E. Pearson. 1979. On the Theory of Central Place Foraging. Ohio State University Press, Columbus.

Pablo-Rodriguez, N., D. Aurioles-Gamboa and J.L. Montero-Muñoz. 2015. Niche overlap and habitat use at distinct temporal scales among the California sea lions (*Zalophus californianus*) and Guadalupe fur seals (*Arctocephalus philippii townsendi*). Mar. Mamm. Sci. DOI: 10.1111/mms.12274.

Paez-Rosas, D. and D. Aurioles-Gamboa. 2010. Alimentary niche partitioning in the Galapagos sea lion, *Zalophus wollebaeki*. Mar. Biol. 157: 2769–2781.

Paez-Rosas, D., D. Aurioles-Gamboa, J.J. Alava and D.M. Palacios. 2012. Stable isotopes indicate differing foraging strategies in two sympatric otariids of the Galapagos Islands. J. Exp. Mar. Biol. Ecol. 424: 44–52.

Pak, H. and J.R.V. Zanveld. 1973. The cromwell current on the east side of the Galapagos Islands. J. Geophys. Res. 78: 4845–7859.

Ponganis, P.J., L.N. Starke, M. Horning and G.L. Kooyman. 1999. Development of diving capacity in emperor penguins. J. Exp. Biol. 202: 781–786.

Ponganis, P.J. and G.L. Kooyman. 2000. Diving physiology of birds, a history of studies on polar species. Comp. Biochem. Physiol. A: Mol. Integr. Physiol. 126: 143–151.

Raeside, J.I. and K. Ronald. 1981. Plasma-concentrations of estrone, progesterone and corticosteroids during late pregnancy and after parturition in the harbor seal, phoca-vitulina. J. Reprod. Fertil. 61: 135–139.

Reijnders, P.J.H. 1990. Progesterone and estradiol-17-Beta concentration profiles throughout the reproductive-cycle in harbor seals (*Phoca-Vitulina*). J. Reprod. Fertil. 90: 403–409.
Rosenzweig, M.L. 1978. Competitive speciation. Biol. J. Linn. Soc. 10: 275–289.
Roughgarden, J. 1972. Evolution of niche width. Am. Nat. 106: 683–718.
Sakamoto, C.M., F.J. Millero, W.S. Yao, G.E. Friederich and F.P. Chavez. 1998. Surface seawater distributions of inorganic carbon and nutrients around the Galapagos Islands: Results from the PlumEx experiment using automated chemical mapping. Deep-Sea Res. Pt. II 45: 1055–1071.
Salazar-Aldás, D. 2007. Ecologia alimentaria del lobo marino de Galapagos (*Zalophus wollebaeki*) en el islote Caamaño. B.S. Thesis, Universidad Central del Ecuador, Quito, Ecuador.
Salazar, S.K. 2005. Variación temporal y espacial del espectro trófico del lobo marino de Galápagos. Masters Thesis, Instituto Politécnico Nacional (IPN), La Paz, BCS, México.
Schooley, R.L., C.R. Mclaughlin, G.J. Matula and W.B. Krohn. 1994. Denning chronology of female black bears—effects of food, weather, and reproduction. J. Mammal. 75: 466–477.
Stephenson, R., D.L. Turner and P.J. Butler. 1989. The relationship between diving activity and oxygen storage capacity in the Tufted duck (*Aythya-Fuligula*). J. Exp. Biol. 141: 265–275.
Szteren, D., D.E. Naya and M. Arim. 2004. Overlap between pinniped summer diet and artisanal fishery catches in Uruguay. Lat. Am. J. Aquat. Mamm. (LAJAM) 3: 119–125.
Tanaka, H. 2006. Winter hibernation and body temperature fluctuation in the Japanese badger, *Meles meles anakuma*. Zool. Sci. 23: 991–997.
Thompson, D., C.D. Dick, J. Mcconell and J. Garrett. 1998. Foraging behavior and diet of lactating female southern sea lions (*Otaria flavescens*) in the Falkland Islands. J. Zool. 246: 135–146.
Tracy, C.R., S.J. Reynolds, L. Mcarthur, C.R. Tracy and K.A. Christian. 2007. Ecology of aestivation in a cocoon-forming frog, *Cyclorana australis* (Hylidae). Copeia 4: 901–912.
Tremblay, Y., S.A. Shaffer, S.L. Fowler, C.E. Kuhn, B.I. Mcdonald, M.J. Weise et al. 2006. Interpolation of animal tracking data in a fluid environment. J. Exp. Biol. 209: 128–140.
Trillmich, F. 1979. Galapagos sea lions and fur seals. Noticias de Galapagos 29: 8–14.
Trillmich, F. and P. Majluf. 1981. First observations on colony structure, behaviour, and vocal repertoire of the South American fur seal (*Arctocephalus australis* Zimmermann, 1783) in Peru. Z. Säugetierkd 46: 310–322.
Trillmich, F. and D. Limberger. 1985. Drastic effects of El Niño on Galapagos pinnipeds. Oecologia 67: 19–22.
Trillmich, F. 1986. Attendance behavior of Galapagos sea lions. pp. 196–208. *In*: R.L. Gentry and G.L. Kooyman [eds.]. Fur Seals: Maternal Strategies on Land and at Sea. Princeton University Press, Princeton, New Jersey.
Trillmich, F., L. Rea, M. Castellini and J.B.W. Wolf. 2008. Age-related changes in hematocrit in the Galapagos sea lion (*Zalophus wollebaeki*) and the Weddell seal (*Leptonychotes weddellii*). Mar. Mamm. Sci. 24: 303–314.
Trillmich, F. 2015a. *Arctocephalus galapagoensis*. The IUCN Red List of Threatened Species 2015: e.T2057A45223722. http://dx.doi.org/10.2305/IUCN.UK.2015-2.RLTS.T2057A45223722.en, Downloaded on 23 September 2015.
Trillmich, F. 2015b. *Zalophus wollebaeki*. The IUCN Red List of Threatened Species 2015: e.T41668A45230540. http://dx.doi.org/10.2305/IUCN.UK.2015-2.RLTS.T41668A45230540.en Downloaded on 23 September 2015.
Trivelpiece, W.Z., S.G. Trivelpiece and N.J. Volkman. 1987. Ecological segregation of Adelie, Gentoo, and Chinstrap Penguins at King-George-Island, Antarctica. Ecology 68: 351–361.
Villegas-Amtmann, S., D.P. Costa, Y. Tremblay, S. Salazar and D. Aurioles-Gamboa. 2008. Multiple foraging strategies in a marine apex predator, the Galapagos sea lion *Zalophus wollebaeki*. Mar. Ecol. Prog. Ser. 363: 299–309.
Villegas-Amtmann, S., S. Atkinson and D.P. Costa. 2009. Low synchrony in the breeding cycle of Galapagos sea lions revealed by seasonal progesterone concentrations. J. Mammal. 90: 1232–1237.

Villegas-Amtmann, S. and D.P. Costa. 2010. Oxygen stores plasticity linked to foraging behaviour and pregnancy in a diving predator, the Galapagos sea lion. Funct. Ecol. 24: 785–795.

Villegas-Amtmann, S., S.E. Simmons, C.E. Kuhn, L.A. Huckstadt and D.P. Costa. 2011. Latitudinal range influences the seasonal variation in the foraging behavior of marine top predators. Plos One 6(8): e23166. doi: 10.1371/journal.pone.0023166.

Villegas-Amtmann, S., J.W.E. Jeglinski, D.P. Costa, P.W. Robinson and F. Trillmich. 2013. Individual foraging strategies reveal niche overlap between endangered Galapagos pinnipeds. Plos One 8(8): e70748. doi: 70710.71371/journal.pone.0070748.

Waite, J.N., V.N. Burkanov and R.D. Andrews. 2012. Prey competition between sympatric Steller sea lions (*Eumetopias jubatus*) and northern fur seals (*Callorhinus ursinus*) on Lovushki Island, Russia. Can. J. Zool./Rev. Can. Zool. 90: 110–127.

Weise, M.J. and D.P. Costa. 2007. Total body oxygen stores and physiological diving capacity of California sea lions as a function of sex and age. J. Exp. Biol. 210: 278–289.

Werner, E.E., G.G. Mittelbach and D.J. Hall. 1981. The role of foraging profitability and experience in habitat use by the bluegill sunfish. Ecology 62: 116–125.

Widdowson, E.M. 1981. The Role of Nutrition in Mammalian Reproduction. University Park Press, Baltimore, Maryland.

Wilson, R.P. 2010. Resource partitioning and niche hyper-volume overlap in free-living Pygoscelid penguins. Funct. Ecol. 24: 646–657.

Wolf, J.B.W. and F. Trillmich. 2007. Beyond habitat requirements: Individual fine-scale site fidelity in a colony of the Galapagos sea lion (*Zalophus wollebaeki*) creates conditions for social structuring. Oecologia 152: 553–567.

Wolf, J.B.W., C. Harrod, S. Brunner, S. Salazar, F. Trillmich and D. Tautz. 2008. Tracing early stages of species differentiation: Ecological, morphological and genetic divergence of Galapagos sea lion populations. BMC Evol. Biol. 8: 150, DOI: 10.1186/1471-2148-8-150.

Young, H.S., D.J. Mccauley, R. Dirzo, R.B. Dunbar and S.A. Shaffer. 2010. Niche partitioning among and within sympatric tropical seabirds revealed by stable isotope analysis. Mar. Ecol. Prog. Ser. 416: 285–294.

9

Management Strategies and Conservation Status of Galapagos Sea Lion Populations at San Cristobal Island, Galapagos, Ecuador

Diego Páez-Rosas[1,2,*] *and Nataly Guevara*[1]

Introduction

In recent years, public opinion, scientists and environmental groups have played an important role in the environmental and marine policies of all countries. The relevance of marine mammals in general and pinnipeds in particular is not merely based on their biological and economic importance, but also on their role as "charismatic species" (Kovacs et al. 2012). This designation has a far-reaching social impact, manifested in the management, which has been granted to their populations. On this basis, the Galapagos National Park (GNP), the institution responsible for the management and conservation of Galapagos sea lions (*Zalophus wollebaeki*), recognized the

---

[1] Universidad San Francisco de Quito, Galapagos Science Center. Av. Alsacio Northia s/n frente a Playa Mann EC200150, Isla San Cristóbal, Galápagos Ecuador.
[2] Dirección Parque Nacional Galápagos, Oficina Técnica San Cristóbal. Av. Perimetral s/n EC200150, Isla San Cristóbal, Galápagos Ecuador.
* Corresponding author: dpaez@usfq.edu.ec

need to coordinate the efforts of various actors (governmental, academic, non-government organizations and the organized civil society) with the purpose of generating research, monitoring and conservation actions. Thus, the establishment of a vision of regional management for the long-term conservation of the species throughout its entire geographic range is of paramount importance.

In this context, a specific management plan focused on the conservation of Galapagos sea lions from San Cristobal Island, a semi-urbanized island with an expected projected human population of more than 8000 people (Alava et al. 2014), was developed. The following management and conservation criteria were set up: the Galapagos sea lion (i) is an "endangered" species (Salazar et al. 2011, Kovacs et al. 2012, Trillmich 2015) listed under the category "special protection" in the Law of Forestry and Conservation of Natural Areas and Wildlife of Ecuador (Páez-Rosas and Guevara 2012); (ii) is a priority species in the Galapagos National Park Management Plan due to its significant role in maintaining the functional biodiversity of the marine ecosystems of the region (Alava and Salazar 2006, Páez-Rosas et al. 2012); (iii) has one of its most important breeding rookeries located within the urban limits of Puerto Baquerizo Moreno-Isla San Cristobal, one of the most populated areas of the archipelago (Páez-Rosas and Aurioles-Gamboa 2010); and, (iv) is a conspicuous and charismatic species, features which make it one of the main tourist attractions of the region (Lorden et al. 2012).

The environmental changes, which can be observed in the quality of the environment throughout the world as well as the limited information available on the structure and functioning of ecosystems, attract attention and reflect the urgency of finding mechanisms and methodological approaches, which in the short term may help us to identify environmental issues at a stage where they still can be mitigated at relatively low cost or when their impacts are still low. Marine mammals are among the species responding quickly to changes in environmental conditions and quality. Due to their evolutionary life history, sea lions depend on both the terrestrial habitat for breeding and rearing and on marine habitat for feeding and local movements; therefore, they are exposed to a greater number of environmental drivers affecting their survival (Montero-Serra et al. 2014, Alava et al. 2014).

Several studies have demonstrated the critical ecological role of this species in the ecosystem. Given the magnitude of their foraging movements and the seasonal variation in abundance of their main prey species (Villegas-Amtmann et al. 2008), these marine mammals function as sentinels of the state of health and conservation in the areas they inhabit (Páez-Rosas 2011). Therefore, the conservation efforts for this species must be based on a solid scientific knowledge. The management of free ranging sea lion populations must include not only aspects of research and science, but also the economic, social and cultural dimensions around them, and potential opportunities for the sustainable use of this resource by the community (e.g., Tourism Sector) (Lorden et al. 2012). Concerning the scientific knowledge, it is necessary to more fully understand the ecology, behavior, foraging, reproduction and health of these populations.

The Galapagos sea lion (*Zalophus wollebaeki*) is endemic to the archipelago and at the same time the most abundant marine mammal in the region (Fig. 1). Nonetheless, the population has been showing an alarming decline in numbers over the last 30 years as a result of their high vulnerability to environmental disturbances such as El Niño and ocean warming (Alava and Salazar 2006, Páez-Rosas 2008, Trillmich 2015), the deterioration of their habitat (Denkinger et al. 2014, Alava et al. 2014) and a potential competition for fish resources with other species and fishers (Salazar 2005). These conservation concerns have placed this tropical pinniped on the Red List of the International Union for Conservation of Nature IUCN as an "endangered species" (Trillmich 2015). Despite this conservation initiative, the species is still indirectly affected by various anthropogenic stressors, including fishing activities, the presence of introduced animals, the use or deterioration of their reproductive habitats and marine pollution (Denkinger et al. 2014, Alava et al. 2014).

Galapagos sea lions are highly gregarious; they aggregate on islands, islets and sites protected from predators, forming so called haul outs

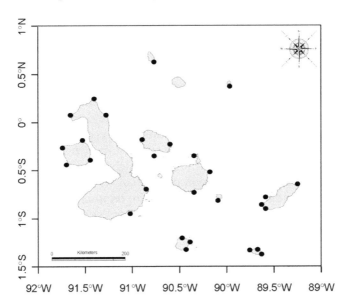

Figure 1. Map of the main breeding rookeries of Galápagos sea lion found in the Galapagos Islands. Black dots along the coastline of islands and islets indicate rookeries.

(Peterson and Bartholomew 1967). Their territories include an aquatic habitat used for thermoregulation and copulation, and a terrestrial habitat used by females to give birth, nursery, lactation and resting (Riedman 1990). The local dispersal behavior and inter-island movements of the species shows a sexual differentiation of distributional patterns, where adult and sub-adult males perform regional movements around the entire archipelago in search of food, resting in various places before returning to their original rookeries at the commencement of the next breeding season (Trillmich 1986). On the other hand, most of the females and juveniles remain in their same rookeries performing small migrations primarily focused on foraging (Trillmich 1986, Páez-Rosas 2008).

The reproductive period comprises the months from August to December, depending on the geographic location of the rookery (Trillmich 1986). The rookeries are composed of approximately 55% females, 31% juveniles, 14% adult and sub-adult males during this period (Páez-Rosas 2008). Within a breeding rookery, there are generally three zones: a juvenile zone, a bachelor's zone and a zone for breeding and mating (Páez-Rosas 2008). Territorial adult males arrive at the breeding grounds early during the reproductive period and begin to define their territories which forces other males (adults and sub-adults) to occupy marginal areas of the rookery (i.e., bachelor's zone) (Kunc and Wolf 2008). Because of their strong sexual selection, the females are distributed irregularly (i.e., in random clusters/aggregations), forming groups of different sizes along the beaches, while areas that are not used to establish territories are commonly occupied by the highly gregarious juveniles (Wolf et al. 2005).

Currently, there are no accurate data on the population size; however, censuses conducted in 2001 indicate that the population varies between 14,000 and 16,000 individuals (Alava and Salazar 2006), with the southern and eastern islands (Floreana, Espanola and San Cristobal) having the higher concentration of individuals (Salazar 2005, Páez-Rosas 2008). In the Galapagos Islands, about 55 rookeries have been registered, of which 31 are of reproductive character (57%), and 24 are non-breeding sites or temporary haul outs (43%) (Salazar et al. 2011). Specifically, six reproductive sites have been established in San Cristobal, among which the rookery of El Malecón (Fig. 2) stands out as the largest reproductive colony in the entire archipelago (D. Páez-Rosas, unpublished data). This site requires particular attention since it is located within the human settlement of Puerto Baquerizo Moreno. This situation has generated a high level of negative interactions between the species and the local human community (Denkinger et al. 2014).

In order to define the main conservation issues affecting the Galapagos sea lions, a research and participatory process were performed to characterize and assess the social, economic, environmental and institutional matters brought forth by the sea lion as a natural resource on San Cristobal Island.

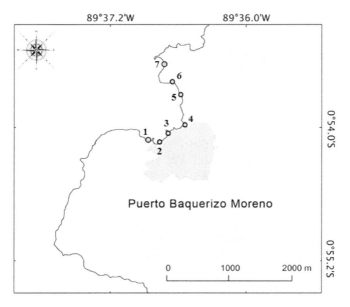

Figure 2. Map of Puerto Baquerizo Moreno, highlighting the seven most important settlements of sea lions (circles) embracing the rookery El Malecón. (1) Naval base, (2) Marines beach, (3) Malecón, (4) Gold beach, (5) Mann beach, (6) Predial, and (7) Carola beach.

In order to establish lines of action framed by a sustainable development strategy aimed at the conservation of the species in the Galapagos, this work relied on the participation of various sectors related to this resource.

Review and creation of a bibliographic database

A comprehensive literature review was the foundation for the development of the theoretical, legal and scientific framework of the management plan for the conservation of this species. By studying the local legislation related to the conservation of sea lions, we sought to develop proposals to improve or implement command and control strategies, which may facilitate achieving the established objectives. Another cornerstone on which the work was based on was the collection of scientific information on the species, published over the past 40 years. We identified records of about 50 manuscripts of peer-reviewed scientific publications and theses (Páez-Rosas and Guevara 2012).

The aim was to create a scientific and technical database to provide managers and decision makers of the Galapagos National Park with qualitative and quantitative information when establishing control and conservation measures. Acquiring this information was essential for the development of management activities aimed at minimizing potential threats (environmental and anthropogenic) since there is lack of scientific basis to generate quantitative conservation models that would improve management of this population.

Participatory work with the fisheries

A survey about the perception of fishers whose main economic activity is artisanal fishing in the region was performed. The survey consisted of personal interviews of about thirty minutes; the format contained 10 multiple-choice questions with specific answers, from which they consulted general aspects of the species, issues related to the interaction between sea lions and their fishing activities, as well as perceptions that the artisanal fisheries sector had concerning to management actions and conservation of this species. Based on this work, we could learn in detail about the perception fishers have about sea lions, and how this sector considers itself to be affected by the behavior of the species. We surveyed 100 fishers, of which the 90% felt that their economic activity is affected by sea lions since the animals generate significant economic losses mainly due to the damage caused to fishing vessels (Fig. 3A). Other adverse effects are negative interactions during fishing activities at sea, where sea lions steal the fishers' catch, causing damage to hooks and gillnets. This situation has led fishers to opt for actions, which threaten the species, and it is common to see boats protected with barbed wire and wood with nails as strategies against the species impact (Fig. 3B).

It has been shown that the negative impact on fishing vessels is directly associated with the sea lion behavior. During the day, the animals choose to rest on floating artificial substrates around their rookeries (Fig. 3A) to facilitate thermoregulation (Montero-Serra et al. 2014). One of initiative to mitigate the problem was the installation of floating platforms in order to provide an artificial resting area for sea lions (Fig. 3C), and thus achieve a reduction of the amount of negative interaction between sea lions and fisheries sector. This project was initiated mainly by the Spanish Agency of International Cooperation for Development and implemented by the Galapagos National Park.

Subsequently, to gauge the appreciation of the fishing sector for this alternative solution, the active fishers were asked again if they considered the installation of these floating rafts effective, generating a positive response of 47%. It is important to note that at the time of the survey, there was only one floating raft functioning. Therefore, many of the interviewed mentioned that it would be necessary increase the number of rafts to make this idea work (Páez-Rosas and Guevara 2012).

Figure 3. Images of interactions between sea lions of El Malecón rookery (San Cristobal Island) and fishing boats of Puerto Baquerizo Moreno. A) Sea lions resting on fishing boats; B) boats with barbed wire and nails; and C) rafts floating in the bay of San Cristobal. Photo credits: I. Montero.

Participatory work with the tourism sector

Ecotourism plays an important role in the conservation of natural reserves of Ecuador and the world. Moreover, tourism is the main source of income for the inhabitants of Galapagos. However, mishandling this activity may cause negative impacts on the natural resources of the region (Lorden et al. 2012). Part of the aim of this work was to identify the impacts tourism has on the sea lions of the region. Despite the issue being evident, it had been ignored or overlooked by the community (Fietz 2012, Lorden et al. 2012). Moreover, burgeoning tourism leads to the increase of people living in the islands, causing more contamination.

A suggestion from this sector was to increase research on the behavior of the species in order to have a technical basis for the generation of preventive measures. Additionally, establishing an appropriate minimum distance to be maintained between visitors and the animals was proposed. However, this latter point was widely debated since at present the Galapagos National Park rules require a minimum distance of 2 meters between the tourist and wildlife of Galapagos; which was insufficient in the opinion of most respondents (Fig. 4A). Preliminary studies based on the behavior of

Figure 4. Images of interactions between sea lions of El Malecón rookery (San Cristobal Island), tourist activities and the introduced species in the archipelago. Sea lions resting on tourist beaches in close proximity to people (A); and, domestic dogs walking in the rookeries of sea lions (B). Photo credits: N. Guevara.

Galapagos sea lion on the beaches of San Cristobal suggest that the proper distance to minimize impacts on the species should be at least 5 meters (Fietz 2012), which should be implemented in the future in order to reduce the anthropogenic pressure caused to the species by this sector (Páez-Rosas and Guevara 2012).

Participatory work with the social sector

San Cristobal Island is characterized by the close relationship between the human population and the sea lion rookeries, a situation that causes a high rate of negative interactions. Therefore, it is important to generate lines of action to reduce the impact. One of the factors which is of most concern in the Galapagos Islands is the negative interaction between the native wildlife and introduced animals (Fig. 4B). This situation is particularly evident around Puerto Baquerizo Moreno where domestic and feral dogs and cats cause damage to native flora and fauna become hosts or reservoirs of several emerging infectious diseases for sea lion populations (Brock et al. 2012, Guevara 2012).

Given the complexity of the subject, the only action for control, which has been achieved, is to implement a ban on animals on the streets and beaches of Puerto Baquerizo Moreno. According to 73% of the 100 people interviewed, this rule is not being followed since it is common to see dogs and cats freely wandering in the urban area and in the sea lions' reproductive zones. This situation not only involves direct disturbances such as attacks on sea lions and physiological stress, but may also generate more serious consequences, including the possible transmission of zoonotic diseases such as canine distemper virus (CDV), leptospirosis, toxoplasmosis and intestinal parasites. Viral and bacterial diseases can be transmitted by air or by contact with the feces and urine of infected animals (Levy et al. 2008, Guevara 2011).

Resource managers need to invest more attention to this issue. The Special Law of the Galapagos states that it is responsibility of municipalities to establish the conditions under which pets must be kept through the Municipal Ordinance on animal husbandry, but these institutions have not taken actions to address this problem (Páez-Rosas and Guevara 2012).

Systematization of results

Due to the close relationship between sea lions and these three social sectors, their active participation in the development of the management plan for the conservation of Galapagos sea lions was imperative. Of concern was the level of knowledge and lack of social mobilization of the sectors involved, which were often unaware of basic information about the species' taxonomy (endemism), or its conservation status (being classified as "endangered").

Based on the results of the surveys, the discussion sessions and workshops conducted with all sectors directly involved to the living resource Galapagos sea lion, the main threats to the species were identified (Table 1), and contingency (long-term) actions to mitigate these impacts were proposed.

168 Tropical Pinnipeds: Bio-Ecology, Threats and Conservation

Table 1. Major threats and management issues of Galapagos sea lions that inhabit populated areas. Identification of the responsible institutions and possible solutions. GNP = Galapagos National Park, ABG = Biosecurity and Quarantine Agency for Galapagos, MINTUR = Ministry of Tourism of Ecuador.

SUBJECT	THREATS	HOW IT AFFECTS	WHO IS RESPOSIBLE	WHO CAN TAKE ACTION?
SCIENTIFIC INVESTIGATION	CENSUS DATA OUTDATED	Unknown current number of sea lions in the population	GNP	GNP, NGOs, Universities
	UNKNOWN STATE OF HEALTH	Inability to implement prevention-oriented actions	GNP	GNP, NGOs, Universities
CONTAMINATION AND MARINE POLLUTION	WASTEWATER AND BIOLOGICAL POLLUTION	Increase of fecal and total coliform bacteria in the coastal-marine environment, habitat deterioration, illness	Community, Municipality, Government	Municipality, Government, Universities
	CHEMICAL ASSAULTS AND POLLUTION	Potential oil spills from vessels and contamination by persistent organic pollutants (POPs) and metals	Community, Municipality, Government	GNP, Municipality, Government, Universities
	SOLID WATER (GARBAGE)	Habitat degradation, entanglement and injury, illness or death	Community, Municipality, Government	Municipality, Government Fishing Sector

Management Strategies for Galapagos Sea Lion   169

INTRODUCED ANIMALS	ENTRY OF INVASIVE SPECIES	Increased animals introduced	ABG, GNP, Government	ABG, GNP, Municipality, Government
	DOGS, CATS AND RATS	Potential emerging infectious diseases, carriers, attacks	ABG, GNP, Government	ABG, GNP, Municipality, Government
EDUCATION	LACK OF SOCIAL MOBILIZATION AND MISINFORMATION ABOUT THE SPECIES	Ignorance of the ecological and socio-economic importance of sea lion	GNP, Government, Naturalist Guides Association	GNP, Government, Universities, NGOs, Naturalist Guides Association
ECOLOGICAL IMPACTS	DISTURBANCES	Alteration of the natural behavior of sea lions. Impacts on the population dynamics, and the environmental stressors	GNP, Government, MINTUR, Municipality	GNP, Government, Universities, NGOs, MINTUR

Management and conservation strategies

In order to institute conservation efforts, it is first necessary to develop a conceptual framework that identifies goals and objectives, and the direct and indirect threats that interfere with the accomplishment of the desired outcome. This will prioritize activities and/or resources aimed to monitor and evaluate the success of the proposed actions.

The management strategies proposed in this paper are based on the main threats to the conservation of the Galapagos sea lion (Table 1). The main goal is the establishment of lines of actions to coordinate and promote the efforts of the Galapagos National Park to ensure the maintenance and recovery of the population of the Galapagos sea lions by preventing and mitigating both direct and indirect anthropogenic impacts on the species.

Strategy 1

Stop the deterioration of the health status of Galapagos sea lions with the aim to ensure the maintenance and recovery of their populations in the long term through in-depth health analysis, which will implement technical solutions to improve the current conditions of this species (Fig. 5).

Strategy 2

Mitigate the impacts of introduced animals and increase control at points of entry to the Galapagos Islands for total restriction of introduction of exotic species to the archipelago (Fig. 6). Pursuing the support of the Control and Monitoring Unit of the ABG (Biosecurity and Quarantine Agency for Galapagos) to better accomplish this strategy.

Strategy 3

Improve the quality of the marine and terrestrial habitat occupied by the populations of Galapagos sea lions in urban areas in order to reduce the level of anthropogenic impacts resulting from the daily activities of the inhabitants of the islands (Fig. 7).

Strategy 4

Provide and promote scientific investigation in order to increase the biological and ecological knowledge and monitor the population dynamics

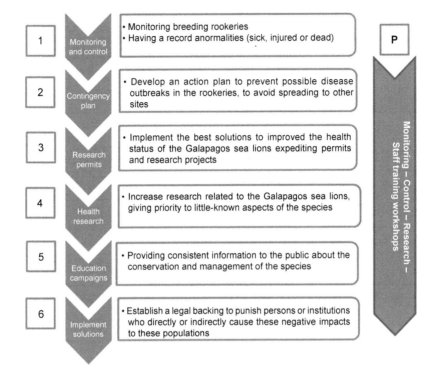

Figure 5. Conceptual model diagram for Strategy 1 to address conservation threats impairing the health of Galapagos sea lions at San Cristobal, Galapagos.

of the species, a basic requirement for generating priority conservation lines for the management and conservation of these populations (Fig. 8).

General conclusions and recommendations

The successful implementation of the Management Plan for the Conservation of Galapagos sea lion will depend on the involvement not only of the authorities but also of the local population of stakeholders and all industries that interact with the sea lions in this region. It is necessary to include social mobilization activities, including education, communication, and outreach, under which the benefits of conserving the species and ecosystems to the society are further fostered and emphasized, not only in economic terms but also as part of the intangible and priceless heritage of this region.

Another key focus of this work is the generation of scientific knowledge, information that is essential when developing management procedures. It is therefore a priority to conduct applied research to generate the required

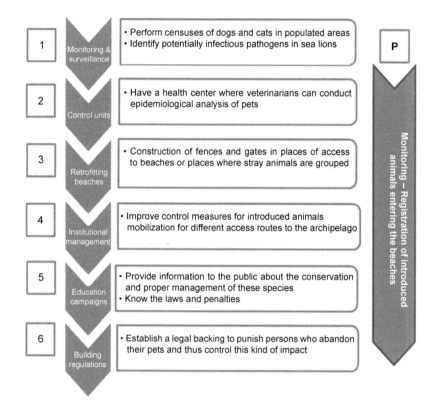

Figure 6. Conceptual model diagram for Strategy 2 to minimize the impact of introduced species affecting the population health and habitat of Galapagos sea lion at San Cristobal, Galapagos.

information to help plan specific activities to support the conservation of sea lions in the Galapagos Islands.

The Galapagos National Park, as administrator of the biodiversity in both the Galapagos National Park and in the Galapagos Marine Reserve, is the institution responsible for leading these proposals. However, one must take into account the degree of susceptibility of this species that faces interactions with the human population of the islands. The success of all management and conservation actions will require the cooperation of the local community, the autonomous governments (municipalities) and non-government organizations.

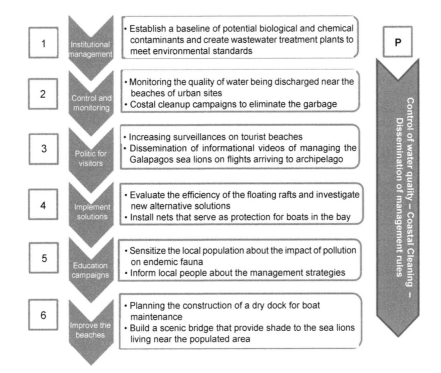

Figure 7. Conceptual model diagram for Strategy 3 to improve and recover the human impacted habitat of Galapagos sea lion at San Cristobal, Galapagos.

Acknowledgement

We want to thank the Galapagos National Park and the Spanish Agency for International Cooperation for Development for providing all logistical and financial facilities for the preparation of this work. We would like to thank the artisanal fishers' cooperatives: COPESPROMAR COPESAN Galapagos, the Association of Naturalist Guides Galapagos, the Universidad San Francisco de Quito and all social sectors who collaborated in the process of interviews, surveys and dissemination workshops. Special acknowledgement to Maximilian Hirschfeld and the park rangers of San Cristobal Technical Office for all their assistance given during the preparation of this work. Finally, thanks to the editor of this book for all the comments that helped improve our manuscript.

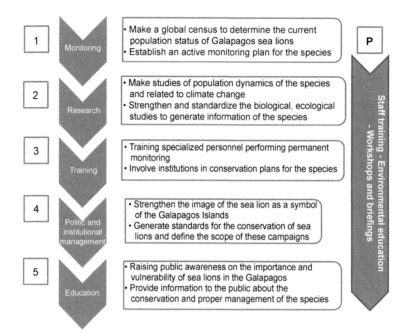

Figure 8. Conceptual model diagram for Strategy 4 to continue fostering scientific research and increase population data of Galapagos sea lion to support conservation and management of the species at San Cristobal, Galapagos.

Keywords: Galapagos sea lions, endangered species, conservation strategies, charismatic species, anthropogenic threats, Management Plan, Galapagos Islands

References

Alava, J.J. and S. Salazar. 2006. Status and conservation of Otariids in Ecuador and the Galapagos Islands. pp. 495–520. *In*: A.W. Trites, S.K. Atkinson, D.P. De Master, L.W. Fritz, T.S. Gelatt, L.D. Rea and K.M. Wynne [eds.]. Sea Lions of the World. Alaska Sea Grant College Program: Fairbanks, Alaska, USA.

Alava, J.J., C. Palomera, l. Bendell and P.S. Ross. 2014. Pollution as a threat for the conservation of the Galapagos marine reserve: Environmental impacts and management perspectives. pp. 247–283. *In*: J. Denkinger and L. Vinueza [eds.]. The Galapagos Marine Reserve: A Dynamic Socio-ecological System. Social and Ecological Interactions in the Galapagos Islands. Springer Science+Business Media, New York.

Brock, P., A.J. Hall, S. Goodman, M. Cruz and K. Acevedo-Whitehouse. 2012. Applying the tools of ecological immunology to conservation: A test case in the Galapagos sea lion. Anim. Conser. 16: 19–31.

Denkinger, J., D. Quiroga and J. Murillo. 2014. Assessing human–wildlife conflicts and benefits of Galapagos sea lions on San Cristobal Island, Galapagos. pp. 285–305. *In*: J. Denkinger and L. Vinueza [eds.]. The Galapagos Marine Reserve: A Dynamic Socio-

ecological System. Social and Ecological Interactions in the Galapagos Islands. Springer Science+Business Media, New York.
Fietz, K. 2012. General Behavioral Patterns and Human Impact on Behavior of the Galápagos Sea Lion (*Zalophus wollebaeki*) on San Cristóbal, Galápagos. Master Dissertation. University of Hamburg, Germany.
Guevara, N. 2011. Línea base del estado de salud y detección de *Leptospira* patógena por PCR en lobos marinos de Galápagos (*Z. wollebaeki*) de la Isla San Cristóbal. Bachellor Dissertation. Universidad San Francisco de Quito, Ecuador.
Kovacs, K., A. Aguilar, D. Aurioles, V. Burkanov, C. Campagna, N. Gales et al. 2012. Global threats to pinnipeds. Mar. Mamm. Sci. 28: 414–436.
Kunc, H. and W. Wolf. 2008. Seasonal changes of vocal rates and their relation to territorial status in Male Galápagos sea lions (*Zalophus wollebaeki*). Ethology 114: 381–388.
Levy, J., P. Crawford, M. Lappin, E. Dubovi, M. Levy, R. Alleman et al. 2008. Infectious diseases of dogs and cats on Isabela Island, Galapagos. J. Vet. Intern. Med. 22(1): 60–65.
Lorden, R., S. Sambrook and R. Mitchell. 2012. Residents' and Tourists' knowledge of sea lions in the Galápagos. Society & Animals 20: 342–363.
Montero-Serra, I., D. Páez-Rosas, J. Murillo, T. Vegas-Vilarrúbia, K. Fietz and J. Denkinger. 2014. Environment-driven changes in terrestrial habitat use and distribution of the Galapagos sea lion. Endang. Species Res. 24: 9–19.
Páez-Rosas, D. 2008. Diversificación de dietas en tres colonias de lobo marino de Galápagos, *Zalophus wollebaeki*, evaluada con análisis de excretas e isótopos estables de C y N. Master Dissertation, Instituto Politécnico Nacional, México.
Páez-Rosas, D. and D. Aurioles-Gamboa. 2010. Alimentary niche partitioning in the Galapagos sea lion, *Zalophus wollebaeki*. Mar. Biol. 157(12): 2769–2781.
Páez-Rosas, D. 2011. Ecología trófica de los pinnípedos de las Islas Galápagos: Análisis espacial y temporal. PhD Dissertation. Instituto Politécnico Nacional, México.
Páez-Rosas, D., D. Aurioles-Gamboa, J.J. Alava and D. Palacios. 2012. Stable isotopes indicate differing foraging strategies in two sympatric otariids of the Galápagos Islands. J. Exp. Mar. Biol. Ecol. 424: 44–52.
Páez-Rosas, D. and N. Guevara. 2012. Plan de Manejo para la Conservación de los Lobos Marinos de la Isla San Cristóbal. Dirección Parque Nacional Galápagos DPNG. Galápagos, Ecuador. 96 pp.
Peterson, R. and G. Bartholomew. 1967. The natural history and behavior of the California sea lion. Am. Soc. Mammal. Spec. Pub. No. 1 79 pp.
Riedman, M. 1990. The Pinnipeds: Seals, Sea Lions and Walruses. University of California Press, CA, USA.
Salazar, S. 2005. Variación temporal y espacial del espectro trófico del lobo marino de Galápagos. Master Disertation, Instituto Politécnico Nacional, México.
Salazar, S., J.J. Alava, V. Utreras and D.G. Tirira. 2011. Lobo Marino de Galapagos (*Zalophus wollebaeki*) [Galapagos sea lion *Zalophus wollebaeki*]. pp. 136–137. *In*: D.G. Tirira [ed.]. Libro Rojo de los mamíferos del Ecuador [The Red Book of Ecuadorian mammals]. Fundación Mamíferos y Conservación, Pontificia Universidad Católica del Ecuador y Ministerio del Ambiente del Ecuador. Quito, Ecuador.
Trillmich, F. 1986. Maternal investment and sex-allocation in the Galapagos fur seal, *Arctocephalus galapagoensis*. Behav. Ecol. Sociobiol. 19: 157–164.
Trillmich, F. 2015. *Zalophus wollebaeki* (Galapagos sea lion). The IUCN Red List of Threatened Species.557 Version 2015.2. www.iucnredlist.org.
Villegas-Amtmann, S., D. Costa, Y. Tremblay, D. Aurioles-Gamboa and S. Salazar. 2008. Multiple foraging strategies in a marine apex predator, the Galapaos sea lion. Mar. Ecol. Prog. Ser. 24: 785–795.
Wolf, J., G. Kauermann and F. Trillmich. 2005. Males in the shade: Habitat use and sexual segregation in the Galápagos sea lion (*Zalophus californianus wollebaeki*). Behav. Ecol. Sociobiol. 59: 293–302.

10

Population Ecology, Trends and Distribution of the Juan Fernandez Fur Seal, *Arctocephalus philippii* (Peters 1866) in Chile

Layla P. Osman[1,2,]* and *Carlos A. Moreno*[3]

Introduction

Theoretical population ecology deals with the study of changes in population size through time. In other words, it is the study of why populations of living organisms increase, decrease or remain unchanged, i.e., population dynamics (Berryman 1999). Under favorable environmental conditions, populations will begin to increase in numbers. The maximum rate at which a population can increase is determined by the intrinsic life history characteristics of the species. Nevertheless, many extrinsic factors can influence the dynamics of a population such as environmental

---
[1] Fisheries, Marine Conservation and Environmental Sustainability, International Consultant.
[2] Programa de investigación Pesquera, Instituto de Acuicultura, Universidad Austral de Chile, Valdivia, Chile.
[3] Instituto de Ciencias Ambientales y Evolutivas, Universidad Austral de Chile, Valdivia, Chile.
  Email: cmoreno@uach.cl
* Corresponding author: laylaosman@gmail.com

A Poem (in Spanish) by Dr. Layla P. Osman appears on the last page of the book. This is on the Selkirk Island where the major population of Juan Fernandez fur seal are found.

variability, disease, competition and predation (e.g., Akcakaya et al. 1997, Berryman 1999). Additionally, no population grows indefinitely, leading to population regulation. A similar set of external factors can determine the maximum population size.

Marine mammals, as with most long-lived animals, have relatively slow intrinsic rates of increase. Modest population growth rates are the consequence of their life history characteristics. Characteristics such as age at which females start reproducing, the number of years between births and how many years a female will live and reproduce determine how quickly a population can increase. Most marine mammal species take many years to reach sexual maturity and have long gestation periods that result in the production of mostly only one young a year. In general, low rates of population growth make nearly all species of marine mammals vulnerable to overexploitation (Wade 2002).

The populations of many species of seals (Family Phocidae) and sea lions (Family Otariidae) were substantially reduced by commercial sealers and whalers in the 18th and 19th centuries, and some may have been reduced centuries earlier by coastal aboriginal people (e.g., Bonner 2000, Reeves 2009). In fact, most species of fur seals were presumed to be extinct by the late 19th century (Weber et al. 2004). The Juan Fernandez fur seal *Arctocephalus philippii* is the only endemic pinniped of Chile (Fig. 1), being limited in range to the islands that make up the Juan Fernandez Archipelago (i.e., Robinson Crusoe, Santa Clara and Alejandro Selkirk), and

Figure 1. Adult individuals of Juan Fernandez fur seal (*A. philippii*) resting on rocky shores at the Juan Fernandez Archipelago. Photo Credit: L. Osman.

also Desventuradas Islands (i.e., San Felix and San Ambrosio), located in the Southeastern Pacific off the coast of Chile (Torres 1987a) (Fig. 2). *A. philippii* had a long history of exploitation, which began at Robinson Crusoe Island before the end of the seventeenth century. From 1687 until 1898, at least 3,870,170 fur seals were estimated to have been taken in commercial endeavors (see Hubbs and Norris 1971, Torres 1987b). By 1900, *A. philippii* was believed to be extinct (King 1964, Maxwell et al.

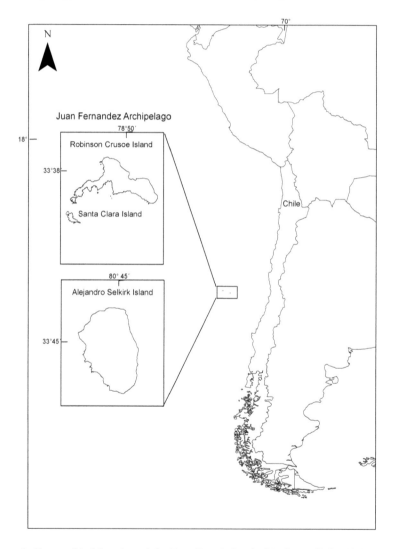

Figure 2. Geographical location of the Juan Fernández Archipelago off the Chilean coast (small rectangle). Detailed positions of the two major islands of the Archipelago are found within the left hand squares.

1967, Hubbs and Norris 1971, Torres 1987a) as a consequence of the vast exploitation regime their populations were subject to. It was not until 1965, when Bahamonde (1966) reported the presence of small colonies on Alejandro Selkirk and Robinson Crusoe Islands and that the species was rediscovered again. Harvesting of Juan Fernández fur seals has been banned since 1965 (Aguayo 1979). The status of total protection was extended to all *Arctocephalus* species in Chile in 1978 (Torres 1987a).

Currently, *A. philippii* is one of the least known species of fur seals although its conservation status is defined as "Least Concern", according to the IUCN Red List of Threatened Animals (Aurioles-Gamboa 2015). In view of the history of exploitation and the lack of information and uncertainty about its population size and local distribution, the objectives of this study were (a) to estimate the current population size and trends; (b) to compare the per capita rates of change (R) between island populations; and, (c) to described the distribution of *A. philippii* at the Juan Fernandez Archipelago, where the majority of the population inhabits. Hence, this work provides the results of the first analysis of *A. philippii* population process recovery.

Material and Methods

Study area

The Juan Fernandez Archipelago (Fig. 2) is located in the Eastern South Pacific, between 670 km and 860 km from the coast of Chile. The Archipelago contains three islands: Robinson Crusoe (33°37' S; 78°51' W), Santa Clara (33°42' S; 79°01' W) and Alejandro Selkirk (33°45' S; 80°45' W). Robinson Crusoe (47.9 km^2) is the closest island to the continent (670 km), while Santa Clara (2.2 km^2) is located at 1.2 km southwest from Robinson Crusoe Island. Alejandro Selkirk (49.5 km^2) is the most isolated island, being 860 km away from the coast of Chile and 187 km west of Robinson Crusoe Island (Fig. 2).

Data collection, censuses and survey methods

To estimate the current population size of *A. philippii* at Juan Fernandez Archipelago three censuses of the fur seal population were conducted during the 2003/2004, 2004/2005 and 2005/2006 (hereafter referred to as the year in which summer season ended; e.g., 2005/2006 = 2005) breeding seasons in the austral summer from the end of December to January. At Robinson Crusoe and Santa Clara Islands, access to fur seal colonies was possible by zodiac, and direct counts of fur seals were made on land, from high vantage points and/or walking close to the shore by three observers. Each observer independently recorded numbers of fur seals in the following six categories when possible: pup, juvenile male, juvenile female, unidentified juvenile, adult female and adult male. The counts were made with the aid of binoculars and manual counters. Triple counts were performed by each observer on every category to asses inter- and intra-observer variability. If individual counts differed by more than 5%, further counts were undertaken until the estimates were within 5% of each other. On Alejandro Selkirk Island the counts were made only on land from high vantage points and/or walking close to the shore. We classified each colony visited as "reproductive" by the presence of pups or "non-reproductive" by the lack of them, and noted its geographical position using a handheld GPS.

These datasets, together with available published census data and additional data kindly provided by CONAF (Corporación Nacional Forestal), were used in concert to analyse the population trajectory of *A. philippii*. Since the data was not evenly distributed throughout years and total counts are only available for some islands in some years, different sets of data were used to evaluate population growth for each specific island individually.

Estimates of pup numbers are often used as an index of abundance to describe trends in pinniped populations because pups form a readily recognizable age class. Further, most of them are ashore at one time (at least at an early age); it is more difficult to estimate what proportion of other age classes are ashore and hence to make accurate corrections to visual counts (Berkson and DeMaster 1985, Shaughnessy et al. 2002). Even though the counts of pups are one the best estimates of population size, the available pup time series are very short for this species; thus, total population counts were used instead. The total population reported here includes adult males, juveniles, females and pup counts. The reported values provide a minimum population estimate because of the inherent bias of excluding non-reproductive females and juveniles that possibly do not remain long in the area, whereas subadult males arrive later in the season. Since the Robinson Crusoe and Santa Clara populations are very close, these sets of data were combined and analysed as one population, particularly since contrasting changes in population size are due to movements of individuals between islands rather than the populations of these islands behaving differently.

Population analysis

Both the raw time series and log-transformed data were investigated for trends to evaluate the pattern of population growth (Berryman 1999). Changes in populations of living organisms are caused by factors that affect the birth and death rates of individual organism. Hence, we looked for associations between the per capita birth and death rates and specific suspected causal variables (s) (Royama 1977, Berryman 1999, Münster-Swendsen and Berryman 2005). Even though birth and death rates may not be directly measured, we can estimate their net effect by calculating the realized per capita rate of change from a time series (Berryman 1999). Thus the intrinsic (r_i) (equations 2 and 3; Caughley 1977) and finite (λ_t) (equation 4; Caughley 1977) rates of increase were calculated for the *A. philippii* populations by using the following equations:

1) $e^{r_i} = \lambda$
2) $r_i = \ln e^{r_i}$
3) $\lambda_t = \dfrac{N_{t+1}}{N_t}$

And then, the realized *per capita* rate of change or R-function (R) (Berryman 1999) was calculated using the following model which takes into account the time between censuses:

4) $R = \ln \left(\dfrac{N_{t+1}}{N_t} \right)^{1/t}$

The R function describes how the net reproductive rate of the average individual changes with population density. The R-function is an extremely important concept in population dynamics theory, since it defines how the well-being or fitness of an average individual changes with the density of the population within which it lives (Berryman 1999).

Results

Logistic limitations and bad weather conditions did not allow us to perform the counts on all islands every year. In 2003, we counted Robinson Crusoe, Santa Clara and Alejandro Selkirk Islands. During 2004, we counted Robinson Crusoe only and in 2005 we counted the entire Archipelago again. At Robinson Crusoe Island, 19 colonies were visited, from which "Tierras Blancas", "Tres Puntas", "Punta O'Higgins" and "Bahía del Padre" are the most important breeding colonies (Fig. 3). At Santa Clara Island, there are no real boundaries between colonies, but the highest population numbers and pup production are in the side of "Morro Spartan" and "Weste" (Fig. 2). Additionally, one important breeding colony, "La Matriz", is located in a small bay on the west side of the Island separated from the rest (Fig. 3). In Alejandro Selkirk Island, the fur seals are mostly congregated at "Lobería Vieja" (Fig. 4), but it is important to note the existence of two additional rookeries to which we did not have access (and hence are not included in the census).

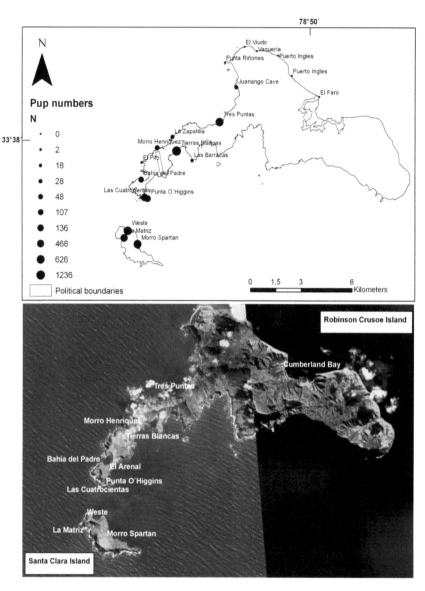

Figure 3. Pup numbers and breeding and non-breeding colony distribution of *A. philippii* at Robinson Crusoe and Santa Clara Islands (Map above was made and taken from Google earth, http://earth.google.com). The smallest dots (i.e., points = 0 pups) correspond to non-breeding colonies.

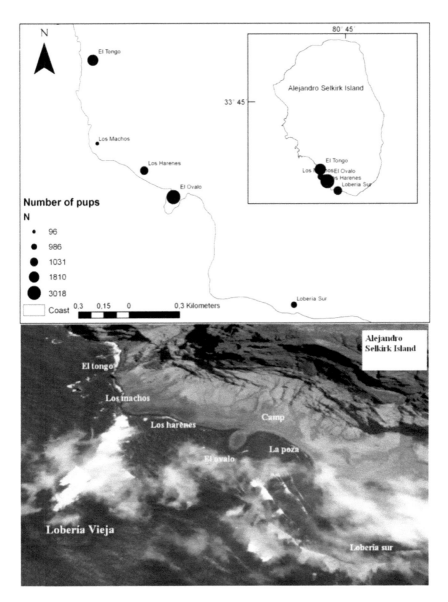

Figure 4. Pup numbers and distribution of *A. philippii* at Lobería Vieja, Alejandro Selkirk Island (image taken from Google earth, http://earth.google.com).

Tables 1, 2, and 3 show the data gathered to assess the *A. philippii* population growth at Robinson Crusoe (Table 1), Santa Clara (Table 2) and Alejandro Selkirk (Table 3) Islands. The data for combined time series from Robinson Crusoe and Santa Clara Islands is shown in Table 4.

As shown in Figs. 2 and 3, *A. philippii* populations are distributed on the three islands of the Juan Fernandez Archipelago and have experienced an increase in population size through time (Tables 1–3). In Robinson Crusoe Island, the number of pups increased from 125 individuals in 1983 to 2732 in 2005 (Table 1). In Santa Clara, the population increased from

Table 1. *A. philippii* total population (1968–2005) and pup production (1983–2005) at Robinson Crusoe Island.

Island	Year	Population size	Pup numbers	Reference
RC	1968	192	-	Torres 1987a, 1987b
RC	1969	246	-	Torres 1987a, 1987b
RC	1978	512	-	Torres 1987a, 1987b
RC	1983	1544	125	Torres 1987a, 1987b
RC	1984	1206	-	Torres 1987a, 1987b
RC	1998	5837	1481	Conaf, Juan Fernández
RC	1999	7142	1220	Conaf, Juan Fernández
RC	2000	8503	2230	Conaf, Juan Fernández
RC	2001	8729	2377	Conaf, Juan Fernández
RC	2003	8080	2880	This study
RC	2004	9165	3498	This study
RC	2005	7253	2732	This study

Table 2. *A. philippii* total population (1969–2005) and pup production (1998–2005) at Santa Clara Island.

Island	Year	Population size	Pup numbers	Reference
SC	1969	4	-	Torres 1987a, 1987b
SC	1978	84	-	Torres 1987a, 1987b
SC	1983	497	-	Torres 1987a, 1987b
SC	1984	368	-	Torres 1987a, 1987b
SC	1998	2898	575	CONAF, Juan Fernández
SC	1999	2595	482	CONAF, Juan Fernández
SC	2000	2890	583	CONAF, Juan Fernández
SC	2001	4114	508	CONAF, Juan Fernández
SC	2003	5113	1176	This study
SC	2005	5110	1325	This study

Table 3. *A. philippii* population counts (1969–2005) and pup production (1987–2005) at Alejandro Selkirk Island.

Island	Year	Population size	Pup numbers	Reference
AS	1968	267	-	Torres 1987a, 1987b
AS	1969	500	-	Torres 1987a, 1987b
AS	1978	1820	-	Torres 1987a, 1987b
AS	1982	3480	-	Torres 1987a, 1987b
AS	1983	4318	548	Torres 1987a, 1987b
AS	1987	4510	721	Torres *pers. com.* 2006
AS	2003	13941	4821	This study
AS	2005	19270	6941	This study

Table 4. *A. philippii* total population counts (1969–2005) at Robinson Crusoe (RC) and Santa Clara Islands (SC).

Island	Year	N
RC & SC	1969	250
RC & SC	1978	596
RC & SC	1983	2041
RC & SC	1984	1574
RC & SC	1998	8735
RC & SC	1999	9737
RC & SC	2000	11393
RC & SC	2001	12843
RC & SC	2003	13193
RC & SC	2005	12363

575 pups in 1998 to 1325 in 2005 (Table 2) and in Alejandro Selkirk island it increased from 548 pups in 1983 to 6941 in 2005 (Table 3). The times series for both Robinson Crusoe-Santa Clara Islands shows that the population growth had an increasing trend through time (Fig. 5), but in the last years the rate of population increase diminished, reaching a negative growth rate at the end of the period. The R-function shows that the population passed through two equilibrium points (Point A and B, Fig. 6A), and currently, the population is reaching equilibrium as it is evidenced by point B (Fig. 6). The Alejandro Selkirk time series shows that the population has a trend of continued rapid increase (Fig. 7) reflective of positive per capita population growth rate, despite a possible past equilibrium point sometime between 1984–88 (Point C, Fig. 6B).

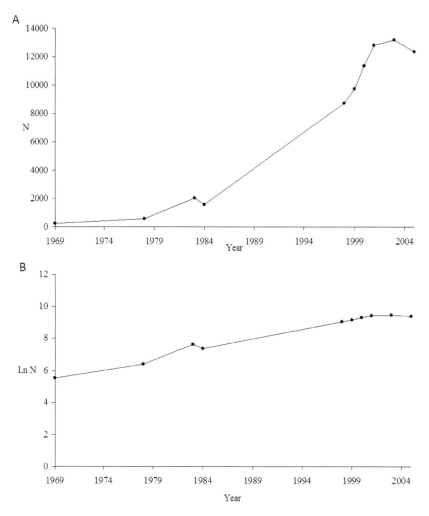

Figure 5. Population growth of *A. philippii* plotted on arithmetic (A) and logarithmic (B) scales at Robinson Crusoe-Santa Clara Islands.

Discussion

Direct harvesting of wildlife populations often results in substantial reductions in population size and overexploitation is the prime suspect in the global extinction of many mammalian species (Purvis 2001). Endemic island species are more prone to extinction than non-endemic ones (Frankham et al. 2002). For nearly two centuries, *A. philippii* was under an exploitation regime that almost exterminated this species (Hubbs and Norris

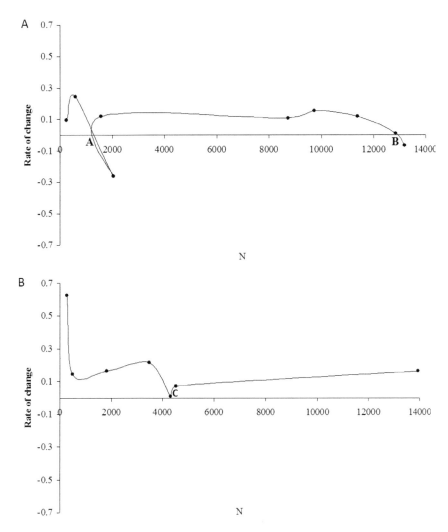

Figure 6. R-function of population growth of *A. philippii* at Robinson Crusoe-Santa Clara Islands (A) and at Lobería Vieja, Alejandro Selkirk Island (B). A and B are equilibrium points at Robinson Crusoe-Santa Clara Islands (A), while C is an equilibrium point at Lobería Vieja, Alejandro Selkirk Island (B).

1971, Torres 1987b). During the end of the 18th century and beginning of the 19th century, millions of fur seals were harvested from the Juan Fernández islands (Hubbs and Norris 1971). In 1872, trade with China was initiated with the harvest of approximately 37,000–38,000 animals, while "a single ship carried one million skins to the London Market in 1801" (Busch 1985). Between 1793 and 1807, an estimated of 3,500,000 animals were harvested (Hubbs and Norris 1971). By 1891, the population was estimated at

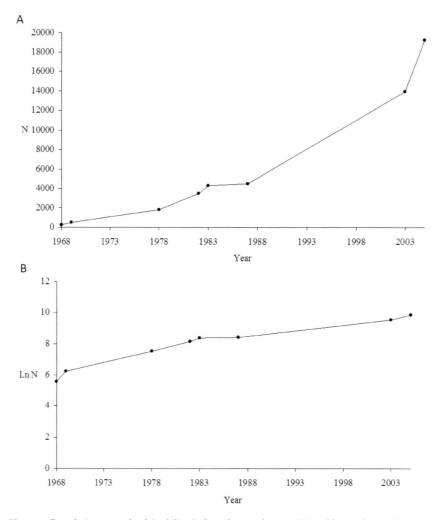

Figure 7. Population growth of *A. philippii* plotted on arithmetic (A) and logarithmic (B) scales at Lobería Vieja, Alejandro Selkirk Island.

approximately 300–400 animals and by 1900 the population was presumed extinct (Hubbs and Norris 1971, Gerber and Hilborn 2001).

The results of this study show that currently *A. philippii* populations have recovered from past exploitation, increasing from an initial population of 750 individuals in 1969 to 32,278 in 2005. This population estimate should be considered as a minimum population size since it excludes the individuals who cannot be registered during the census such as females performing foraging trips and juveniles. Although all known rookeries were identified, one minor site could not be counted. Observer error was minimized by comparing the counts of the three observers, and if a difference of more than 5% was found, the count was performed again.

Even though the populations of *A. philippii* have been increasing, the Robinson Crusoe-Santa Clara population demonstrated signs that it may have reached equilibrium (Fig. 6A), such that the population is oscillating between negative and positive per capita population growth rates. The population time series in Robinson Crusoe-Santa Clara islands showed a continuing increase. We suspect the mathematically-identified equilibrium point in year must have been due to a count bias rather than a population growth decrease, particularly given that different research teams collected the data on different dates. The population trend from Alejandro Selkirk Island shows that this population is still growing, and is likely to continue to do so in the near future since it stills has a high per capita population growth rate (Fig. 6B).

The last census conducted around the Juan Fernández Archipelago yielded a total pup production of 10,998 individuals, from which Lobería Vieja contributed 63% of the animals, with 25% from Robinson Crusoe 25% and 12% from Santa Clara. The differential size of the islands might account for the different population sizes; Santa Clara only contributes with 12% of the pup production since it is the smallest island. However, this is not the sole explanation for the observed differences. Robinson Crusoe (47.9 km^2) and Alejandro Selkirk (49.5 km^2) are similar in size, but Alejandro Selkirk with only one important colony - Lobería Vieja – which accounts for half of the population of *A. philippii*. In contrast, Robinson Crusoe, which has several colonies, contributes a smaller fraction to the total and may be at maximum capacity.

Of course, total surface area is not a salient feature to the number of sea lions an island can support. Fur seals are colonial, and when they are ashore, they form groups along rocky coastlines to rest and breed (e.g., Riedman 1990). Sites used for breeding must be accessible from the sea and must contain habitat suitable for mating, birth, lactation, shelter from heavy seas, and in some cases, escape routes from predators (e.g., Limberger et al. 1986, Riedman 1990, Ryan et al. 1997, Bradshaw et al. 1999). For example, high density colonies of New Zealand fur seals (*Arctocephalus forsteri*) contain more and smaller rocks, more crevices and ledges which provide shelter for pups, less-pronounced slopes, higher cliffs, and more westerly exposure than that found in low-density colonies (Bradshaw et al. 1999). Semi-tropical and temperate-zone fur seals species, including *A. philippii*, have lactation periods lasting between 8 and 12 months (Gentry and Kooyman 1986) and pups are largely restricted to land and require suitable terrestrial habitat throughout most of their first year of life (Limberger et al. 1986). Pups are the age-class most vulnerable to thermal stress when ashore (Limberger et al. 1986, Trites and Antonelis 1994) and high rates of pup mortality can occur during periods of extremely high or low temperatures (Trites 1990, Trites and Antonelis 1994). Thus, optimal fur seal breeding habitat must contain terrain that facilitates behavioral thermoregulation by pups. Additionally, fur seal and sea lion social behavior in hotter climes is constrained by the form and availability of cool substrate provided by the rookery environment (Francis and Boness 1991). Female *A. philippii* at Lobería Vieja made daily movements from inland pupping and rest sites to the shoreline and into the water in response to rapid increases in solar radiation to levels exceeding 1.3 cal/cm^2/min. Females float and groom offshore in the afternoon in areas protected from the surf by offshore islets and rocky reefs. As with other Otariids, *A. philippii* males held territories on land either along the shoreline or in land-locked areas as is typical for fur seals, or held completely aquatic territories encompassed with the site where females floats (Francis and Boness 1991). The existence of aquatic territoriality as a successful mating strategy has been only reported for *A. philippii* and the occurrence of this behavioral strategy is likely a product of the interaction of thermoregulatory constrains and topographical features which together promote predictable aggregations of females offshore (Francis and Bonness 1991).

The importance of Lobería Vieja on Alejandro Selkirk Island as breeding habitat is likely related to the characteristics of its terrain. Robinson Crusoe and Santa Clara islands both have narrow coastlines, and the fur seal colonies are located mostly on the few available rocky shores. In addition, the access to many of this rocky shores is difficult since they are elevated from the sea, which may be a constraint for pups. On Robinson Crusoe Island, there are two larger and wider places that could had been occupied as a breeding area, one is a sandy beach named "El Arenal" (Fig. 3), on which juveniles can occasionally be observed. Here, the high tide covers almost the whole beach, which is probably why it is not used as a breeding site and only sporadically as a resting place. The other place is in Cumberland Bay (Fig. 3), which has similar conditions as Lobería Vieja, but is the site of San Juan Bautista town. Santa Clara Island has similar topography as Robinson Crusoe Island, dominated by a narrow rocky-shore coastline. In contrast, Lobería Vieja is a wide and long rookery of near 3 km in length, facing south and westward. This colony has different sectors called Lobería Sur, El Ovalo, Los Harenes, Los Machos and El Tongo (Torres 1987b, Torres et al. 1984; Fig. 4), with plenty of available space, mostly pebble beaches but also rocky shores. The main breeding sites are El Ovalo, Los Harenes and El Tongo (Fig. 4). Los Machos is used mainly by juveniles and subadult males. Lobería Vieja has the characteristics apparently favoured by *A. forsteri* (i.e., more and smaller rocks, more crevices and ledges which provide shelter for pups, less-pronounced slopes, higher cliffs, and more westerly exposure than all the other breeding and non-breeding colonies on the Archipelago). Therefore, the most plausible explanation of why the Robinson Crusoe-Santa

Clara population is reaching equilibrium and Alejandro Selkirk is not, is the availability of breeding habitat, for which Lobería Vieja still has plenty. Another hypothesis could be that the populations may be differentially regulated by food availability. However, there is no reason to assume that the fur seals from the islands are not foraging in similar areas.

Gathering sufficient time-series data to undertake accurate analyses and detect trends is difficult, but essential for the implementation of management and conservation practices (Van den Hoff et al. 2007), especially on vulnerable species such as *A. philippii*. The small amount of available population data prevents the use of more sophisticated analyses, and certainly continued efforts are necessary to continue building the population time series for *A. philippii* on the Juan Fernandez Archipelago. Our analysis indicates that Robinson Crusoe-Santa Clara population is not growing, and has reached an equilibrium state. The only population which shows an upward trend is Alejandro Selkirk Island, where Lobería Vieja is the most important breeding colony for both its current population size and available breeding habitat, which currently, seems to be the most important factor regulating *A. philippii* population growth. However, variation in the intrinsic growth rates within this population could be also driven by changes in the conditions of the marine ecosystem of the Southeastern Pacific. However, greater numbers of observations are required to identify any underlying causal relations between the reproductive success and oceanographic events of the marine ecosystem in which this species lives and forages most of the year in the pelagic environment.

Acknowledgements

Special thanks to Dr. David Rosen for revising and providing insights to this paper.

Keywords: Juan Fernández fur seal, population dynamics, exploitation, recovery, intrinsic growth rate, Juan Fernandez Archipelago, Chile

References

Aguayo, L.A. 1979. Juan Fernández fur seal. pp. 28–30. *In*: FAO, Mammals in the Seas, Volume II: Pinniped Species Summaries and Report on Sirenians. Food and Agriculture Organization of the United Nations, Rome, Italy.

Akcakaya, H.R., M.A. Burgman and L.R. Ginzburg. 1997. Applied Population Ecology. Applied Biomathematics. Setauket, New York, USA.

Aurioles-Gamboa, D. 2015. *Arctocephalus philippii*. The IUCN Red List of Threatened Species 2015.

Bahamonde, N. 1966. El Mar y sus recursos. pp. 81–90. *In*: N. Bahamonde [ed.]. Geografía Económica de Chile. Primer Apéndice, CORFO.

Berkson, J.M. and D.P. DeMaster. 1985. Use of pup counts in indexing population changes in pinnipeds. Can. J. Fish. Aquat. Sci. 42: 873–879.

Berryman, A. 1999. Principles of Population Dynamics and their Application. Stanley Thornes (Publishers) Ltd., Cheltenham, UK.

Bonner, N. 2000. Seals of the World. Cassell plc Blandford, UK.

Bradshaw, C.J.A., C.M. Thompson, L.S. Davis and C. Lalas. 1999. Pup density related to terrestrial habitat use by New Zealand fur seals. Can. J. Zool. 77: 1579–1586.

Busch, B.C. 1985. The War Against the Seals. A History of the North American Seal Fishery. McGill-Queen´s University Press, Kingston and Montreal, Canada.

Cuevas, J.G. and G. Van Leersum. 2001. Project "Conservation, Restoration, and Development of the Juan Fernández islands, Chile". Rev. Chil. Hist. Nat. 74: 899–910.

Francis, J.M. and D.J. Boness. 1991. The effect of thermoregulatory behaviour on the mating system of the Juan Fernández fur seal, *Arctocephalus philippii*. Behavior 119(1-2): 104–126.

Frankham, R., J.D. Ballou and D.A. Briscoe. 2002. Introduction to Conservation Genetics. Cambrigde University Press, Cambridge, UK.

Gentry, R.L. and G.L. Kooyman. 1986. Fur Seals Maternal Strategies on Land and at Sea. Princeton University Press, Princeton, NJ, USA.

Gerber, L.R. and R. Hilborn. 2001. Catastrophic events and recovery from low densities in populations of otariids: Implications for risk of extinction. Mammal. Rev. 31(2): 131–150.

Hubbs, C.L. and K.S. Norris. 1971. Original teeming abundance, supposed extinction, and survival on the Juan Fernández fur seal. pp. 35–52. *In*: W.H. Burt [ed.]. Antarctic Pinnipedia. Antarctic Research Series Volume 18, American Geophysical Union, Washington, USA.

King, J. 1964. Seals of the World. Br. Mus. (Nat. Hist.). London.

Limberger, D., F. Trillmich, H. Biebach and R.D. Stevenson. 1986. Temperature regulation and microhabitat choice by free-ranging Galapagos fur seal pups (*Arctocephalus galapagoensis*). Oecologia 69: 53–59.

Maxwell, G., J. Stidworthy and D. Williams. 1967. Seals of the World. World Wildlife Series Series 2. Constable, London, UK.

Münster-Swendsen, M. and A. Berryman. 2005. Detecting the causes of population cycles by analyses of R-functions: The spruce needle-miner, *Epinotia tedella*, and its parasitoids in Danish spruce plantations. Oikos 108: 495–502.

Purvis, A. 2001. Mammalian life histories and responses of population to exploitation. pp. 169–181. *In*: J.D. Reynolds, G.M. Mace, K.H. Redford and J.G. Robinson [eds.]. Conservation of Exploited Species. Cambridge University Press, Cambridge.

Reeves, R.R. 2009. Hunting of marine mammals. pp. 585–588. *In*: W.F. Perrin, B. Würsig and J.G.M. Thewissen [eds.]. Encyclopedia of Marine Mammals, 2nd Ed. Academic Press, San Diego, CA, USA.

Riedman, M. 1990. The Pinnipeds: Seals, Sea Lions and Walruses. University of California Press, Ltd., London, England.

Royama, T. 1977. Population persistence and density dependence. Ecol. Monogr. 47: 1–35.

Ryan, C.J., G.J. Hickling and K.J. Wilson. 1997. Breeding habitat preferences of the New Zealand Fur seal (*Arctocephalus forsteri*) on Banks Peninsula. Wild. Res. 24: 225–235.

Shaughnessy, P.D., R.J. Kirkwood and R.M. Warneke. 2002. Australian fur seals, *Arctocephalus pusillus doriferus*: Pup numbers at Lady Julia Percy Island, Victoria, and a synthesis of the species´ population status. Wild. Res. 29: 185–192.

Torres, D., C. Guerra and J.C. Cárdenas. 1984. Primeros registros de *Arctocephalus gazella* y nuevos hallazgos de *Arctocephalus tropicalis* y *Leptonychotes weddelli* en el Archipiélago de Juan Fernández. Ser. Cient. INACH 31: 115–148.

Torres, D. 1987a. Juan Fernández fur seal, *Arctocephalus philippii*. Status, biology, and ecology of fur seals. pp. 37–41. *In*: J.P. Croxall and R.L. Gentry [eds.]. Proceedings of an International Symposium and Workshop, 23–27 April 1984. NOAA Tech. Rep. NMFS 51. Cambridge, England.

Torres, D. 1987b. Antecedentes sobre el lobo fino de Juan Fernández *Arctocephalus philippii* y proyecciones para su estudio. pp. 287–317. *In*: J.C. Castilla [ed.]. Islas Oceánicas Chilenas: conocimiento Científico y Necesidades de Investigaciones. Ediciones Universidad Católica de Chile, Santiago, Chile.

Trites, A.W. 1990. Thermal budgets and climate spaces: The impact of weather on the survival of Galapagos (*Arctocephalus galapagoensis* Heller) and northern fur seals pups (*Callorhinus ursinus* L.). Funct. Ecol. 4: 753–768.

Trites, A.W. and G.A. Antonelis. 1994. The influence of climatic seasonality on the life cycle of the Pribilof northern fur seal. Mar. Mamm. Sci. 10: 311–324.

Van den Hoff, J., H. Burton and B. Raymond. 2007. The population trend of Southern elephant seals (*Mirounga leonina* L.) at Macquarie Island (1952–2004). Polar Biol. 30: 1275–1283.

Wade, P.R. 2002. Population dynamics. pp. 974–79. *In*: W.F. Perrin, B. Wursig and J. Thewissen [eds.]. Encyclopedia of Marine Mammals. Academic Press, San Diego, CA, USA.

Weber, D.S., B.S. Stewart and N. Lehman. 2004. Genetic consequences of a severe population bottleneck in the Guadalupe fur seal (*Arctocephalus townsendi*). J. Hered. 95(2): 144–153.

11

Population Ecology and Conservation Status of the South American Sea Lion in Uruguay

Valentina Franco-Trecu,[1,*] *Massimiliano Drago,*[2,a]
Diana Szteren[1,b] *and Federico G. Riet-Sapriza*[3,c]

Introduction

South American sea lion (*Otaria flavescens*, SASL) is distributed along the South American coast from southern Brazil in the Atlantic Ocean to northern Peru in the Pacific Ocean (Vaz-Ferreira 1982, Rosas et al. 1994). Throughout its entire geographical range, the SASL history of abundance has been tightly associated to human activities such as intense sealing from the 1950s until recent days (Crovetto et al. 1979, Crespo and Pedraza 1991, Sielfeld 1999, Ponce de León 2000). The post-harvesting dynamics of local SASL populations in the Pacific and Atlantic Ocean has shown a remarkable

---

[1] Departamento de Ecología y Evolución, Facultad de Ciencias, Universidad de la República, Montevideo, Uruguay.
[2] Departamento de Ecología y Evolución, Centro Universitario Regional Este (CURE), Universidad de la República, Maldonado, Uruguay.
[3] Laboratorio de Ecología Molecular de Vertebrados Acuáticos (LEMVA), Departamento de Ciencias Biológicas, Facultad de Ciencias, Universidad de los Andes, Bogotá, Colombia.
[a] Email: m.drago@ub.edu
[b] Email: dszteren@gmail.com
[c] Email: frietsapriza@gmail.com
* Corresponding author: pinnipedosuy@gmail.com

variation among rookeries. In the Pacific Ocean, the SASL population in Peru has slowly recovered since the large decline after the 1997 El Niño event (Olivera and Majluf 2012). While in Chile, the SASL population has showed different trends according to local environmental variation, ranging from slow recovery in the North (Bartheld et al. 2008) to stable in Central and Southern Chile (Sepúlveda et al. 2011). In the Atlantic Ocean, for instance, in Argentina, a range of local SASL populations are recovering with an annual growth rates from 5 to 8% (Reyes et al. 1999, Dans et al. 2004, Schiavini et al. 2004, Grandi et al. 2014). In North Patagonia, the recovering of some SASL populations has been associated with colonization events of new areas and to changes in the local age structure (Grandi et al. 2008, 2014). The Malvinas Island (Falkland) SASL population has been slowly recovering since 1995 (Thompson et al. 2005) and its current annual pup production is estimated at ca. 4000 individuals (Baylis et al. 2015). In Uruguay, the SASL breeds mainly in two rookeries (Isla—Islote de Lobos and Islas de Torres—del Marco, see Fig. 1), and at Isla Verde-Islote La Coronilla with a low reproductive activity

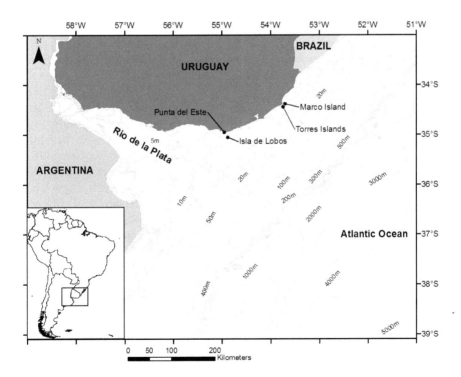

Figure 1. Locations of the two main breeding rookeries, Isla de lobos and Cabo Polonio (Torres and Marco Island) of South American sea lions *Otaria flavescens*, in Uruguay. Depth (isobaths, light grey lines) data was collected in 2002 by FREPLATA-Proyecto de Protección Ambiental del Río de la Plata y su Frente Marítimo (www.freplata.org).

where a few number of births has been recorded (Szteren 2015). However, no new breeding areas have been observed for at least 60 years. Direct estimates of the population size of SASLs in Uruguay were lacking until 2011. However, an indirect estimate of population size (12,000 individuals in 2005) was obtained from a matrix population model using demographic parameters from other species of otariids (Páez 2006). Unlike the observed trend of population increase in the post-harvesting SASL population in northern Patagonia (Dans et al. 2004), the aforementioned model predicted a decreasing trend in the SASL population in Uruguay between 1993 and 2005 (Páez 2006). This decreasing trend has continued almost 20 years after the end of their exploitation in Uruguay, in which ~109,000 adults between 1910 and 1942 (Ximénez and Langguth 2002) and ~50,000 pups between 1959–1986 (Ponce de León 2000) were harvested.

Reproductive Behaviour

Polygynous mating systems in mammals are classified as resource-defence polygyny, female-defence polygyny or male dominance polygyny - leks (Emlen and Oring 1977, Davies 1991, Shuster and Wade 2003). The combination of behavioural and molecular data indicated that the SASL reproductive behaviour in Isla de Lobos - Uruguay actually involves the coexistence of two types of polygyny, each occurring in different locations of the same rookery (Franco-Trecu et al. 2015a). Males at the tide line monopolize relatively stable groups of females (female-defence polygyny) into territories (Fig. 2), whose locations change with the random micro-tidal variation at the study site (Wilson 1975, Alcock et al. 1978, Barrows 1983). Alternatively, males at the internal pools defend fixed territories (stable location during the tenure) (Dewsbury 1978) and establish a resource-defence polygyny (Emlen and Oring 1977). The coexistence of different types of polygyny in a single population has been observed in different taxa (Alvarez et al. 1990, Thirgood et al. 1999), but never in pinnipeds. Other successful reproductive tactics (e.g., wandering or sneaking males) were not observed at Isla de Lobos due paternity was assigned to a much higher percentage of pups (80% Franco-Trecu et al. 2015a) than it had been commonly possible in other studies with otariids (~15–30%) (Gemmell et al. 2001, Pörschmann et al. 2010, Flatz et al. 2012, Franco-Trecu et al. 2014a).

Ecological factors such as the magnitude of heat stress, the topography and type of substratum of the breeding site, the degree and frequency of female movements and the rookery population density, are key to determine the extent of the spatial variation in the polygynous mating system of SASL populations (Campagna and Le Boeuf 1988, Paves et al. 2005, Fernandez-Juricic and Cassini 2007, Soto and Trites 2011). For instance, resource- and female-defence polygyny have been described for SASL populations in

Figure 2. A Male and females of South American sea lion at the territorial tidal line on a rocky beach of coastal Uruguay. Photo credit: V. Franco-Trecu.

Argentinean rookeries (Campagna and Le Boeuf 1988, Cassini and Vila 1990, Cappozzo and Perrin 2009) and in Chile (Paves et al. 2005), and a lek-like mating system has been reported for the tropical islands of Peru (Soto and Trites 2011). However, each population had a single or "pure" mating system, in contrast to the coexistence of different types of polygyny that we found at Isla de Lobos. This coexistence generated one of the highest values of statistical variability for reproductive success ever estimated in otariids (Franco-Trecu et al. 2014a, Gonzalez-Suarez and Cassini 2014).

We suggest that the constraints imposed by the topography of the rookery on the availability of sites that facilitate thermoregulation during the breeding season, determines the SASL males' spatial use and the individual mating behaviour in Isla de Lobos. To our knowledge, this is the first time that the coexistence of two types of polygyny in the same breeding rookery is reported for otariids.

Foraging Behaviour and Ecology

The SASL are central-place-foragers (Orians and Pearson 1979) whose females alternate foraging trips to the sea with suckling bouts on land to nurse their pup. As a result, foraging and nursing are separated in time and space (Bonner 1984, Oftedal et al. 1987). Most information about the foraging and diving behaviour of the Uruguayan SASL population comes from studies conducted at Isla de Lobos (see Fig. 1). The SASL is a predator that feeds on a variety of prey including Argentine shortfin squid (*Illex argentinus*), Whitemouth croaker (*Micropogonias furnieri*), Stripped weakfish (*Cynoscion guatucupa*) and Largehead hairtail (*Trichiurus lepturus*) (Franco-Trecu et al. 2013). Sea lion females instrumented with time-depth-recorders, satellite and GPS tags have shown that they forage exclusively over the shallow continental shelf covering large areas (2,224 km^2 to 9,577 km^2) and make short foraging trips to sea at a maximum distance around 100 km from the rookeries (Riet Sapriza et al. 2013, Rodriguez et al. 2013). During their short (1.9 min) and shallow benthic diving bouts (average depth of 21 m, maximum 78m), SASL search prey near or at the bottom of the sea (Fig. 3) (Riet-Sapriza et al. 2013).

The SASL is a highly sexually dimorphic species, with adult males doubling females in weight (Cappozzo and Perrin 2009). Body size differences clearly influence diving physiology capacity in pinnipeds since adults of large-bodied species often exploit deep, benthic habitats, whereas those of small-bodied species typically forage in epipelagic habitats (Kooyman and Trillmich 1986, Costa 1993, Costa et al. 2004). Likewise, small-bodied females of dimorphic species often have a more pelagic diet than males throughout their life (Le Boeuf et al. 2000, Meynier et al. 2008). Nevertheless, the relationships among body size, diving performance and habitat use are not necessarily univocal, and physiological traits other than size, such as the mass-specific total oxygen stores, may also play a fundamental role on determining resource partitioning between the two sexes (Weise and Costa 2007).

As expected from differences in body mass, sea lion males from northern Patagonia had been reported to exploit benthic and deeper foraging grounds than females (Campagna et al. 2001, Drago et al. 2009a); however, dissimilarities in foraging habits between the sexes are not constant over time (Drago et al. 2009b). In Uruguay, $\delta^{13}C$ and $\delta^{15}N$ stable isotope values of skin and bone were used to infer the trophic relationships between the sexes during the pre-breeding period and year round respectively. They revealed that male and female sea lions used a wide diversity of foraging strategies throughout the year and that no sex-diferences existed (Drago et al. 2015). However, the diversity of foraging strategies was strongly reduced in both sexes during the pre-breeding period, when all individuals increased their consumption of pelagic prey over benthic prey, but the two sexes differing considerably in trophic level, being significantly higher in males, and the isotopic niche space of males and females did not overlap at all (Drago et al. 2015). These results would indicate that sexual foraging segregation only takes place during the pre-breeding season, when crowding in the areas surrounding the breeding rookeries increases and

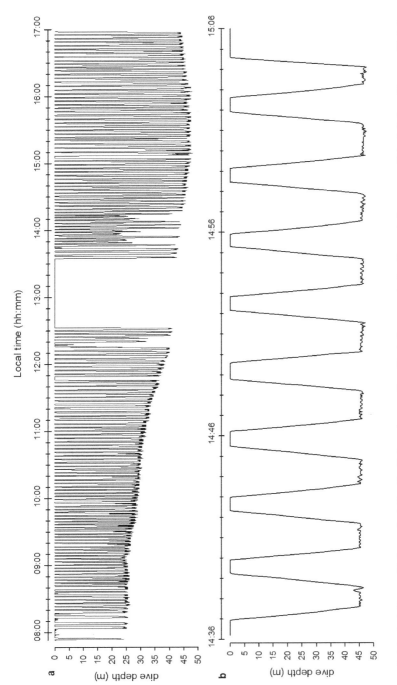

Figure 3. A typical dive profile from a lactating South American sea lion *Otaria flavescens* during the austral summer at Isla de Lobos, Uruguay (data from Riet-Sapriza et al. 2013), (a) the dive profile from 8 am to 17:00 pm indicating intensive diving during this time period and period of resting or recovery time (between 12:30 and 13:30), and (b) is a close up of the dive profile in (a) and shows a typical pattern of benthic divers (U shaped dive pattern).

per-capita resource declines. Therefore, niche divergence is the most likely explanation for the trophic segregation between SASL females and males. Trophic segregation and a dietary shift favouring pelagic species with higher energy density at this time of year may provide a means of reducing intraspecific competition related to high niche overlap and would allow individuals of both sexes to increase their energy reserves needed to face the energetic demands during the breeding period (Beck et al. 2007). On the other hand, intersexual differences largely vanish during the rest of the year, since individuals of both sexes spread over a huge area and increase both their dietary niche breadth and overlap. This suggests that SASL of both sexes could exploit a greater range of habitats and prey when they are more scattered. Furthermore, stable isotopes values from whiskers (which include information from ~2–3 years), show that SASL have a smaller niche area with a higher individual trophic specialization compared to their sympatric species, the South American fur seal, *Arctocephalus australis* (Franco-Trecu et al. 2014b).

Anthropogenic Threats

Conflicts between marine mammals and fisheries can be categorized as direct and indirect interactions (Beverton 1985). The former may be known as operational interaction and implies direct interaction with fisheries operations (i.e., fishing gear such as nets, lines) that results in bycatch of marine mammals, depredation by marine mammals (remove caught fish from nets or lines) and gear damage (Gulland 1986, Wickens 1994, 1995, Dans et al. 2003, Shaughnessy et al. 2003, Wilkinson et al. 2003, Read 2005, 2008, Read et al. 2006, Hamer et al. 2012). The later, also named "ecological, biological, trophic" interactions, could imply competition by depletion of food resources, either directly or indirectly via the food chain (Abrams et al. 1996). The overlap in food resources between marine mammals and fisheries are complex, poorly understood, and largely unmanaged (Lowry and Frost 1985, Trites et al. 1997, Goldsworthy et al. 2003, Planganyi and Butterworth 2005, Cornick et al. 2006). Populations of marine mammals can have significant effect on the abundance of prey populations and reduce fisheries target species catch (Trites et al. 1997), whereas fisheries may be depriving marine mammals of food resources, and consequently may be impeding the growth and the recovery of their populations (Gulland 1987, Butterworth et al. 1988, Butterworth 1992, Wickens et al. 1992, Trites et al. 1997, DeMaster et al. 2001). In Uruguay, the direct operational interaction with the coastal bottom trawl fishery (industrial) has negative effects for SASL by means of bycatch (Franco-Trecu et al. unpublished data). While their interaction with artisanal fisheries, affects both parts and has become a conflict with a strong socio-ecological component.

Direct and indirect interactions with industrial fisheries

The coastal bottom trawl fishery effort (e.g., catch per unit effort, CPUE) and the number of SASL bycatch as a result of the operational interaction, registered over a period of 3 years (2009 to 2011) accounted for an average annual mortality of ca. 50 individuals (I.C. 15–80) (Franco-Trecu et al. unpublished data). The potential biological interactions between coastal bottom trawl fisheries and SASL, is a complex ecological conflict that is difficult to measure. However, it has been shown that the diet and the foraging areas of lactating SASL females from Isla de Lobos, overlap with industrial fish species and fisheries operational areas, respectively (Riet-Sapriza et al. 2013). In terms of landings, the main catch of the coastal bottom trawl fisheries are Whitemouth croaker and Stripped weakfish which are important species in sea lions diet (Riet Sapriza et al. 2013). The magnitude of the fisheries landing and sea lions consumption differ substantially. Annual prey consumption of female SASL from Isla de Lobos was estimated at 1,005 ton, approximately 97% less than the bottom trawl fishery landings (35,511 ton) (Riet Sapriza et al. 2013). Given the benthic diving behaviour of SASL and their exclusively-restricted foraging areas to shallow continental shelf (Riet Sapriza et al. 2013, Rodriguez et al. 2013), if food resource are scarce, it is likely that competition with fisheries could create a sub-optimal foraging condition for lactating females. In this context lactating females may have to work harder to find food, i.e., make longer and farther foraging trips and dive near or at their physiological limits. This may affect the reproductive success of females and the recovery of the population in the long term.

Direct and indirect interactions with artisanal fisheries

Operational interactions are a persistent problem in most fishing communities along the Uruguayan coast. As other pinnipeds (Wickens 1995, Northridge and Hofman 1999), the SASL interacts with coastal fisheries preying upon fish caught in gillnets and long-lines, damaging fishing gear, entanglements and sometimes being incidentally caught (Szteren and Páez 2002, Szteren 2006, Franco-Trecu et al. 2009, Franco-Trecu et al. 2017). Artisanal fishermen consider this a major problem for their activity, a threat for their sustainable way of living and frequently complain about sea lion predation and gear damage. Artisanal fisheries target mainly whitemouth croaker, stripped weakfish, brazilian codling (*Urophycis brasiliensis*) and narrownose smoothhound (*Mustelus schmitti*), depending on the area and the season.

From 1997 to 2004 at 4 fishing ports on the Rio de la Plata estuary and the Atlantic coast, damages to fisheries catches ranged from 0.25 to 46.2% of the total potential landings considering minimum and maximum scenarios (Szteren and Lezama 2006). In this period, the frequency of direct interactions between SASL and the artisanal fisheries decreased in Piriápolis, a fishing port located on the Rio de la Plata estuary (Szteren and Lezama 2006), but in 2010 the frequency of these interactions increased again (De Maria et al. 2014). There were no reported seasonal differences in fisheries catches damages, except in 2010 when they were significantly higher in Spring and Autumn (De Maria et al. 2014). Interannual variability of fish populations and the reproductive cycle could explain this pattern, because higher damages coincided with the pre-breeding and post-breeding seasons in which males need to recover their body condition and females are progressively making longer foraging trips, hence interacting with fisheries. In the studied years and fishing ports, catch per unit of effort was never significantly different in the presence or absence of interactions. Furthermore, from 1997 to 2004, the mean number of sea lions involved in fisheries interactions did not show any clear trend in time, between different types of fishing gear, or among locations (Szteren and Páez 2002, Szteren and Lezama 2006). However, in 2010 the number of pinnipeds interacting with artisanal fisheries was significantly higher in winter and spring (De Maria et al. 2014), outside the breeding season.

The SASL prey species preferences while interacting with fisheries, vary depending on their location. At the Rio de la Plata estuary SASL exhibit an opportunistic strategy: the most consumed prey are the same most caught by the artisanal fishery (*Macrodon ancylodon, Urophycis brasiliensis* in 2007-2008; *Micropogonias furnieri, Cynoscion guatucupa* in 2012) (Szteren 2006, De Maria et al. 2014). In contrast, on the Atlantic coast, SASL consume secondary-target fisheries species such as *Cynoscion guatucupa* while the most fished was *Mustelus schmitti* (Szteren 2006).

In conclusion, in Autumn, when interactions between artisanal fisheries in the coast intensify, sea lions predation increases, and landings of the main species (*M. furnieri*) is at its lowest (Horta and Defeo 2012). Two factors may be influencing the magnitude of the interactions: distance to the closest colony and magnitude of fisheries landings. In fishing ports located relatively far from a sea lion colony, the presence of large schools of fish is most likely what attracted the sea lions. On the other hand, the fishing ports with the highest percentage of interactions are near to a haul out site, showing that the closeness to a rookery is also an important factor (De Maria et al. 2014). According to our studies, along the Uruguayan coast, sea lions did not cause a significant loss to artisanal gillnets. However, small damages should be analyzed in the context in which they take place, because a small quantity of fish being lost may represent a great percentage in relation with the daily total catch. It would be important to consider this conflict as a component of socio-ecological systems under an ecosystem point of view.

In order to effectively reduce these conflicts, management strategies should involve participative research with an integrative approach to facilitate co-management (Trimble and Berkes 2013). In this context, a participatory research initiative has been formed since 2011 to investigate the problems of the artisanal fishery in Piriápolis, Uruguay. At the moment, a mitigation measure is being assessed by the design and testing of different types of fish traps as a complementary fishing gear, where fish caught is kept alive and damage caused by sea lions have been null. The participatory research approach is based on the involvement of the group members throughout the whole research process, so that the perception, experience and know-how of each individual are acknowledged and taken into account. Because fishers are carrying out the experiments on board their boats, they are able to propose solutions to the problems that may arise, watching the performance at first hand.

Although, the mayor concern is direct interaction because of fishing gear damage caused by sea lions; indirect interactions between artisanal fishermen and sea lions also occur and should be taken into account. The artisanal fisheries and sea lion foraging areas overlap, as well as the species landed by the fisheries and consumed by sea lions. In contrast with the industrial fisheries, 99% of the total landing by artisanal fisheries in the Uruguayan Atlantic coast overlapped with the species consumed by sea lions (Riet Sapriza et al. 2013). Furthermore, artisanal fisheries total annual catches represented 59% more than the total annual prey consumed by lactating SASL (Riet-Sapriza et al. 2013). Evidently, from the point of the artisanal fisheries the trophic interaction for food resources with sea lion is much more significant than with coastal bottom trawl.

Sealing and Population Dynamics

An aerial survey of the total SASL abundance in Uruguay in 2013, estimated about 10,000 individuals, with pups representing a ~30% of the total population (Franco-Trecu 2015). Both pups and total abundance were higher at Cabo Polonio than at Isla de Lobos. Considering only the breeding areas, male/pup ratio was 0.66 at Cabo Polonio while at Isla de Lobos the ratio accounted for 1.44. The SASL birth rate in Uruguay exhibited a negative trend between 1956 and 2013 with a 2% annual decline rate (Franco-Trecu et al. 2015b). However, births trend at Cabo Polonio rookery did not differ significantly from zero and thus, this local population has been stable.

While pup specific growth rate was significantly higher at Isla de Lobos (the smallest colony), pup survival in the second month after birth was lower in this colony compared to Cabo Polonio. Furthermore, differences in the social structure between the colonies were also found, with a higher proportion of subadult and adult males in the smallest colony. Antisocial behaviour (abduction and eventual killing of pups) by peripheral adult and subadult SASL males has been reported in both Uruguayan

(Vaz-Ferreira 1965, 1976) and Argentinean colonies (Campagna et al. 1988). Given the differences in the social structures found between the colonies, the antisocial behaviour could point to infanticide as a likely cause for the higher pup mortality rate found in the smallest colony. In the total Uruguayan population (considering Isla de Lobos and Cabo Polonio), the pup birth rate has shown a historical decline since 1956 (Franco-Trecu 2015) and until the 1980s, likely as a result of the impact of pup harvesting. The recovery of the Isla de Lobos population would have been expected to occur 30 yrs after the end of harvesting; however, the Isla de Lobos colony had a lower early pup survival and a higher male/pup ratio than Cabo Polonio (Franco-Trecu et al. 2015b).

Conservation and Management

The decline of the SASL population in Uruguay may be attributed to the lack or inappropriate management decisions taken over the last decades. The cumulative effect of population extractions, such as: (i) pup harvesting (~50,000), (ii) zoo and marine aquariums sales (144 and 285 young males and females, respectively), and (iii) fisheries incidental mortalities (i.e., direct interactions with industrial and artisanal fisheries may result in bycatch and illegal shooting of sea lions, respectively), may be responsible for the lack of recovery of the species in Uruguay. These extractions and mortalities combined, not only reduced the local population size, but also could have disrupted its social structure to a degree that the Allee effect (decline in individual fitness at low population size or density) could have become so significant (Franco-Trecu et al. 2015b) as to hamper the SASL post-harvesting population recovery at Isla de Lobos. Coastal ecosystems may be subject to environmental impacts associated with high human population densities such as overfishing (Milessi et al. 2005, Defeo et al. 2009), pollution (Clausen and York 2008, Bulleri and Chapman 2010) and perturbation due to recreational activities and ports (Davenport and Davenport 2006). These activities may have also further affected the food resource availability and the quality of SASL foraging areas in Uruguay, and contributed to the lack of the population recovery after the end of the SASL commercial harvest in the 80s.

Acknowledgements

Special thanks to Dr. S. Villegas-Amtmann for revising and providing insights to this paper.

Keywords: South American sea lion, *Otaria flavescens,* diet composition, diving, fishery interaction, foraging behaviour, harvesting, reproductive behaviour, Uruguay

References

Abrams, P.A., B.A. Menge, G.G. Mittelbach, D.A. Spiller and P. Yodzis. 1996. The role of indirect effects in food webs. pp. 371–395. *In*: G.A. Polis and K.O. Winemiller [eds.]. Food Webs: Integration of Patterns and Dynamics. Chapman & Hall, New York, USA.

Alcock, J., E.M. Barrows, G. Gordh, L.J. Hubbard, L. Kirkendall, D.W. Pyle et al. 1978. The ecology and evolution of male reproductive behaviour in the bees and wasps. Zoological Journal of the Linnean Society 64: 293–326.

Alvarez, F., F. Braza and C. San Jose. 1990. Coexistence of territoriality and harem defense in a rutting fallow deer population. J. Mammal. 71: 692–695.

Barrows, E.M. 1983. Male territoriality in the carpenter bee *Xylocopa virginica*. Anim. Behav. 31: 806–813.

Bartheld, J., H. Pavés, F. Contreras, C. Vera, C. Manque, D. Miranda et al. 2008. Cuantificación poblacional de lobos marinos en el litoral de la I a IV región. Informe Final Proyecto FIP 2006-50.

Baylis, A.M.M., R.A. Orben, J.P.Y. Arnould, F. Christiansen, G.C. Hays and I.J. Staniland. 2015. Disentangling the cause of a catastrophic population decline in a large marine mammal. Ecology 96(10): 2834–2847.

Beck, C.A., S.J. Iverson, W.D. Bowen and W. Blanchard. 2007. Sex differences in grey seal diet reflect seasonal variation in foraging behaviour and reproductive expenditure: Evidence from quantitative fatty acid signature analysis. J. Anim. Ecol. 76(3): 490–502.

Beverton, R.J.H. 1985. Analysis of marine mammal–fisheries interactions. pp. 3–33. *In*: J.R. Beddington, R.J.H. Beverto and D. Lavigne [eds.]. Marine Mammals and Fisheries. George Allen & Unwin, London, United Kingdom.

Bonner, Nigel W. 1984. Lactation strategies in pinnipeds: Problems for a marine mammalian group. Symp. Zool. Soc. Lond. 51: 253–272.

Bulleri, F. and M.G. Chapman. 2010. The introduction of coastal infrastructure as a driver of change in marine environments. J. Appl. Ecol. 47(1): 26–35.

Butterworth, D.S., D.C. Duffy, P.B. Best and M.O. Bergh. 1988. On the scientific basis for reducing the South African fur seal population. S. Afr. J. Sci. 84: 179–188.

Butterworth, D.S. 1992. Will more seals result in reduced fishing quotas? S. Afr. J. Sci. 88: 414–416.

Campagna, C. and B.J. Le Boeuf. 1988. Reproductive behaviour of southern sea lions. Behaviour 104(3/4): 233–261.

Campagna, C., Le Boeuf, B.J. and Cappozzo, H.L. 1988. Pup abduction and infanticide in southern sea lions. Behaviour 107: 44–60.

Campagna, C., R. Werner, W. Karesh, M.R. Marin, F. Koontz, R. Cook et al. 2001. Movements and locations at sea of South American sea lions (*Otaria flavescens*). J. Zool. (Lond.) 257: 205–220.

Cappozzo, H.L. and W.P. Perrin. 2009. South American sea lion (*Otaria flavescens*). pp. 1076–1079. *In*: W.F. Perrin, B. Würsing and J.G.M. Thewissen [eds.]. Encyclopedia of Marine Mammals, 2nd Ed. Academic Press, San Diego, CA, USA.

Cassini, M.H. and B.L. Vila. 1990. Male mating behavior of the Southern sea lion. Bull. Mar. Sci. 46(2): 555–559.

Clausen, R. and R. York. 2008. Global biodiversity decline of marine and freshwater fish: A cross-national analysis of economic, demographic, and ecological influences. Soc. Sci. Res. 37(4): 1310–1320.

Campagna, C., B.J. Le Boeuf and H.L. Cappozzo. 1988. Pup abduction and infanticide in southern sea lions. Behaviour 107: 44–60.
Cornick, L.A., W. Neill and W.E. Grant. 2006. Assessing competition between Steller sea lions and the commercial groundfishery in Alaska: A bioenergetics modelling approach. Ecol. Model. 199(1): 107–114.
Costa, D.P. 1993. The relationship between reproductive and foraging energetics and the evolution of the Pinnipedia. pp. 293–314. *In*: I.L. Boyd [ed.]. Marine Mammals: Advances in Behavioural and Population Biology. Symposium Zoological Society of London No. 66. Oxford University Press, Oxford.
Costa, D.P., C.E. Kuhn, M.J. Weise, S.A. Shaffer and J.P.Y. Arnould. 2004. When does physiology limit the foraging behaviour of freely diving mammals? Int. Congr. Ser. 1275: 359–366.
Crespo, E.A. and S.N. Pedraza. 1991. Estado actual y tendencia de la población de lobos marinos de un pelo (*Otaria flavescens*) en el litoral norpatagónico. Ecología Austral 2(1): 87–95.
Crovetto, A., L. Duran, D. Oliva and J. Torres. 1979. Sobre la Explotación de *Otaria flavescens* (Shaw 1800), en la Localidad de Punta Lobos (Carnivora, Otariidae). Anales del Museo de Historia Natural (Valparaíso, Chile) 12: 241–243.
Dans, S., M.F. Alonso, E.A. Crespo, S.N. Pedraza and N. García. 2003. Interactions between marine mammals and high seas fisheries in Patagonia: An integrated approach. pp. 100–115. *In*: N. Gales, M. Hindell and R. Kirkwood [eds.]. Marine Mammals: Fisheries, Tourism and Management Issues. CSIRO Publishing, Collingwood, Australia.
Dans, S.L., E.A. Crespo, S.N. Pedraza and M.K. Alonso. 2004. Recovery of the South American sea lion population in northern Patagonia. Can. J. Fish. Aquat. Sci. 61(9): 1681–1690.
Davenport, J. and J.L. Davenport. 2006. The impact of tourism and personal leisure transport on coastal environments: A review. Estuar. Coast. Shelf Sci. 67(1-2): 280–292.
Davies, N.B. 1991. Mating systems. pp. 263–294. *In*: J.R. Krebs and N.B. Davies [eds.]. Behavioural Ecology: An Evolutionary Approach, 3rd Ed. Blackwell Science Publications, Oxford.
De Maria, M., F.R. Barboza and D. Szteren. 2014. Predation of South American sea lions (*Otaria flavescens*) on artisanal fisheries in the Rio de la Plata estuary. Fish Res. 149: 69–73.
Defeo, O., S. Horta, A. Carranza, D. Lercari, A. De Alava, J. Gómez et al. 2009. Hacia un manejo ecosistémico de pesquerías: Áreas Marinas Protegidas en Uruguay. Facultad de Ciencias - DINARA. Montevideo, Uruguay.
DeMaster, D.P., C.W. Fowler, S.L. Perry and M.E. Richlen. 2001. Predation and competition: The impact of fisheries on marine-mammal populations over the next one hundred years. J. Mammal. 82(3): 641–651.
Dewsbury, D. 1978. Comparative Animal Behavior. McGraw-Hill Book Company, New York, NY, USA.
Drago, M., L. Cardona, E.A. Crespo and A. Aguilar. 2009a. Ontogenic dietary changes in South American sea lions. J. Zool. 279: 251–261.
Drago, M., E.A. Crespo, A. Aguilar, L. Cardona, N. García, S.L. Dans et al. 2009b. Historic diet change of the South American sea lion in Patagonia as revealed by isotopic analysis. Mar. Ecol. Prog. Ser. 384: 273–286.
Drago, M., V. Franco-Trecu, L. Zenteno, D. Szteren, E.A. Crespo, F.R. Sapriza et al. 2015. Sexual foraging segregation in South American sea lions increases during the pre-breeding period in the La Plata River. Mar. Ecol. Prog. Ser. 525: 261–272.
Emlen, S.T. and L.W. Oring. 1977. Ecology, sexual selection, and the evolution of mating systems. Science 197(4300): 215–223.
Fernandez-Juricic, E. and M.H. Cassini. 2007. Intra-sexual female agonistic behaviour of the South American sea lion (*Otaria flavescens*) in two colonies with different breeding substrates. Acta Ethol. 10(1): 23–28.
Flatz, R., M. González-Suárez, J.K. Young, C.J. Hernández-Camacho, A.J. Immel and L.R. Gerber. 2012. Weak polygyny in California sea lions and the potential for alternative mating tactics. Plos One 7(3): e33654.

Franco-Trecu, V., P. Costa, C. Abud, C. Dimitriadis, P. Laporta, C. Passadore et al. 2009. Bycatch of franciscana *Pontoporia blainvillei* in uruguayan artisanal gillnet fisheries: An evaluation after a twelve-year gap in data collection. Lat. Am. J. Aquat. Mamm. (LAJAM) 7(1-2): 11–22.

Franco-Trecu, V., M. Drago, F.G. Riet-Sapriza, A. Parnell, R. Frau and P. Inchausti. 2013. Bias in diet determination: Incorporating traditional methods in Bayesian mixing models. Plos One 8(11): e80019.

Franco-Trecu, V., P. Costa, Y. Schramm, B. Tassino and P. Inchausti. 2014a. Sex on the rocks: Breeding tactics and reproductive success of the for South American fur seal males. Behav. Ecol. 25(6): 1513–1523.

Franco-Trecu, V., D. Aurioles-Gamboa and P. Inchausti. 2014b. Individual trophic specialisation and niche segregation explain the contrasting population trends of two sympatric otariids. Mar. Biol. 161(3): 609–618.

Franco-Trecu, V. 2015. Tácticas comportamentales de forrajeo y apareamiento y dinámica poblacional de dos especies de otáridos simpátricas con tendencias poblacionales contrastantes. PEDECIBA, Universidad de la República, Montevideo, Uruguay.

Franco-Trecu, V., P. Costa-Urrutia, Y. Schramm, B. Tassino and P. Inchausti. 2015a. Tide line vs internal pools: Mating system and breeding success of South American sea lion males. Behav. Ecol. Sociobiol. 69(12): 1985–1996.

Franco-Trecu, V., M. Drago, C. Baladan, M.D. García-Olazabal, E.A. Crespo, L. Cadona et al. 2015b. Post-harvesting population dynamics of the South American sea lion (*Otaria byronia*) in the Southwestern Atlantic. Mar. Mamm. Sci. 31(3): 963–978.

Franco-Trecu, V., M. Drago, H. Katz, E. Machín and Y. Marín. 2017. With the noose around the neck: marine debris entangling otariid species. Environmental Pollution : 985–989 doi.

Gemmell, N.J., T.M. Burg, I.L. Boyd and W. Amos. 2001. Low reproductive success in territorial male Antarctic fur seals (*Arctocephalus gazella*) suggests the existence of alternative mating strategies. Mol. Ecol. 10(2): 451–460.

Goldsworthy, S.D., C. Bulman, X. He, J. Larcombe and C. Littnan. 2003. Trophic interactions between marine mammals and Australian fisheries: An ecosystem approach. pp. 62–99. *In*: N. Gales, M. Hindell and R. Kirkwood [eds.]. Marine Mammals: Fisheries, Tourism and Management Issues. CSIRO Publishing, Collingwood, Australia.

Gonzalez-Suarez, M. and M.H. Cassini. 2014. Variance in male reproductive success and sexual size dimorphism in pinnipeds: Testing an assumption of sexual selection theory. Mamm. Rev. 44(2): 88–93.

Grandi, M.F., S.L. Dans and E.A. Crespo. 2008. Social composition and spatial distribution of colonies in an expanding population of South American sea lions. J. Mammal. 89(5): 1218–1228.

Grandi, M.F., L.D. Silvana and E.A. Crespo. 2014. The recovery process of a population is not always the same: The case of *Otaria flavescens*. Mar. Biol. Res. 11(3): 225–235.

Gulland, J.A. 1986. Marine mammal - fisheries interactions. Ambio 15(3): 152–154.

Gulland, J.A. 1987. Seals and fisheries: A case for predator control? Trends Ecol. Evol. 2: 102–103.

Hamer, D.J., S. Childerhouse and N.J. Gales. 2012. Odontocete bycatch and depredation in longline fisheries: A review of available literature and of potential solutions. Mar. Mamm. Sci. 28(4): E345–E374.

Horta, S. and O. Defeo. 2012. The spatial dynamics of the whitemouth croaker artisanal fishery in Uruguay and interdependencies with the industrial fleet. Fish. Res. 125: 121–128.

Kooyman, G.L. and F. Trillmich. 1986. Diving behavior of Galapagos sea lions. pp. 209–219. *In*: R.L. Gentry and G.L. Kooyman [eds.]. Fur Seals: Maternal Strategies on Land and at Sea. Princeton University Press, Princeton, NJ, USA.

Le Boeuf, B.J., D.E. Crocker, D.P. Costa, S.B. Blackwell, P.M. Webb and D.S. Houser. 2000. Foraging ecology of northern elephant seals. Ecol. Monogr. 70: 353–382.

Lowry, L.F. and K.J. Frost. 1985. Biological interactions between marine mammals and commercial fisheries in the Bering Sea. pp. 41–61. *In*: J.R. Beddington, R.J.H. Beverton and D.M. Lavigne [eds.]. Marine Mammals and Fisheries. George Allen and Unwin, Hemel Hempsted, UK.

Meynier, L., P.C.H. Morel, B.L. Chilvers, D.D.S. Mackenzie, A. MacGibbon and P.J. Duignan. 2008. Temporal and sex differences in the blubber fatty acid profiles of the New Zealand sea lion Phocarctos hookeri. Mar. Ecol. Prog. Ser. 366: 271–279.

Milessi, A.C., H. Arancibia, S. Neira and O. Defeo. 2005. The mean trophic level of uruguayan landings during the period 1990–2001. Fish Res. 74: 223–231.

Northridge, S.P. and R.J. Hofman. 1999. Marine mammals interactions with fisheries. pp. 99–119. In: J. Twiss and R.R. Reeves [eds.]. Conservation and Management of Marine Mammals. Smithsonian Institution Press, Washington, USA.

Oftedal, O.T., D.J. Boness and R.A. Tedman. 1987. The behavior, physiology, and anatomy of lactation in the pinnipedia. pp. 175–221. In: H.H. Genoways [ed.]. Current Mammalogy. Plenum Press, New York, NY, USA.

Oliveira, L.R., L.D. Fraga and P. Majluf. 2012. Effective population size for South American sea lions along the Peruvian coast: the survivors of the strongest El Niño event in history. Journal of the Marine Biological Association of the United Kingdom 92(8): 1835.

Orians, G.H. and N.E. Pearson. 1979. On the theory of central place foraging. pp. 154–177. In: Horn, D.J., R.D. Mitchell and G.R. Stairs [eds.]. Analysis of Ecological Systems. The Ohio State University Press, Columbus.

Páez, E. 2006. Situación de la administración del recurso lobos y leones marinos en Uruguay. pp. 577–583. In: R. Menafra, L. Rodríguez-Gallego, F. Scarabino and D. Conde [eds.]. Bases para la conservación y el manejo de la costa uruguaya. Vida Silvestre, Sociedad Uruguaya para la Conservación de la Naturaleza. Montevideo, Uruguay.

Paves, H.J., R.P. Schlatter and C.I. Espinoza. 2005. Breeding patterns in southern sea lions, *Otaria flavescens* (Shaw 1800), in south-central Chile. Rev. Chil. Hist. Nat. 78(4): 687–700.

Planganyi, E.E. and D.S. Butterworth. 2005. Indirect fisheries interactions. pp. 19–46. In: J.E. Reynolds III, W.F. Perrin, R.R. Reeves, S. Montgomery and T.J. Ragen [eds.]. Marine Mammal Research: Conservation Beyond Crisis. The Johns Hopkins University Press, Baltimore, MD, USA.

Ponce de León, A. 2000. Taxonomía, sistemática y sinopsis de la biología y ecología de los pinipedios de Uruguay. pp. 6–35. In: M. Rey and F. Amestoy [eds.]. Sinopsis de la biología y ecología de las poblaciones de lobos finos y leones marinos de Uruguay. Pautas para su manejo y Administración. Parte I. Biología de las especies, edited by Montevideo-Uruguay: Proyecto URU/92/003. Instituto Nacional de Pesca-Programa de las Naciones Unidas para el Desarrollo.

Pörschmann, U., F. Trillmich, B. Mueller and J.B.W. Wolf. 2010. Male reproductive success and its behavioural correlates in a polygynous mammal, the Galapagos sea lion (*Zalophus wollebaeki*). Mol. Ecol. 19(12): 2574–2586.

Read, A.J. 2005. Bycatch and depredation. pp. 5–17. In: J.E. Reynolds III, W.F. Perrin, R.R. Reeves, S. Montgomery and T.J. Ragen [eds.]. Marine Mammal Research: Conservation Beyond Crisis. The Johns Hopkins University Press, Baltimore, MD, USA.

Read, A.J., P. Drinker and S. Northridge. 2006. Bycatch of marine mammals in U.S. and global fisheries. Conserv. Biol. 20: 193–169.

Read, A.J. 2008. The looming crisis: Interactions between marine mammals and fisheries. J. Mammal. 89(3): 541–548.

Reyes, L.M., E.A. Crespo and V. Szapkievich. 1999. Distribution and population size of the southern sea lion (*Otaria flavescens*) in central and southern Chubut, Patagonia, Argentina. Mar. Mamm. Sci. 15(2): 478–493.

Riet-Sapriza, F.G., D.P. Costa, V. Franco-Trecu, Y. Marín, Y. Chocca, B. González et al. 2013. Foraging behavior of lactating South American sea lions, *Otaria flavescens* and spatial-resource overlap with the Uruguayan fisheries. Deep Sea Res. Part 2 Top. Stud. Oceanogr. 88-89: 106–119.

Rodriguez, D.H., M. Dassis, A.P. de Leon et al. 2013. Foraging strategies of Southern sea lion females in the La Plata River Estuary (Argentina-Uruguay). Deep Sea Res. Part 2 Top. Stud. Oceanogr. 88-89: 120–130.

Rosas, F.C.W., M.C. Pinedo, M. Marmontel and M. Haimovici. 1994. Seasonal movements of the South American sea lion (*Otaria flavescens*, Shaw) off the Rio Grande do Sul coast, Brazil. Mammalia 58(1): 51–59.

Schiavini, A., E.A. Crespo and V. Szapkievich. 2004. Status of the population of South American sea lion (*Otaria flavescens*) in Santa Cruz and Tierra del Fuego Provinces, Argentina. Mamm. Biol. 69(2): 108–118.

Sepúlveda, M., D. Oliva, A. Urra, M.J. Pérez-Álvarez, R. Moraga, D. Schrader et al. 2011. Distribution and abundance of the South American sea lion *Otaria flavescens* (Carnivora: Otariidae) along the central coast off Chile. Rev. Chil. Hist. Nat. 84(1): 97–106.

Shaughnessy, P., R. Kirkwood, M. Cawthorn, C. Kemper and D. Pemberton. 2003. Pinnipeds, cetaceans and fisheries in Australia: A review of operational interactions. pp. 136–152. *In*: N. Gales, M. Hindell and R. Kirkwood [eds.]. Marine Mammals: Fisheries, Tourism and Management Issues. CSIRO Publishing, Collingwood, Australia.

Shuster, S. and M. Wade. 2003. Mating Systems and Strategies. Princeton University Press, Princeton, NJ, USA.

Sielfeld, W. 1999. Estado del conocimiento sobre conservación y preservación de *Otaria flavescens* (Shaw 1800) y *Arctocephalus australis* (Zimmermann 1783) en las costas de Chile. Estudios Oceanológicos 18: 81–96.

Soto, K.H. and A.W. Trites. 2011. South American sea lions in Peru have a lek-like mating system. Mar. Mamm. Sci. 27(2): 306–333.

Szteren, D. and E. Páez. 2002. Predation by southern sea lions (*Otaria flavescens*) on artisanal fishing catches in Uruguay. Mar. Freshwater Res. 53(8): 1161–1167.

Szteren, D. 2006. Predation of *Otaria flavescens* over artisanal fisheries in Uruguay: Opportunism or prey selectivity? Lat. Am. J. Aquat. Mamm. (LAJAM) 5(1): 29–38.

Szteren, D. and C. Lezama. 2006. Southern sea lions and artisanal fisheries in Piriapolis, Uruguay: Interactions in 1997, 2001, and 2002. pp. 591–604. *In*: A.W. Trites, S.K. Atkinson, D.P. De Master, L.W. Fritz, T.S. Gelatt, L.D. Rea and K.M. Wynne [eds.]. Sea Lions of the World. Alaska Sea Grant College Program. Fairbanks, Alaska, USA.

Szteren, D. 2015. *Otaria flavescens* and *Arctocephalus australis* abundance in poorly known sites: A spatial expansion of colonies? Braz. J. Oceanogr. 63(3): 83–92.

Thirgood, S., J. Langbein and R.J. Putman. 1999. Intraspecific variation in ungulate mating strategies: the case of the fallow deer. Adv. Study Behav. 28: 333–361.

Thompson, D., I. Strange, M. Riddy and C.D. Duck. 2005. The size and status of the population of southern sea lions *Otaria flavescens* in the Falkland Islands. Biol. Conserv. 121: 357–367.

Trimble, M. and F. Berkes. 2013. Participatory research towards co-management: Lessons from artisanal fisheries in coastal Uruguay. J. Environ. Manage. 128: 768–778.

Trites, A.W., D. Pauly and V. Christensen. 1997. Competition between fisheries and marine mammals for prey and primary production in the Pacific Ocean. J. Northwest Atl. Fish. Sci. 22: 173–187.

Vaz-Ferreira, R. 1965. Comportamiento antisocial en machos subadultos de Otaria byronia (de Blainville), ("Lobo manino de un pelo"). Revista de la Facultad de Humanidades y Ciencias pp. 203–207.

Vaz-Ferreira, R. 1976. Otaria flavescens (Shaw) South American sea Lion. Advisory Committee on Marine Resources research pp. 1–20.

Vaz-Ferreira, Raul. 1982. *Otaria flavescens* (Shaw) South American sea lion. pp. 477–495. *In*: FAO Advisory Committee on Marine Resources Research. Working Party on Marine Mammals. Mammals in the Seas, Volume IV: Small Cetaceans, Seals, Sirenians and Otters. Food and Agricultural Organization of the United Nations. FAO Fisheries Series No. 5. Rome, Italy.

Weise, M.J. and D.P. Costa. 2007. Total body oxygen stores and physiological diving capacity of California sea lions as a function of sex and age. J. Exp. Biol. 210: 278–289.

Wickens, P.A., D.W. Japp, P.A. Shelton, F. Kriel, P.C. Goosen, B. Rose et al. 1992. Seals and fisheries in South Africa competition and conflict. Afr. J. Marine Sci. 12: 773–789.

Wickens, P.A. 1994. Operational interactions between seals and fisheries in South Africa. South African Department of Environmental Affairs/Southern African Nature Foundation, Cape Town, South Africa.

Wickens, P.A. 1995. A review of operational interactions between pinnipeds and fisheries. Food and Agriculture Organization of the United Nations. FAO Fisheries Technical Paper I–VII. Rome.

Wilkinson, I.S., J. Burgess and M. Cawthorn. 2003. New Zealand sea lions and squid: Managing fisheries impacts on a threatened marine mammal. pp. 192–207. *In*: N. Gales, M. Hindell and R. Kirkwood [eds.]. Marine Mammals: Fisheries, Tourism and Management Issues. CSIRO Publishing, Collingwood, Australia.

Wilson, E.O. 1975. Sociobiology: The New Synthesis. Harvard University Press, Cambridge, USA.

Ximénez, I. and E. Langguth. 2002. Isla de Lobos. Ed. Graphics. Montevideo, Uruguay.

12

Ecology and Conservation Status of the South American Fur Seal in Uruguay

Valentina Franco-Trecu

Introduction

The South American fur seal (SAFS), *Arctocephalus australis,* is distributed along the Atlantic and Pacific coasts of South America (Vaz-Ferreira 1982), where it lives in sympatry with the South American sea lion (*Otaria flavescens*), forming mixed and dense colonies. SAFS populations were the basis of livelihood for many Pre-Hispanic indigenous populations along the South American coastal areas (Schiavini 1985). Later, the species was intensively exploited from the 18th century until the end of the 20th century in most of its geographic range (Ponce de León 2000). In Uruguay, private concessions killed nearly 527,000 individuals until 1949 (Ponce de León 2000). When the State oversaw the management and exploitation of the species between 1950 and 1991, were culling approximately 280,000 individuals (Ponce de León 2000). In 2013, the population size was ca. 130,000 individuals in Uruguay, according to the last estimate of abundance available. Sixty percent of the Uruguayan SAFS population is found at Isla de Lobos, while the remaining 40% occurs in Torres and Castillo Islands (Cabo Polonio—Valizas), located at an approximate distance of

---

Departamento de Ecología y Evolución, Facultad de Ciencias, Universidad de la República, Montevideo, Uruguay.
  Email: pinnipedosuy@gmail.com

70 km from the former. The annual population growth rate at Isla de Lobos has been estimated at 1.5% (I.C. 0.2 to 2.8), based on historical data of pup counts from 1956 to 2013 (Franco-Trecu 2015).

It is possible that the SAFS population abundance of the Uruguayan colonies has a significant impact on the dynamics of other Atlantic populations. Some individuals that were marked at birth, at Isla de Lobos, Uruguay, were observed at an older age in Brazil (Santa Catarina) and in several Argentinean localities, including Bahía Blanca, Isla Escondida, Pehuen Co, Quequen and Golfo San Matías (Crespo et al. 2015). This evidence indicates that, following weaning, young individuals from Uruguay disperse to different rookeries within the South Atlantic Ocean, often travelling more than 1,000 km from their birthplace.

Maternal Behaviour

SAFS females reach sexual maturity by 5 years of age, and at 10 year of age they achieved an approximate body weight of 42 kg and total length of 130 cm (Lima and Páez 1995). On average, new born pups weight 6 kg and their length is 65 cm (V. Franco-Trecu, unpublished data).

At Isla de Lobos, pregnancy rates differ from year to year, suggesting that environmental factors and demographic variability play an important role (Lima and Páez 1995). Parturitions occur with similar frequencies of breech and cephalic presentation of the foetus, and the presentation of the foetus did not affect the total duration of the parturition process, even when breech presentation implies a longer first phase of parturition, a trend also observed in other Otariid species (Franco-Trecu et al. 2016). SAFS females give birth in December, to a pup conceived during the previous breeding season (Franco-Trecu 2005). A few days after giving birth, females enter into postpartum oestrous (Katz et al. 2013), and mate with territorial males (Franco-Trecu et al. 2014a). For the first 11 months of their pup lives, females switch between periods on land (feeding their offspring) and at sea (foraging), while their pups fast (mean 13 days; Franco-Trecu 2005), as they depend on maternal milk. Given that gestation lasts approximately 12 months, during maternal care, females simultaneously sustain two energetically costly activities, in terms of breeding investment, i.e., pregnancy and lactation (Franco-Trecu 2005). Foetuses from lactating females are smaller than those from non-lactating females, probably because simultaneous lactation and pregnancy affect the amount of energy assigned to embryonic growth (Lima and Páez 1997).

Upon females' return to the colony after a foraging trip, the maternal–offspring recognition system is essential for mother-pup reunion and pup survival (Phillips 2003) (Fig. 1). Pups play an active role in this recognition and, in general, avoid responding to incorrect females, most likely due to the high injury risk of aggressive rejection by unrelated females (Phillips 2003). However, successful allo-suckling events in the SAFS Uruguayan population have been registered, representing around 4% of all suckling events (Franco-Trecu and Soutullo 2010). It appears that in this colony, in contrast to other colonies elsewhere, the putative costs of allo-suckling (increased energy expenditure) for females, and potential risks of aggression by non-kin mothers for pups, have not prevented the persistence of allo-suckling behaviour. It is possible that the persistence of this behaviour is associated to additional benefits for the pups, such as obtaining immune compounds (Roulin and Heeb 1999) and extra energy during the most demanding period, increasing their survival while their mothers are foraging.

The reproductive success of SAFS females depends on the sex of its offspring, given that males have higher mortality rates. In the perinatal period, survival of offspring is influenced by its duration, and in the case of males, is also influenced by their lactating rate (Franco-Trecu 2010). Offspring survival is increased in cases when mothers' perform short foraging trips. In contrast, when a females' foraging trip is longer than the mean, her offspring probability of survival decreases. This variation in

Figure 1. South American fur seal female with a pup on a rocky shore of the Uruguayan coast. Female recognition of pups is an essential maternal behavior for mother-pup interactions, nursing and survival. Photo credit: V. Franco-Trecu.

offspring survival is more pronounced in males (Franco-Trecu 2010), and is probably related to differences in body size and energetic requirement.

Reproductive Behaviour

The SAFS mating system was typically described as a female defence polygyny or a harem, as males were thought to defend groups of females (Vaz-Ferreira 1982). Nevertheless, it has recently been reported that the mating system is consistent with a lek (Franco-Trecu et al. 2014a), given that males cluster and display in areas where potentially receptive females give birth, and subsequently come into postpartum oestrus. In these reproductive areas, males defend very small territories, barely larger than the area occupied by their own bodies, and lack of any resource required for female reproduction.

Females have extensive home ranges that overlap with each other and with the territories of many males, indicating that they move freely around the colony, and that males do not monopolize access to them. Under these circumstances, females can easily assess and choose potential mates. SAFS females aggregate in dense breeding colonies and exhibit a high degree of oestrus synchrony (31 days), compared to other otariid species (e.g., for *Zalophus wollebaeki*, synchrony lasts five months) (Villegas-Amtmann et al. 2009, Pörschmann et al. 2010). The variance of male individual breeding success (i.e., paternity variation) was 8.17 (Franco-Trecu et al. 2014a), being the most extreme estimated for pinniped populations (Gonzalez-Suarez and Cassini 2014). This extreme variance occur because some territorial male have a low or null breeding success. All these conditions meet the criteria for a lek mating system (Bradbury et al. 1986, Wiley 1991, Höglund and Alatalo 1995). The low breeding success of some territorial SAFS males may result from the relatively high female oestrus synchrony and possible sperm production limitation, that prevents the monopolization of mating events by individual males (Franco-Trecu et al. 2014a).

In addition to fasting, territorial males face substantial energetic costs while defending small territories, by continuously interacting with other males, and thermoregulating during the austral summer. Accordingly, the average residence time of males holding a territory, is 13 days (range 1–40, sd = 10 days), significantly shorter than the duration of the breeding season (45 days), thus implying a high male turnover rate (Franco-Trecu et al. 2014a). This suggests that, males in a good body condition will be able to sustain these high energetic costs and hold their territory for longer periods. In contrast, males with a poor body condition, will avoid these energetic costs, and adopt a subordinate social tactic such as satellite males (Isvaran 2005, Bro-Jorgensen et al. 2008, Taborsky et al. 2008, Noble et al. 2013, York et al. 2014).

Overall, the SAFS male breeding success is influenced by a complex interaction of several behavioural traits, which makes it highly variable. Given the high costs of territoriality, it is possible that some satellite males achieve greater average breeding success than territorial ones (Franco-Trecu et al. 2014a).

Foraging Behaviour

SAFS females are pelagic foragers that dive mostly at night. During their first foraging trip postpartum, SAFS females utilize areas that are close to the rookery (ca. 80 km), which allows them to nurse their newborn pups frequently. In subsequent trips, females utilize more distant areas, on the break of the continental shelf, at approximately 500 km from the breeding colony (Franco-Trecu 2015). The duration of the SAFS first foraging trip postpartum, is approximately 7 days, and is significantly shorter than their second one (~13 days). The duration of nearly half of their dives (45%) ranges between 0.08 and 0.50 min, while the other half (45%) ranges between 0.51 and 1.75 min. The remaining 10% dives were greater than 1.75 min and the longest dive duration recorded was 5.25 min. The individual bottom time/dive time ratio vary between 0.3 and 0.5, indicating that the time spent at the bottom is minimal compared to the duration of the whole dive, suggesting that SAFS females are epi or mesopelagic diving. With regard to the depth of their dives, most of them (~60%) are shallower than 20 m, followed by depths of 21–60 m in 31% of their dives, and very few dives (~5%) are deeper than 61 m. The maximum depth recorded was 191 m. On their first foraging trip postpartum, lactating SAFS dive significantly deeper during daytime than night-time, and this difference tends to disappear in subsequent trips. A similar pattern is observed for diving effort. However, for all foraging trips, the duration, and time spent at the bottom of their dives, as well as the ratio of bottom time/dive time, is lower for dives occurring at night than during the day (Franco-Trecu 2015). This suggests that SAFS are feeding on vertically migrating prey, which are located at shallower depths during the night. Accordingly, their diet comprises mainly pelagic species such as Argentine shortfin squid (*Illex argentinus*), Argentine anchovy (*Engraulis anchoita*) and Argentine hake (*Merluccius hubbsi*), which migrate up through the water column at night (Franco-Trecu et al. 2013).

The extent of the foraging home range of SAFS for the female population is approximately 61,000 km^2 (95% Kernel), while for an individual it varies from 6,900 km^2 to 47,000 km^2. However, spatial overlap of individual home-ranges is high (i.e., mean: 95.6%, range: 71%–100%) (Franco-Trecu 2015), and most foraging trips occur toward the south or southwestern part of Isla de Lobos. Because the foraging areas of females overlap spatially to a great extent, the contribution of each female to the population home range size is not highly significant. This is supported by the results of a stable isotopes analysis in whiskers, which demonstrates that the isotopic niche space of SAFS individuals overlaps and is extensive; which suggests that this generalist population is composed by generalist individuals (Franco-Trecu et al. 2014b).

Conservation and Management

Despite the offshore oceanic habits of SAFS, there have been occasional records of individuals by-caught by both artisanal (Franco-Trecu et al. 2009) and commercial coastal fisheries, operating in shallow, nearshore waters. In the latter, mean mortality estimates vary between 16 and 38 individuals per year, which would be nearly only 0.05% of the Uruguayan population, thus not representing a problematic issue (Franco-Trecu et al. submitted). An additional source contributing to the removal of individuals from the population is the species trade performed by the Government in the last two decades, when 239 young males and 339 young females were sold to zoos and marine aquaria between 1999 and 2013 (DINARA, http://www.dinara.gub.uy). Although the Uruguayan SAFS population has no evident conservation risks, the species is listed as a priority for conservation in Uruguay (Soutullo et al. 2013). Furthermore, it has been identified as a conservation target in the Marine Protected Area-Cabo Polonio National Park (Soutullo et al. 2009) because it is a charismatic species and its population in Uruguay can represent a reservoir to the species. At a global level, the South American fur seal is catalogued as Least Concern by UICN, but it is listed on CITES Appendix II.

Acknowledgements

I thank all my colleagues who helped me, especially Pablo Inchausti, Helena Katz, Massimiliano Drago, Paula Costa and Mateo García, for their assistance and logistical support during my field work. Special thanks to Dr. S. Villegas-Amtmann for revising and providing insights to this paper.

Keywords: *Arctocephalus australis,* South American fur seal, maternal behaviour, harvesting, reproductive behaviour, Uruguay

References

Bradbury, J., R. Gibson and I.M. Tsai. 1986. Hotspots and the dispersion of leks. Anim. Behav. 34: 1694–1709.
Bro-Jorgensen, J., M.E. Brown and N. Pettorelli. 2008. Using the satellite-derived normalized difference vegetation index (NDVI) to explain ranging patterns in a lek-breeding antelope: The importance of scale. Oecologia 158(1): 177–182.
Crespo, E.A., A.C.M. Schiavini, N.A. García et al. 2015. Status, population trend and genetic structure of South American fur seals *Arctocephalus australis* in southwestern Atlantic waters. Mar. Mamm. Sci. 31(3): 866–890.
Franco-Trecu, V. 2005. Comportamiento maternal y aspectos reproductivos de *Arctocephalus australis*, en Isla de Lobos, Uruguay. Licenciatura (Bachellor Dissertation), Sección Etología, Universidad de la República, Montevideo, Uruguay.
Franco-Trecu, V., P. Costa, C. Abud, C. Dimitriadis, P. Laporta, C. Passadore et al. 2009. By-catch of franciscana *Pontoporia blainvillei* in uruguayan artisanal gillnet fisheries: An evaluation after a twelve-year gap in data collection. Lat. Am. J. Aquat. Mamm. (LAJAM) 7(1-2): 11–22.
Franco-Trecu, V. 2010. Éxito de crianza y hábitos alimenticios en hembras del lobo fino sudamericano (*Arctocephalus australis*) y su relación trófica con hembras del león marino sudamericano (*Otaria flavescens*). PEDECIBA, Universidad de la República, Montevideo, Uruguay.
Franco-Trecu, V., B. Tassino and A. Soutullo. 2010. Allo-suckling in the South American fur seal (*Arctocephalus australis*) at Isla de Lobos - Uruguay: Cost or benefit of living in group? Ethol. Ecol. Evol. 22: 143–150.
Franco-Trecu, V., M. Drago, F.G. Riet-Sapriza, A. Parnell, R. Frau and P. Inchausti. 2013. Bias in diet determination: Incorporating traditional methods in Bayesian mixing models. Plos One 8(11): e80019.
Franco-Trecu, V., P. Costa, Y. Schramm, B. Tassino and P. Inchausti. 2014a. Sex on the rocks: Breeding tactics and reproductive success of the for South American fur seal males. Behav. Ecol. 25(6): 1513–1523.
Franco-Trecu, V., D. Aurioles-Gamboa and P. Inchausti. 2014b. Individual trophic specialisation and niche segregation explain the contrasting population trends of two sympatric otariids. Mar. Biol. 161(3): 609–618.
Franco-Trecu, V. 2015. Tácticas comportamentales de forrajeo y apareamiento y dinámica poblacional de dos especies de otáridos simpátricas con tendencias poblacionales contrastantes. PEDECIBA, Universidad de la República, Montevideo, Uruguay.
Franco-Trecu, V., M.D. García-Olazabal, B. Tassino and J. Acevedo. 2016. Parturition process in an amphibian mammal species: New evidences in South American fur seal (*Arctocephalus australis*). Anim. Biol. 66(1): 21–29.
Gonzalez-Suarez, M. and M.H. Cassini. 2014. Variance in male reproductive success and sexual size dimorphism in pinnipeds: Testing an assumption of sexual selection theory. Mamm. Rev. 44(2): 88–93.
Höglund, J. and R.V. Alatalo. 1995. Leks. Princeton University Press, Princeton, NJ, USA.
Isvaran, K. 2005. Variation in male mating behaviour within ungulate populations: Patterns and processes. Curr. Sci. 89(7): 1192–1199.
Katz, H., P. Pessina and V. Franco-Trecu. 2013. Serum progesterone concentration in female South American fur seals (*Arctophoca australis*) during the breeding season. Aquat. Mamm. 39(3): 290–295.
Lima, M. and P. Páez. 1995. Growth and reproductive patterns in the South American fur seal. J. Mammal. 76(4): 1249–1255.
Lima, M. and E. Páez. 1997. Demography and population dynamics of South American fur seals. J. Mammal. 78(3): 914–920.

Noble, D.W.A., K. Wechmann, J. Scott Keogh and M.J. Whiting. 2013. Behavioral and morphological traits interact to promote the evolution of alternative reproductive tactics in a lizard. Am. Nat. 182: 726–742.

Phillips, A.V. 2003. Behavioral cues used in reunions between mother and pup south american fur seals (*Arctocephalus australis*). J. Mammal. 84(2): 524–535.

Ponce de León, A. 2000. Taxonomía, sistemática y sinopsis de la biología y ecología de los pinipedios de Uruguay. pp. 6–35. *In*: M. Rey and F. Amestoy [eds.]. Sinopsis de la biología y ecología de las poblaciones de lobos finos y leones marinos de Uruguay. Pautas para su manejo y Administración. Parte I. Biología de las especies, edited by Montevideo-Uruguay: Proyecto URU/92/003. Instituto Nacional de Pesca-Programa de las Naciones Unidas para el Desarrollo.

Pörschmann, U., F. Trillmich, B. Mueller and J.B.W. Wolf. 2010. Male reproductive success and its behavioural correlates in a polygynous mammal, the Galapagos sea lion (*Zalophus wollebaeki*). Mol. Ecol. 19(12): 2574–2586.

Roulin, A. and P. Heeb. 1999. The immunological function of allosuckling. Ecol. Lett. 2(5): 319–324.

Schiavini, A.C.M. 1985. Determinación de las pautas de captura de Pinnipedos del Canal Beagle por los Aborigenes Prehistoricos - Resultados Preliminares. pp. 58. *In*: Primeras Jornadas Argentinas de Mastozoologia, 26–28 September, 1985. Sociedad Argentina para el Estudio de los Mamíferos (SAREM). Mendoza, Argentina.

Soutullo, A., E. Alonso, D. Arrieta, R. Beyhaut, S. Carreira, C. Clavijo et al. 2009. Especies prioritarias para la conservación 2009. Informe del Proyecto Fortalecimiento del Proceso de Implementación del Sistema Nacional de Áreas Protegidas. Serie de Informes 16: 1–95. Uruguay.

Soutullo, A., C. Clavijo and J.A. Martínez-Lanfranco. 2013. Especies prioritarias para la conservación en Uruguay. Vertebrados, moluscos continentales y plantas vasculares. Montevideo: SNAP/DINAMA/MVOTMA y DICYT/MEC. Uruguay.

Taborsky, M., R.F. Oliveira and H.J. Brockmann. 2008. The evolution of alternative reproductive tactics: Concepts and questions. pp. 1–21. *In*: R. Oliveira, M. Taborsky and H.J.E. Brockmamm [eds.]. Alternative Reproductive Tactics. Cambridge University Press, Cambridge.

Vaz-Ferreira, Raul. 1982. *Arctocephalus australis*, Zimmermann, South American fur seal. pp. 497–508. *In*: FAO Advisory Committee on Marine Resoruces Research. Working Party on Marine Mammals. Mammals in the Seas, Volume IV: Small Cetaceans, Seals, Sirenians and Otters. Food and Agricultural Organization of the United Nations. FAO Fisheries Series No. 5. Rome, Italy.

Villegas-Amtmann, S., S. Atkinson and D.P. Costa. 2009. Low synchrony in the breeding cycle of Galapagos sea lions revealed by seasonal progesterone concentrations. J. Mammal. 90(5): 1232–1237.

Wiley, R. 1991. Lekking in birds and mammals: Behavioral and evolutionary issues. Adv. Study Behav. 20: 201–291.

York, J.R., T.A. Baird and M.L. Haynie. 2014. Unexpected high fitness payoff of subordinate social tactics in male collared lizards. Anim. Behav. 91: 17–25.

13

The Uncertain Fate of the Endangered Mediterranean Monk Seal *Monachus monachus* in the 21st Century

Population, Ecology and Conservation Threats

Panagiotis Dendrinos*, Styliani Adamantopoulou, Eleni Tounta and Alexandros A. Karamanlidis

Introduction

> *"The seal... does not take water in but breathes and sleeps and gives birth on land, albeit near the beach, as if it belongs to the animals living on land. On the other hand, it spends most of its life in the water and gets its food from the water, therefore we should examine it along with the marine animals".*
>
> Aristotle, 4 B.C.

MOm/Hellenic Society for the Study and Protection of the Monk Seal, 18 Solomou str, 10682, Athens, Greece.
Email: info@mom.gr
* Corresponding author: p.dendrinos@mom.gr

In his 4th century B.C. seminal work the "History of Animals", Aristotle, the ancient Greek philosopher, penned the first known description of a pinniped in history. Aristotle described the only pinniped species inhabiting the Mediterranean Sea, the Mediterranean monk seal, which he encountered along the coasts of the Aegean Sea in Greece. The first description of this seal in modern times was published in 1779 by the German naturalist Johann Hermann (Hermann 1779). Hermann described the Mediterranean monk seal from an animal in a traveling exhibition in Strasbourg. Actually, Hermann is also responsible for naming the species a "monk seal". He came up with this name by combining the fact that around the area of Marseilles locals seemed to refer to the species with the name "moine" (monk in French), with the fact that the animal he had seen in Strasbourg resembled a monk wearing a cloak. It is unknown to us how many Mediterranean monk seals lived in Aristotle's world, or even Hermann's, but today the species is considered to be one of the rarest marine mammals and the most endangered seal on Earth. It is estimated that no more than 700 individual Mediterranean monk seals survive today (Karamanlidis and Dendrinos 2015, Karamanlidis et al. 2016a).

Taxonomy, Range, Population and Population Structure

It has been thoroughly documented that in the past Mediterranean monk seals lived throughout the Black and Mediterranean Sea, the island complexes of the Azores, Madeira, the Canary and Cape Verde Islands in the Atlantic Ocean, as well as at the northwestern coasts of Africa, from Morocco in the north to Senegal and quite possibly Gambia in the south (Johnson and Lavigne 1999, Johnson 2004, González 2015, Karamanlidis et al. 2016a). The systematic exploitation of the Mediterranean monk seal for its skin and oil during the ancient Greek and Roman times and the Middle Ages, appears to have been the main factor leading to the significant population decreases throughout the species' range (Johnson and Lavigne 1999, Johnson 2004, González 2015). The unabated exploitation of the species during the past two centuries ultimately led to the extinction of the Mediterranean monk seal from most countries in the Mediterranean and Black Sea, as well as from most parts of the Atlantic.

At present, the global range of the Mediterranean monk seal is extremely limited and fragmented. The species survives in four isolated subpopulations (Karamanlidis et al. 2016a): In the Mediterranean Sea, the stronghold of the species remains the islands of the Ionian and Aegean Sea in Greece, where Aristotle first described the species, almost 2,500 years ago. The species is also found along the coasts of mainland Greece, the Mediterranean coast of Turkey, and recently, along the coasts of Cyprus. Some individuals still survive in the Sea of Marmara, while in the Black Sea, monk seals are believed to have gone extinct in the 1990s. It is estimated that the Eastern Mediterranean population may count up to 450 individuals. In the Atlantic, two subpopulations exist: one at the Cabo Blanco peninsula, at the border between Mauritania and Western Sahara, and one at the Archipelago of Madeira. In the mid-1990s, the population at Cabo Blanco was estimated at 317 seals.

However, a mass mortality event in 1997 reduced this population nearly by two thirds. Since then the population has been showing encouraging signs of recovery and in 2015 it was estimated that approximately 220 seals composed the second largest monk seal population in the world. The third largest population of approximately 40 monk seals survives in the Archipelago of Madeira, while an unknown number of monk seals might still survive along the Mediterranean coasts of Algeria and eastern Morocco. Without systematic monitoring and conservation activities however in place the status of the species in this area is uncertain.

According to the most recent observations, the monk seal populations at Cabo Blanco in the Atlantic (Martínez-Jauregui et al. 2012) and at the island of Gyaros in the eastern Mediterranean Sea (Karamanlidis et al. 2013) are the only large extant aggregations of the species that still preserve the structure of a colony; all other subpopulations in the eastern Mediterranean are composed of fragmented breeding groups of reduced size (usually less than 30 individuals) (Karamanlidis et al. 2016a).

During the last decade sporadic sightings of individual Mediterranean monk seals have been recorded in countries where the species has been considered to be extinct for a long time, such as Italy, Croatia and Albania in the central Mediterranean and Syria, Lebanon, Israel, Egypt, and Libya in the southern and southeastern Mediterranean Sea (see Fig. 1). In March 2015, a female monk seal was found trapped in fishing nets on the coast of Beirut, Lebanon. Sadly, the necropsy revealed that the animal was at the final stage of pregnancy. However unfortunate this incident might have been, it is, on the other hand, an encouraging sign of reproductive activity of the species from a region where it was long considered to be extinct.

According to the most recent taxonomic classification, the Mediterranean monk seal (*Monachus monachus*; Fig. 2) is the sole representative of the genus *Monachus* (Scheel et al. 2014). Examination of mitochondrial DNA has indicated genetic differences between Atlantic and eastern Mediterranean seal populations: only one haplotype was found in Mediterranean monk seals in Madeira and Cabo Blanco in the Atlantic in contrast to four different haplotypes found in monk seals in the eastern Mediterranean (Karamanlidis et al. 2016b). Also, a comparison of 24 nuclear microsatellite loci between eastern Mediterranean and Atlantic Mediterranean monk seal populations showed that the first group had 14 unique alleles and the second 18, highly significant differences in allele frequencies between the two subpopulations

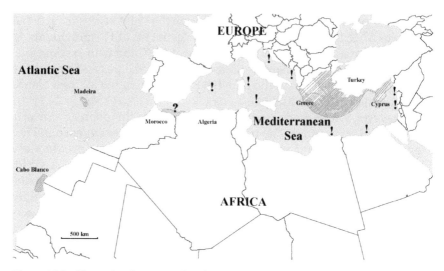

Figure 1. Map illustrating the current distribution and recent sightings of Mediterranean monk seal along the coastal regions of the Mediterranean Sea and North Africa. Cross-hatched areas indicate the geographical range of extant monk seal populations; the question mark indicates an area where the fate of the population is unknown; the exclamation marks indicate areas outside the current range where Mediterranean monk seals have recently been seen. Adapted from Karamanlidis et al. (2016a).

Figure 2. An adult female with her pup resting on an open beach near a pupping cave, in Greece (A); and, an adult male resting on an open beach in Greece (B). Photo credits: P. Dendrinos (Fig. 2A) and A.A. Karamanlidis (Fig. 2B)/MOm/Hellenic Society for the Study and Protection of the Monk Seal.

were found for 14 out of 17 loci (Pastor et al. 2007). Although until now no further taxonomic separation of the species has been suggested, given the genetic differences and the distance separating the two subpopulations it is considered appropriate to treat the Atlantic and the Eastern Mediterranean monk seal subpopulations as two separate management units (Karamanlidis et al. 2016b).

Biology

The Mediterranean monk seal is considered a medium-sized phocid. In the Cabo Blanco population, average lengths of 2.42 and 2.51 m have been recorded in adult females (Fig. 2A) and males (Fig. 2B), respectively. Data on adult body length from Greece suggest that Mediterranean monk seals in this population might be a little bit smaller than their counterparts in Cabo Blanco, while there are no significant differences between the two sexes. The average length of newborn pups is approximately 1 m and they weigh 15–26 kg; adult monk seals weigh 240–300 kg, with anecdotal records of a pregnant female reaching 302 kg and a male reaching 400 kg. Adult males are uniformly black with a white patch on their ventral side that is unique for every individual (Fig. 2B). Adult females are generally brown or gray, with the lower half of their body showing a lighter coloration. Adult Mediterranean monk seals undergo an annual molt, which appears to be a gradual process and therefore the color of the pelage can vary significantly throughout the year. The external appearance of the body includes also areas with multiple scars that are sustained during mating and social interactions. Females carry many scars on the back, while males often have an area of light coloration on the throat as a result of multiple scars from aggressive interactions with other males. Pups are born with a white belly patch on the otherwise black to dark chocolate, woolly coat (Badosa et al. 1998, Dendrinos 2011). The ventral patch is often marked by black spots and varies in shape, size and position between different individuals and according to gender (Badosa et al. 1998). Apart from the pups, which possess a soft and woolly pelt or "lanugo", juvenile and adult Mediterranean monk seals have very short and bristly hair, about 0.5 cm long; the shortest hair amongst pinnipeds (Ling 1970). The neonatal molt is progressive and is completed on average eight weeks postpartum. The completion of the neonatal molt is not associated with weaning, as molted pups have been observed to suckle (Gazo et al. 2006). Molting of juveniles and adults occurs over a very protracted period, extending throughout the year (Androukaki et al. 1999, Güçlüsoy and Savaş 2003b, Pastor and Aguilar 2003); there appears to be no significant difference in the molting period between monk seal populations in the eastern Mediterranean and the Cabo Blanco region. In the Cabo Blanco monk seal population, the intermolt period is close to one year, except in females nursing a pup. Such females have longer intermolt periods and can even molt while still lactating (Pastor and Aguilar 2003). In males, the process of developing the mature pelage pattern of bulls is gradual. It involves at least two annual molts and can be completed by the age of 4 years (Badosa et al. 2006).

Male Mediterranean monk seals are estimated to mate for the first time after their 6th year of age. Females are considered to attain sexual maturity by the 4th year of their life, but most recent data suggests that they may reach sexual maturity as early as the age of three. This is considered to be the lowest known age of sexual maturity for any phocid species. Mating in Mediterranean monk seals occurs in the water (Karamanlidis et al. 2016a). Gestation lasts approximately 9 to 11 months (Marchessaux and Pergent-Martini 1991, Pastor and Aguilar 2003) and each female gives birth to one pup (Fig. 2) during every reproductive cycle (King 1956). Females can give birth in successive years. Before giving birth, females will often retreat to isolated areas within their pupping caves which they actively defend against other approaching seals (Layna et al. 1999). Aggressive interactions in the pupping caves between females and other females and pups have been documented using infrared cameras also in Greece (Karamanlidis et al. 2012).

Habitat and ecology

The life of the Mediterranean monk seal, as well as that of all pinnipeds is inextricably connected to land, where pupping occurs. Throughout their range Mediterranean monk seals use nowadays exclusively marine caves to give birth to their young. In Cabo Blanco, seals use less than five big caves for this task; these caves have large-sized entrances and extensive sandy beaches inside. In contrast, monk seals in the eastern Mediterranean use a wide network of sea caves of much smaller size than those of Cabo Blanco, with entrances that protect the inner beach (or beaches) from the outside environment. In Greece alone, a country with an extensive coastline (> 16,000 km), more than 150 reproductive caves have been identified so far. A long term study of the reproductive biology of the Mediterranean monk seal in the National Marine Park of Alonnisos, Northern Sporades in Greece has concluded that the sea caves preferred by females for pupping share some specific morphological characteristics (Dendrinos et al. 2007b). In the Archipelago of Madeira in the Atlantic, the pupping habitat is very similar to the one in the Eastern Mediterranean (Karamanlidis et al. 2004), with the main difference being however that in Madeira, the intense sea tide greatly affects the use of caves by the seals (Pires et al. 2007). The exclusive use of sea caves by females for pupping is considered to be partially a consequence of the intense persecution by humans over the course of many centuries.

In regard to the use of marine caves by the Mediterranean monk seal and based on the available historical sources, it is still very difficult to draw a conclusion whether the species was using in the past exclusively open beaches or open beaches and caves for resting and pupping. It is interesting to note that in the first actual historical note of the Mediterranean monk seal, Homer's Odyssey, the "cave use" by the species is also mentioned. In Homer's monumental epos the sea nymph Eidothea is reported saying that when emerging from the sea and after entering the *"arching caves,* (the sea god Proteus) *will pass along all the seals and count them; then, having viewed them and made his reckoning, he will lie down among them all like a shepherd among his flock of sheep"* (Odyssey, IV. 398 et seq.). A possible explanation for this is that the species has always used both, open beaches and caves to rest and reproduce and that because of the intense persecution by humans only the harder-to-access Mediterranean monk seal populations living in caves survived. It is also possible that the use of inaccessible habitat for reproduction has been throughout time a defense strategy to avoid predators that threatened the newborn pups, especially during times when mothers were out at sea to feed. In well-protected areas, lactating females have been recorded recently to lead their pups to open beaches close to the pupping caves and nurse them there for hours or even days. In some areas in Greece it is not uncommon nowadays to observe at times lone individuals hauling out on open beaches.

Regarding the use of the marine environment, the most recent data suggests that the Mediterranean monk seal is a coastal species as it searches for its food near the coasts and in depths rarely deeper than 200 meters. Compared to other well-studied pinnipeds, little is known about the diving capacities and behavior of Mediterranean monk seals. The maximum depth and duration of diving for one lactating female were 78 m and 15 minutes, respectively (Gazo et al. 2006); however, diving behavior of monk seals at Cabo Blanco appears to be limited by the topographic features of the marine environment in this area. In the eastern Mediterranean Sea (which is characterized by much deeper near-shore waters than the Cabo Blanco region), monk seals have been recorded to dive much deeper. Maximum dive depths for a rehabilitated male and a female juvenile monk seal were 196 and 205 m respectively (Dendrinos et al. 2007a, MOm/Hellenic Society for the Study and Protection of the Monk seal, unpublished data). Recently, researchers in Greece have documented monk seals sleeping at sea, at the water surface but also at the sea bottom (Karamanlidis et al. 2017).

Stomach content analysis of deceased Mediterranean monk seals indicates that monk seals feed on a large variety of prey including bony fishes, cephalopods, and crustaceans. In Greece, more than 530 prey species have been identified so far; the common octopus (*Octopus vulgaris* ~ 34%) and bony fishes from the family Sparidae (~ 28%) were identified most frequently in the stomachs of monk seals (Pierce et al. 2011). In the Archipelago of Madeira in the Atlantic, visual observations of individuals with prey at the surface included seals eating golden-grey mullet (*Liza aurata*), parrot fish (*Sparisoma cretense*), barred hogfish (*Bodianus scrofa*), salema (*Sarpa salpa*), cuttlefish (*Sepia officinalis*) and crabs (*Pachygrapsus* spp.) (Neves 1998). In Cabo Blanco, stomach content analyses indicated that 71.3% of prey items by weight were cephalopods, from which 68.3% were octopuses. Fish species were mainly from the families Sparidae, Scianidae and Haemulidae (Muñoz Cañas et al. 2012). Collectively, results from stomach content and stable isotope analyses studies (Pinela et al. 2010, Karamanlidis et al. 2014) confirm that monk seals forage primarily on the continental shelf along the coast.

Threats

The Mediterranean monk seal is one of the most endangered pinniped species (Karamanlidis and Dendrinos 2015), and one of the most endangered Evolutionarily Distinct and Globally Endangered (EDGE) mammals on Earth (Isaac et al. 2007). The species has a long history of negative interactions with humans. The main reasons for the recent, dramatic population declines include increased human pressure displacing seals from their habitat, destruction/alteration of suitable habitat, continued mortality due to deliberate killing by fishermen, fisheries by-catch, and a mass die-off at the Cabo Blanco monk seal colony (Karamanlidis et al. 2016a). Potentially, limited availability of food sources, genetic inbreeding and pollution could also constitute a threat to the survival of the Mediterranean monk seal.

Habitat perturbation

Habitat deterioration, destruction, and fragmentation have played a significant role in the plight of the Mediterranean monk seal. Once a (partially?) open beach dweller, the species has been persecuted by humans for centuries and forced to occupy increasingly marginal habitat. The gradual process from occupying open beaches to using increasingly marginal habitat (i.e., smaller and more unsuitable marine caves) has been thoroughly documented (Johnson and Lavigne 1999). This threat is still in place today, particularly in the eastern Mediterranean (Kiraç et al. 2013). Critical Mediterranean monk seal habitat has been affected by increased tourism activities throughout Turkey, even in protected areas such as the Olympos Beydagları National Park, and the Kas, Kekova Specially Protected Area. Tourists and scuba divers in these areas appear to frequently visit important monk seal shelters. Although some resting activity of Mediterranean monk seals still continues, no pupping activity has been recorded in these caves recently (Gücü et al. 2009). With human populations and coastal activities increasing around the Mediterranean, there are potential threats to the species' habitat.

Fisheries interactions

Interactions with fisheries throughout the species range are of great conservation concern (Güçlüsoy and Savaş 2003a, Güçlüsoy 2008, Karamanlidis et al. 2008, Hale et al. 2011, González and Fernandez de Larrinoa 2012). Deliberate killing of Mediterranean monk seals mainly by fishermen was responsible for one-third of 79 mortalities investigated in Greece (1991–1995) and is considered the single most important source of mortality for this species in the eastern Mediterranean (Androukaki et al. 1999). Mediterranean monk seals have been entangled in a wide variety of fishing gear including set-nets, trawl nets, and long-lines (Johnson and Karamanlidis 2000) and entanglement remains a major source of mortality in the eastern Mediterranean Sea, especially for sub-adult animals (Karamanlidis et al. 2008, Kiraç et al. 2013). Adverse fishing interactions are also considered as one of the probable causes for the lack of recovery of the Cabo Blanco population after commercial sealing ended in the region. Currently, illegal industrial and artisanal fishing is one of the main threats to the survival of the colony, mainly for sub-adult seals (González and Fernandez de Larrinoa 2012). In comparison, the effect of negative seal-fisheries interactions in the Archipelago of Madeira is considered to be lower. Traps, purse seines, and illegally used gill nets are the main fishing gear posing a threat to the species in the region (Hale et al. 2011).

In southern Turkey, an important monk seal colony almost disappeared in the 1990s, when industrial-scale fishing in the area reduced the available fish sources and negative interactions of artisanal fishermen with monk seals (i.e., deliberate killings) increased. However, a series of regulations enforced to protect fish sources alleviated the problems and helped the local monk seal population to resume pupping in the area (Gücü et al. 2004).

Diseases and loss of genetic variability

Following the mass mortality event at Cabo Blanco in 1997, a morbillivirus was isolated from the deceased Mediterranean monk seals. The virus most closely resembled a dolphin morbillivirus that was previously implicated in the 1991 mass mortality of striped dolphins in the Mediterranean Sea (Osterhaus et al. 1992, van de Bildt et al. 1999). However, although this virus was already circulating in Mediterranean monk seals prior to the mass mortality, there is some doubt as to whether it was responsible for the deaths that occurred. Indeed, the active virus was found in pups that went into a rehabilitation center because their mothers had died, and none of them showed clinical signs and all survived the event without specific treatment.

Dinoflagellate-produced saxitoxins were found in tissues from animals that died during the mass mortality and the suddenness of death of the animals and the general clinical symptoms suggest that the cause of death was from these toxins (Hernández et al. 1998). Toxic algal blooms (i.e., red tides) are favored by oceanographic conditions near Cabo Blanco and were reported from nearby Morocco during a 25-year period leading up to the mass mortality. Toxic algal blooms are unpredictable and following the catastrophic loss of Mediterranean monk seals in 1997 must be considered a serious threat to the species in the region (Reyero et al. 2000, UNEP 2005).

Genetic analyses of mitochondrial and nuclear DNA (Pastor et al. 2004, 2007, Karamanlidis et al. 2016b) have shown that, as a consequence of severe population bottlenecks and population/habitat fragmentation, all Mediterranean monk seal sub-populations have suffered a dramatic decrease in genetic variability. The genetic diversity of the species is among the lowest found in pinnipeds; it is comparable to Hawaiian monk seals and Northern elephant seals. The potential consequences of the loss of genetic variability and genetic inbreeding are still hard to evaluate for the Mediterranean monk seal; however, potential consequences of genetic inbreeding may include congenital defects leading to stillborn pups, something that has been recorded in several small monk seal populations (Bareham and Furreddu 1975, Pastor et al. 2004, MOm, unpublished data). Additionally, low fitness and increased susceptibility to disease may be an effect of genetic erosion that can compromise a population and lead to extinction.

Anthropogenic pollutants

Contaminant burdens have always been suspected to be a threat to the Mediterranean monk seal and thus monitoring pollutants has been considered a high conservation priority (Reijnders et al. 1993). However, information is only available on organochlorine pollutants and heavy metals. Organochlorine pollutants were analyzed in the 1990s in the Cabo Blanco and Greek monk seal subpopulations: Residue levels were found to be very low in the former subpopulation and moderate to high in the latter (Yediler et al. 1993, Borrell et al. 1997, Borrell et al. 2007). Recent research efforts in the eastern Mediterranean indicate that monk seals in Greece are not acutely threatened by heavy metals but that this threat should be closely monitored (Formigaro et al. 2016). Mediterranean monk seals are at an unknown, but suspected high level of risk from oil tanker and other ship accidents, spills, and groundings. This results from increased vessel traffic throughout the range of the species and a greater chance for accidents, disturbance, and collisions near important habitat. Four accidents or spills have occurred near important monk seal habitat in the recent past (Isräels 1992, Kiraç 1998, UNEP 2005). None of these spills or accidents had any known impacts on monk seals, but they highlight the threat from a major maritime accident near an important monk seal site (UNEP 2005).

Other threats

Although there is not enough information available to fully evaluate the magnitude of limited availability of food sources, genetic inbreeding and pollution as threats to the Mediterranean monk seal there is no indication that they are significantly affecting the Mediterranean monk seal at present (Karamanlidis et al. 2016a). More recently, the arrival of Lessepsian fishes in the eastern Mediterranean Sea, such as the toxic Pufferfish (*Lagocephalus sceleratus*), could also have a negative impact on Mediterranean monk seals in the region. The same applies to climate change (i.e., sea level rise) which could also have a negative impact, of unknown magnitude, on the habitat of the Mediterranean monk seal. Additional risks to the species come from the challenge of implementing effective conservation measures for a species in a complex, multi-national environment and the weak enforcement of agreements and international laws (Aguilar 1999).

Conservation

Nowadays the Mediterranean monk seal is legally protected throughout its range through national legislations, regional and international conventions, as well as European Union regulations. According to the European Council's Directive 92/43EEC, the Mediterranean monk seal is considered a species of community importance. Especially designed measures and regulations for the protection of important Mediterranean monk seal populations are currently in place in the following areas: the Desertas Islands Nature Reserve in the Madeira Archipelago, the National Marine Park of Alonnisos, Northern Sporades, the marine protected area in Northern Karpathos–Saria and the 3-mile no-take zone at the Island of Gyaros in Greece. With the same rationale and in order to protect the monk seal population and its pupping caves, a no-fishing area and a participative reserve have been created at the Cabo Blanco peninsula. In Turkey, similar measures have been put in place in five coastal locations in the country: Foça, Karaburun, Alaçatı-Sigacik, the Bodrum Peninsula, and the Cilician coast. Finally, the species is explicitly mentioned in 102 Natura 2000 sites within the European Union (82 sites in Greece, 10 in Italy, five in Spain, three in Portugal, and two in Cyprus) (Karamanlidis et al. 2016a). Based on the European legislation, Natura 2000 sites are legally considered by EU member states as Protected Areas and appropriate management actions should be implemented within their boundaries.

During the last two decades and throughout the range of the species, widespread action has been taken for the conservation of the Mediterranean monk seal. Especially, in areas where important monk seal populations live and breed (namely Greece, Turkey, Archipelago of Madeira and Cabo

Blanco), long-term initiatives have been carried out in order to sensitize the local human population towards monk seal conservation, to protect pupping sites, to restrict fishing gear use and relocate the most adverse fishing practices, to develop monitoring programs and intervention protocols, and to increase on-site capability to rehabilitate sick and injured individuals, particularly pups.

In addition, many workshops and conferences have brought together scientists and managers to discuss Mediterranean monk seal conservation issues and problems. Furthermore, numerous international bodies and fora, including the Regional Activity Center for Specially Protected Areas and the General Fisheries Commission for the Mediterranean, have put forward initiatives and proposals in order to ameliorate existing threats and mitigate pressures from relevant sectors (i.e., fisheries, bycatch, etc.). In Greece, the "National Strategy and Action Plan for the Conservation of the Mediterranean Monk Seal in Greece, 2009–2015" (Notarbartolo di Sciara et al. 2009) described in detail actions that had to be carried out in the country by 2015 in order to safeguard the future of the species. Similarly, in Turkey the National Monk Seal Committee has drafted a "National Action Plan for the Conservation of the Mediterranean monk seal *Monachus monachus* in Turkey" that has been approved by the Turkish Ministry of Forest and Water Works (Kiraç et al. 2013). Recently, a new Regional Strategy for the Conservation of the Mediterranean monk seal has also been adopted by the parties of the United Nations Environmental Program (Notarbartolo di Sciara 2013). Similarly, an Action Plan has been adopted for the recovery of the Mediterranean monk seal in the eastern Atlantic (UNEP 2005).

It appears that the management and protection actions that have been carried out so far have had some positive effects on Mediterranean monk seals, both, in the eastern Mediterranean as well in the Atlantic. In both areas, the populations of the species are showing now encouraging signs of recovery, giving hope for the future of the species. A reflection of the above, together with other factors, is the recent de-listing of the species in the IUCN Red Data list from "critically endangered" to "endangered" (Karamanlidis and Dendrinos 2015). Naturally, this change in IUCN category is no assurance for the species' future, since the populations that remain are still very small in size, while all threats described previously are still in place. The de-listing merely notes that we are moving in the right direction; systematic efforts to properly and efficiently manage and protect the species are obviously still necessary in order to secure the future of the species.

Keywords: Mediterranean monk seal, endangered, population, genetic inbreeding, habitat deterioration, fisheries interaction, pollution, conservation, Cabo Blanco, Atlantic Ocean, Island of Gyaros, eastern Mediterranean Sea

References

Aguilar, A. 1999. Status of Mediterranean Monk Seal Populations. Aloes Editions, Tunis.

Androukaki, E., S. Adamantopoulou, P. Dendrinos, E. Tounta and S. Kotomatas. 1999. Causes of mortality in the Mediterranean monk seal (*Monachus monachus*) in Greece. Contributions to the Zoogeography and Ecology of the Eastern Mediterranean Region 1: 405–411.

Badosa, E., E. Grau, F. Aparicio, J.F. Layna and M.A. Cedenilla. 1998. Individual variation and sexual dimorphism of coloration in Mediterranean monk seal pups (*Monachus monachus*). Mar. Mamm. Sci. 14: 390–393.

Badosa, E., T. Pastor, M. Gazo and A. Aguilar. 2006. Moult in the Mediterranean monk seal from Cap Blanc, western Sahara. Afr. Zool. 41: 183–192.

Bareham, J.R. and A. Furreddu. 1975. Observations on the use of grottos by Mediterranean monk seals (*Monachus monachus*). J. Zool. 175: 291–298.

Borrell, A., A. Aguilar and T. Pastor. 1997. Organochlorine pollutant levels in Mediterranean monk seals from the western Mediterranean and the Sahara Coast. Mar. Pollut. Bull. 34: 505–510.

Borrell, A., G. Cantos, A. Aguilar, E. Androukaki and P. Dendrinos. 2007. Concentrations and patterns of organochlorine pesticides and PCBs in Mediterranean monk seals (*Monachus monachus*) from Western Sahara and Greece. Sci. Total. Environ. 381: 316–325.

Dendrinos, P., A.A. Karamanlidis, E. Androukaki and B.J. McConnell. 2007a. Diving development and behavior of a rehabilitated Mediterranean monk seal (*Monachus monachus*). Mar. Mamm. Sci. 23: 387–397.

Dendrinos, P., A.A. Karamanlidis, S. Kotomatas, A. Legakis, E. Tounta and J. Matthiopoulos. 2007b. Pupping habitat use in the Mediterranean monk seal: A long-term study. Mar. Mamm. Sci. 23: 615–628.

Dendrinos, P. 2011. Contribution to the study of the Mediterranean monk seal's (*Monachus monachus*) ecology and biology at the island complex of Northern Sporades, Greece. National and Kapodistrian University of Athens, Athens, Greece.

Formigaro, C., A.A. Karamanlidis, P. Dendrinos, L. Marsili, M. Silvi and A. Zaccaroni. 2016. Trace element concentrations in the Mediterranean monk seal (*Monachus monachus*) in the Eastern Mediterranean Sea. Science of the Total Environment 576: 528–537.

Gazo, M., C. Lydersen and A. Aguilar. 2006. Diving behaviour of Mediterranean monk seal pups during lactation and post weaning. Mar. Ecol. Prog. Ser. 308: 303–309.

González, L.M. and P. Fernandez de Larrinoa. 2012. Mediterranean monk seal *Monachus monachus* distribution and fisheries interactions in the Atlantic Sahara during the second half of the 20th century. Mammalia 77: 41–49.

González, L.M. 2015. Prehistoric and historic distributions of the critically endangered Mediterranean monk seal (*Monachus monachus*) in the eastern Atlantic. Mar. Mamm. Sci. 31: 1168–1192.

Güçlüsoy, H. and Y. Savaş. 2003a. Interaction between monk seals *Monachus monachus* (Hermann, 1779) and marine fish farms in the Turkish Aegean and the management of the problem. Aquacult. Res. 34: 777–783.

Güçlüsoy, H. and Y. Savaş. 2003b. Status of the Mediterranean monk seal (*Monachus monachus*) in the Foça Pilot monk seal conservation area, Turkey. Zool. Middle East 28: 5–16.

Güçlüsoy, H. 2008. Damage by monk seals to gear of the artisanal fishery in the Foça monk seal pilot conservation area, Turkey. Fish Res. 90: 70–77.

Gücü, A.C., G. Gücü and H. Orek. 2004. Habitat use and preliminary demographic evaluation of the critically endangered Mediterranean monk seal (*Monachus monachus*) in the Cilician Basin (Eastern Mediterranean). Biol. Conserv. 116: 417–431.

Gücü, A.C., S. Sakinan and M. Ok. 2009. Occurence of the critically endangered Mediterranean monk seal, *Monachus monachus*, at Olympos-Beydaglari National Park, Turkey. Zool. Middle East 46: 3–8.

Hale, R., R. Pires, P. Santos and A.A. Karamanlidis. 2011. Mediterranean monk seal (*Monachus monachus*): Fishery interactions in the Archipelago of Madeira. Aquat. Mamm. 37: 298–304.

Hermann, J. 1779. Beschreibung der Moenchs-Robbe. Beschaeftigungen der Berlinischen Gesellschaft Naturforschender Freunde 4: 456–509, pl. 412–413.
Hernández, M., I. Robinson, A. Aguilar, L.M. Gonzalez, L.F. Lopez-Jurado, M.I. Reyero et al. 1998. Did algal toxins cause monk seal mortality? Nature 393: 28–29.
Isaac, N.J.B., S.T. Turvey, B. Collen, C. Waterman and J.E.M. Baillie. 2007. Mammals on the EDGE: Conservation priorities based on threat and phylogeny. PLoS One 2: e296.
Isräels, L.D.E. 1992. Thirty years of Mediterranean monk seal protection—a review. Mededelingen 28: 1–65.
Johnson, W.M. and D.M. Lavigne. 1999. Monk seals in antiquity. The Mediterranean monk seal (*Monachus monachus*) in ancient history and literature. Mededelingen 35: 1–101.
Johnson, W.M. and A.A. Karamanlidis. 2000. When fishermen save seals. The Monachus Guardian 3: 18–22.
Johnson, W.M. 2004. Monk seals in post-classical history. The role of the Mediterranean monk seal (*Monachus monachus*) in European history and culture, from the fall of Rome to the 20th century. Mededelingen 39: 1–91.
Karamanlidis, A.A., R. Pires, N.C. Silva and H.C. Neves. 2004. The availability of resting and pupping habitat for the critically endangered Mediterranean monk seal *Monachus monachus* in the archipelago of Madeira. Oryx 38: 180–185.
Karamanlidis, A.A., E. Androukaki, S. Adamantopoulou, A. Chatzispyrou, W.M. Johnson, S. Kotomatas et al. 2008. Assessing accidental entanglement as a threat to the Mediterranean monk seal *Monachus monachus*. Endang. Species Res. 5: 205–213.
Karamanlidis, A.A., M. Psaradellis and P. Dendrinos. 2012. Social structure and behavior of the unique Mediterranean monk seal colony of the island of Gyaros. 18–22/06/2012.
Karamanlidis, A.A., S. Adamantopoulou, V. Paravas, M. Psaradellis and P. Dendrinos. 2013. Demographic structure and social behavior of the unique Mediterranean monk seal colony of the island of Gyaros. In: 20th Biennial Conference on the Biology of Marine Mammals. 9–13 December 2013. Dunedin, New Zealand.
Karamanlidis, A.A., J.P. Curtis, A.C. Hirons, M. Psaradellis, D. Dendrinos and J.B. Hopkins III. 2014. Stable isotopes confirm a coastal diet for critically endangered Mediterranean monk seals. Isot. Environ. Health Stud. 50: 332–342.
Karamanlidis, A.A. and P. Dendrinos. 2015. *Monachus monachus*. In: The IUCN Red List of Threatened Species 2015: e.T13653A45227543.
Karamanlidis, A.A., P. Dendrinos, P. Fernández de Larrinoa, A.C. Gücü, W.M. Johnson et al. 2016a. The Mediterranean monk seal *Monachus monachus*: Status, biology, threats, and conservation priorities. Mamm. Rev. 46: 92–105.
Karamanlidis, A.A., S. Gaughran, A. Aguilar, P. Dendrinos, D. Huber, R. Pires et al. 2016b. Shaping species conservation strategies using mtDNA analysis: The case of the elusive Mediterranean monk seal (*Monachus monachus*). Biol. Conserv. 193: 71–79.
Karamanlidis, A.A., O. Lyamin, S. Adamantopoulou and P. Dendrinos. 2017. First observations of aquatic sleep in the Mediterranean monk seal (*Monachus monachus*). Aquatic Mammals 43(1): 82–86.
King, J.E. 1956. The monk seals (Genus *Monachus*). Bull. Br. Mus. Nat. Hist. Zool., London 3: 201–256, 208 pls.
Kiraç, C.O. 1998. Oil Spill at Cavus Island. The Monachus Guardian 1: 16–18.
Kiraç, C.O., N.O. Veryeri, H. Güçlüsoy and Y. Savaş. 2013. National action plan for the conservation of the Mediterranean monk seal *Monachus monachus* in Turkey. UNEP MAP RAC/SPA and Republic of Türkiye Ministry of Forest and Water Works.
Layna, J.E., M.A. Cedenilla, F. Aparicio and L.M. Gonzalez. 1999. Observations of parturition in the Mediterranean monk seal (*Monachus monachus*). Mar. Mamm. Sci. 15: 879–882.
Ling, J.K. 1970. Pelage and molting in wild mammals with special reference to aquatic forms. Q. Rev. Biol. 45: 16–54.
Marchessaux, D. and C. Pergent-Martini. 1991. Biologie de la production et developpement des nouveaux nes chez le phoque moine *Monachus monachus*. pp. 349–358. In: C.F.

Boudouresque, M. Avon and V. Gravez [eds.]. Les Especes Marines a Proteger en Mediterranee. GIS Posidonie Publ., Marseille, France.

Martínez-Jauregui, M., G. Tavecchia, M.A. Cedenilla, T. Coulson, P. Fernández de Larrinoa, M. Muñoz and L.M. González. 2012. Population resilience of the Mediterranean monk seal *Monachus monachus* at Cabo Blanco peninsula. Mar. Ecol. Prog. Ser. 461: 273–281.

Muñoz Cañas, M., G. Hernández-Millian, J. Más, P.F. de Larrinoa and G.J. Pierce. 2012. Diet of the Mediterranean monk seal in Mauritanian waters. *In*: 9th MBA Postgraduate Conference. 21–24 May 2012, Cork, Ireland.

Neves, H.C. 1998. Preliminary findings on the feeding strategy of the Monk seal *Monachus monachus* (Pinnipedia: Monachinae) on the Desertas islands. Boletim do Museum Municipal do Funchal Suppl. No. 5: 263–271.

Notarbartolo di Sciara, G., S. Adamantopoulou, E. Androukaki, P. Dendrinos, A.A. Karamanlidis, V. Paravas et al. 2009. National strategy and action plan for the conservation of the Mediterranean monk seal in Greece, 2009–2015. MOm/ /Hellenic Society for the Study and Protection of the Monk seal. Athens, Greece.

Notarbartolo di Sciara, G. 2013. Draft regional strategy for the conservation of monk seals in the Mediterranean (2014–2019). Tethys Research Institute, Milano, Italy.

Osterhaus, A.D.M.E., I.K.G. Visser, R.L. de Swart, M.F. van Bressem, M.W.G. van de Bildt, C. Oervell et al. 1992. Morbillivirus threat to Mediterranean monk seals? Vet. Rec. 130: 141–142.

Pastor, T. and A. Aguilar. 2003. Reproductive cycle of the female Mediterranean monk seal in the western Sahara. Mar. Mamm. Sci. 19: 318–330.

Pastor, T., J.C. Garza, P. Allen, W. Amos and A. Aguilar. 2004. Low genetic variability in the highly endangered Mediterranean monk seal. J. Hered. 95: 291–300.

Pastor, T., J.C. Garza, A. Aguilar, E. Tounta and E. Androukaki. 2007. Genetic diversity and differentiation between the two remaining populations of the critically endangered Mediterranean monk seal. Anim. Conserv. 10: 461–469.

Pierce, G.J., G. Hernandez-Milian, M.B. Santos, P. Dendrinos, M. Psaradellis, E. Tounta et al. 2011. Diet of the Monk seal (*Monachus monachus*) in Greek waters. Aquat. Mamm. 37: 284–297.

Pinela, A.M., A. Borrell, L. Cardona and A. Aguilar. 2010. Stable isotope analysis reveals habitat partitioning among marine mammals off the NW African coast and unique trophic niches for two globally threatened species. Mar. Ecol. Prog. Ser. 416: 295–306.

Pires, R., H.C. Neves and A.A. Karamanlidis. 2007. Activity patterns of the Mediterranean monk seal (*Monachus monachus*) in the Archipelago of Madeira. Aquat. Mamm. 33: 327–336.

Reijnders, P., S. Brasseur, J. van der Torn, P. van der Wolf, I. Boyd, J. Harwood, D. Lavigne and L. Lowry [eds.]. 1993. Seals, Fur Seals, Sea Lions and Walrus. Status Survey and Conservation Action Plan. IUCN, Gland, Switzerland.

Reyero, M., E. Cacho, A. Martinez, J. Vazquez, A. Marina, S. Fraga et al. 2000. Evidence of saxitoxin derivatives as causative agents in the 1997 mass mortality of monk seals in the Cape Blanc Peninsula. Nat. Toxins 8: 1–5.

Scheel, D.-M., G.J. Slater, S.-O. Kolokotronis, C.W. Potter, D.S. Rotstein, K. Tsangaras et al. 2014. Biogeography and taxonomy of extinct and endangered monk seals illuminated by ancient DNA and skull morphology. ZooKeys 409: 1–33.

UNEP. 2005. Action plan for the recovery of the Mediterranean monk seal (*Monachus monachus*) in the eastern Atlantic. Convention on the Conservation of Migratory Species of Wild Animals (Bonn Convention) CMS/ScC.13/Inf.3.

van de Bildt, M.W.G., E.J. Vedder, B.E.E. Martina, B. Abou Sidib, A.B. Jiddou, M.E.O. Barham et al. 1999. Morbilliviruses in Mediterranean monk seals. Vet. Microbiol. 69: 19–21.

Yediler, A., A. Panou and P. Schramel. 1993. Heavy metals in hair samples of the Mediterranean monk seal (*Monachus monachus*). Mar. Pollut. Bull. 26: 156–159.

14

Bioecology and Conservation Threats of the Cape Fur Seal *Arctocephalus pusillus pusillus*

G.J. Greg Hofmeyr

Introduction

Cape fur seals *Arctocephalus pusillus pusillus* (Fig. 1) inhabit the African continental shelf. While their distribution is primarily temperate or subtropical, their range extends into tropical regions. The northward flowing waters of the cold Benguela Current create a suitable and productive marine habitat for them to the north of the Tropic of Capricorn, on the west coast of southern Africa. Large breeding subpopulations are found in low latitudes in northern Namibia and southern Angola.

Cape fur seals are pinnipeds of the family Otariidae. Their common name reflects their area of discovery, although they have a much wider range. They have also been known in the past as the South African fur seal despite the preponderance of their population being distributed in Namibia. Cape fur seals are one of two subspecies of *Arctocephalus pusillus*, the other being the Australian fur seal *A. p. doriferus*. The two subspecies are very similar anatomically and behaviourally, differing primarily in range (Repenning et al. 1971).

Port Elizabeth Museum at Bayworld, Humewood, Port Elizabeth, South Africa; Department of Zoology, Nelson Mandela Metropolitan University, Port Elizabeth, South Africa.
Email: greghofmeyr@gmail.com

Adult male Cape fur seals reach masses of 247 kg and 2.3 m in length while adult females only reach 57 kg in mass and 1.6 m in length. Adult females have a more gracile build with a proportionally smaller head and heavier abdomen compared to the larger head and heavier forequarters of the adult males. Pups weigh some 6 kg at birth (Shaughnessy 1979). In colour, Cape fur seals are dark brown to grey (Fig. 2). Individuals are largely uniform in colour but are slightly darker ventrally between the fore flippers and on the belly. The lanugo is black and is moulted to adult colouration at some four months of age (Fig. 3). They are born with black whiskers that start to become white while they are subadult and are completely white in older adults. Cape fur seals are readily distinguishable from the only other otariid that regularly occurs along the southern African coast, the Subantarctic fur seal *Arctocephalus tropicalis* (Shaughnessy and Ross 1980, Bester 1989) by a combination of size, pelage colour and head shape.

This chapter reviews the bio-ecology of the species and anthropogenic threats to their conservation.

Figure 1. Cape fur seal rookery. Photo credit: Greg Hofmeyr, Black Rocks, Algoa Bay, South Africa 2011.

Figure 2. Adult male Cape fur seals showing variation in pelage colour. Photo credits: Greg Hofmeyr, Black Rocks, Algoa Bay, South Africa 2012.

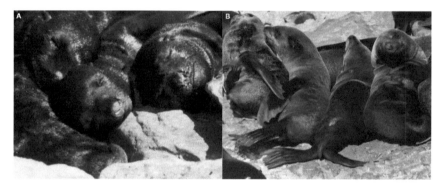

Figure 3. Cape fur seals pups showing colour of lanugo (A) and coat after some four months of age (B). Photo credits: Greg Hofmeyr, Atlas Bay, Namibia 1999; Greg Hofmeyr, Black Rocks, Algoa Bay, South Africa 2012.

Life History

While males become sexually mature at a younger age, they are socially mature at 9–12 years only. Females become sexually mature at 3–6 years (Wickens and York 1997). The estimated annual pregnancy rate is 71% (Wickens and York 1997). The duration of gestation is 51 weeks. Implantation is delayed for three months following conception (Reijnders et al. 1993, Butterworth et al. 1995).

Adults haul out during an annual breeding season at traditional rookeries (Fig. 4). Breeding takes place from late October to early January with a peak of pupping in the first week of December (David 1987b). Adult

Figure 4. Cape fur seal breeding colony with adult male (centre) and adult females with neonate pups. Photo credit: Greg Hofmeyr, Robberg, Plettenberg Bay, South Africa 2012.

males arrive first at the rookeries, establishing territories which they defend by fighting, and by postural and vocal displays (David 1987b, De Villiers and Roux 1992). The breeding aggregations are polygynous with successful males monopolising an average of 7.5 females (Rand 1967). Adult females give birth to a single pup within 2 days of their arrival. They come into oestrus 6–7 days later. Mating takes place before they depart on their first foraging trip (Rand 1955). This is followed by a regular cycle of attending pups ashore and foraging at sea. Initial foraging trips average 5.2 days, and attendance periods, 1.8 days. Foraging trips increase in duration over time (Gamel et al. 2005). While pups are weaned at 10–12 months of age, some may start foraging at 7 months. However, rare instances of suckling of up to 3 years have been recorded (Warneke and Shaughnessy 1985, David and Rand 1986).

Foraging Ecology

Cape fur seals are primarily pelagic foragers but do take benthic prey in parts of their range (David 1987a, Connan et al. 2014). They feed opportunistically on a wide variety of species including pilchards *Sardinops sagax*, Cape hakes *Merluccius* spp., horse mackerel *Trachurus capensis*, pelagic goby *Sufflogobius bibarbatus*, anchovy *Engraulis encrasicolus*, squid *Loligo* spp. and rock lobster

Jasus lalandii (David 1987a, Punt et al. 1995, de Bruyn et al. 2003, Mecenero et al. 2006). Individuals occasionally take seabirds (David et al. 2003, du Toit et al. 2004, Makhado et al. 2006).

While the maximum recorded dive depth is 204 m (Kooyman and Gentry 1986), typical foraging dives are less than 60 m on the south east coast of South Africa (Stewardson 2001), and 50 m on the west coast (Kooyman and Gentry 1986). Mean dive durations of up to 2.1 minutes have been recorded (Kooyman and Gentry 1986). A diurnal pattern is evident with most dives taking place at dusk and during the early night, and again shortly after dawn (Kooyman and Gentry 1986, Stewardson 2001).

Cape fur seal pups face predation by black-backed jackals *Canis mesomelas* and brown hyaenas *Hyaena brunnea* on mainland colonies (Skinner et al. 1995, Oosthuizen et al. 1997, Kuhn et al. 2008). At sea all age classes are preyed on by great white sharks *Carcharodon carcharias* (Martin et al. 2005). They are also taken by killer whales *Orcinus orca* (Rice and Saayman 1987).

Distribution and Demography

Cape fur seal distribution encompasses three countries: Angola, Namibia and South Africa (Fig. 5). They haul out at some 40 sites from Ilha dos Tigres in the south of Angola to Algoa Bay on the south eastern coast of South Africa (Oosthuizen 1991, Kirkman et al. 2013). Some individuals move further eastwards in winter following the annual migration of the pilchard reaching the KwaZulu-Natal coast (O'Donoghue et al. 2010). As a whole, however, the species is not migratory. At sea they are limited to the

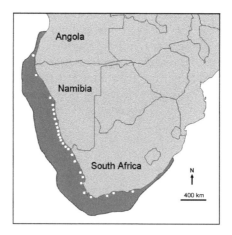

Figure 5. Map showing distribution of Cape fur seals along the coast of southern Africa. White circles indicate the locations of rookeries, or aggregations of rookeries. Following Kirkman et al. (2007) and Hofmeyr (2015).

continental shelf and have been recorded a maximum of 220 km offshore (Shaughnessy 1979). Sightings of vagrants are very rare only having been recorded from Gabon (Thibault 1999) and the Prince Edward Islands (Kerley 1983).

Abundance is estimated from counts of pups ashore at the end of the breeding season every few years. Some 60% of pup production takes place at 23 rookeries in Namibia, 40% at 16 rookeries in South Africa, and a small number at an incipient but growing rookery in Angola (Kirkman et al. 2007, 2013). The most recent estimate for the entire population is approximately two million animals (Kirkman et al. 2013). This indicates relatively little change in the size of the total population since 1993 (Butterworth et al. 1995, Kirkman et al. 2007, 2013). However, the distribution of the population has changed substantially over this time. An increase in the number of colonies has been accompanied by an increase in abundance of populations in northern Namibia and north-western South Africa at the expense of populations in southern Namibia (Kirkman et al. 2013). While three quarters of all pups were born at the three major haulout sites in 2004, these have all experienced small declines (Kirkman et al. 2013). Although the abundance of these larger populations is largely stable, those in southern Namibia have been subject to major mortality events (Kirkman et al. 2013) as a result of the impact of unusual environmental conditions on prey populations (Gammelsrød et al. 1998, Gerber and Hilborn 2001). The smaller rookeries tend to experience greater interannual fluctuations in size (Kirkman et al. 2007, 2013). Few rookeries are smaller than 1000 individuals.

No distinct subpopulations are evident despite some rookeries being separated from others by several hundred kilometres (Matthee et al. 2006). The movement of tagged individuals indicates considerable dispersal between sites (Oosthuizen 1991). While Subantarctic fur seals regularly visit southern Africa in winter (Shaughnessy and Ross 1980, Bester 1989), hybridization is unknown.

Anthropogenic Interactions and Threats

Although Cape fur seals were hunted by indigenous people, levels of exploitation are thought to have been low. Intensive commercial harvesting began in the mid 17th century and reduced abundance to very low numbers (Warneke and Shaughnessy 1985, David 1987b). Exploitation was first controlled in South Africa in 1893, and in the then South West Africa (now Namibia) in 1922, since which times the population has increased substantially (Wickens et al. 1991). Since the pre-exploitation population size is unknown, the extent of recovery to those levels cannot be determined (Kirkman et al. 2007). However, the recent establishment of mainland colonies and the lack of recolonization of certain islands, has likely affected the distribution of the species (Stewardson 1999, Kirkman et al. 2013). Harvesting in South Africa ceased in 1990 (Wickens et al. 1991) and is now prohibited by the Marine Living Resources Act (MLRA 2007). However, harvesting continues in Namibia at the large mainland colonies (Japp et al. 2012). Numbers harvested have not decreased even during periods of high natural mortality (Kirkman et al. 2007, Japp et al. 2012). While Cape fur seals are instinctively afraid of humans and avoid areas of high human population, some adult males become habituated and haul out on the mainland close to settlements (Fig. 6A).

Figure 6. Adult male Cape fur seal on a public city beach (A) and Cape fur seal entangled in fishing debris (B). Photo credits: Greg Hofmeyr, Swartkops River mouth, Port Elizabeth, South Africa 2011 (A); Greg Hofmeyr, Sardinia Bay, Port Elizabeth, South Africa 2008 (B).

Currently, Cape fur seals interact with commercial fisheries, both indirectly via competition, and directly via operational interactions. While some prey species are commercially exploited (David 1987a, Wickens et al. 1992), the complexity of the marine food web and the opportunistic and broad diet of Cape fur seals make it difficult to assess the extent of these interactions (David 1987a, Punt and Butterworth 1995). Nevertheless, changes in fishing effort may reduce prey populations (Barange et al. 1999, Roy et al. 2007, Moloney et al. 2013, Roux et al. 2013). Although seals are taken directly during fishing operations, the levels of mortality are not well known. Levels of direct mortality in the late 20th century are thought to have been low through most of the range, but relatively high in the Eastern Cape (Wickens et al. 1992, David and Wickens 2003). Cape fur seals may be illegally shot during fishing operations and may become entangled in marine debris (Fig. 6B) (Wickens et al. 1992).

Anthropogenic climate change is likely a factor responsible for recent changes in the distribution of population numbers (Kirkman et al. 2013) through changes in the distribution of prey species (Barange et al. 1999, Roy et al. 2007, Moloney et al. 2013, Roux et al. 2013). Episodes of high mortality experienced by Namibian rookeries are likely related to climate change (Gammelsrød et al. 1998, Gerber and Hilborn 2001) illustrating the vulnerability of this species. Further impacts of climate change on pup mortality are possible. Pups frequently die of heat stroke during periods of high temperatures (De Villiers and Roux 1992) and therefore may be susceptible to increases in ambient temperatures and decreases in the number of windy days (Kovacs et al. 2012). Furthermore, pups born on offshore islands are vulnerable to drowning during summer storms (Hofmeyr et al. 2011). A climate change mediated increase in the frequency of such storms will increase rates of pup mortality.

Keywords: Cape fur seal, South African fur seal, foraging ecology, breeding, distribution, anthropogenic interactions, climate change, Namibia, southern Angola, Africa

References

Barange, M., I. Hampton and B.A. Roel. 1999. Trends in the abundance and distribution of anchovy and sardine on the South African continental shelf in the 1990s, deduced from acoustic surveys. S. Afr. J. Mar. Sci. 21: 367–391.

Bester, M.N. 1989. Movements of southern elephant seals and Subantarctic fur seals in relation to Marion Island. Mar. Mamm. Sci. 5: 257–265.

Butterworth, D.S., A.E. Punt, W.H. Oosthuizen and P.A. Wickens. 1995. The effects of future consumption by the Cape fur seal on catches and catch rates of the Cape hakes. 3. Modelling the dynamics of the Cape fur seal *Arctocephalus pusillus pusillus*. S. Afr. J. Mar. Sci. 16: 161–183.

Connan, M.A., G.J.G. Hofmeyr, M.J. Smale and P.A. Pistorius. 2014. Trophic investigations of Cape fur seals at the easternmost extreme of their distribution. Afr. J. Mar. Sci. 36: 331–344.

David, J. and P. Wickens. 2003. Management of Cape fur seals and fisheries in South Africa. pp. 116–131. *In*: N.J. Gales, M.A. Hindell and R. Kirkwood [eds.]. Marine Mammals: Fisheries, Tourism and Management Issues. CSIRO, Collingwood, Australia.

David, J.H.M. and R.W. Rand. 1986. Attendance behaviour of South African fur seals. pp. 126–141. *In*: R.L. Gentry and G.L. Kooyman [eds.]. Fur Seals: Maternal Strategies on Land and at Sea. Princeton University Press, Princeton, New Jersey.

David, J.H.M. 1987a. Diet of the South African (Cape) fur seal (1974–1985) and an assessment of competition with fisheries in southern Africa. S. Afr. J. Mar. Sci. 5: 693–713.

David, J.H.M. 1987b. South African fur seal, *Arctocephalus pusillus pusillus*. *In*: J.P. Croxall and R.L. Gentry [eds.]. Status, Biology and Ecology of Fur Seals. Proceedings of an International Symposium and Workshop, Cambridge, England, 23–27 April 1984. NOAA Technical report NMFS 51: 65–71.

David, J.H.M., P. Cury, R.J.M. Crawford, R.M. Randall, L.G. Underhill and M.A. Meyer. 2003. Assessing conservation priorities in the Benguela ecosystem, South Africa: Analysing predation by seals on threatened seabirds. Biol. Conserv. 114: 289–292.

de Bruyn, P.J.N., M.N. Bester, S. Mecenero, S.P. Kirkman, J.-P. Roux and N.T.W. Klages. 2003. Temporal variation of cephalopods in the diet of Cape fur seals in Namibia. S. Afr. J. Wildl. Res. 33: 85–96.

De Villiers, D. and J.-P. Roux. 1992. Mortality of newborn pups of the South African fur seal *Arctocephalus pusillus pusillus* in Namibia. S. Afr. J. Mar. Sci. 12: 881–889.

du Toit, M., P.A. Bartlett, M.N. Bester and J.-P. Roux. 2004. Seabird predation by individual seals at Ichaboe Island, Namibia. S. Afr. J. Wildl. Res. 34: 45–54.

Gamel, C.M., R.W. Davis, J.H.M. David and M.A. Meyer. 2005. Reproductive energetics and female attendance patterns of Cape fur seals (*Arctocephalus pusillus pusillus*) during early lactation. Am. Midl. Nat. 153: 152–170.

Gammelsrød, T., C.H. Bartholomae, D.C. Boyer, V.L.L. Filipe and M.J. O'Toole. 1998. Intrusion of warm surface water along the Angolan-Namibian coast in February-March 1995: The 1995 Benguela Niño. S. Afr. J. Mar. Sci. 19: 41–56.

Gerber, L.R. and R. Hilborn. 2001. Catastrophic events and recovery from low densities in populations of otariids: Implications for risk of extinction. Mammal. Rev. 31: 131–150.

Hofmeyr, G.J.G., M. Du Toit and S.P. Kirkman. 2011. Early post-release survival of stranded Cape fur seal pups at Black Rocks, Algoa Bay. Afr. J. Mar. Sci. 33: 453–461.

Hofmeyr, G.J.G. 2015. *Arctocephalus pusillus*. The IUCN Red List of Threatened Species 2015: e.T2060A45224212. http://dx.doi.org/10.2305/IUCN.UK.2015-4.RLTS.T2066A66991045.en. Downloaded on 03 August 2016.

Japp, D.W., M.G. Purves and S. Wilkinson. 2012. Benguela Current Large Marine Ecosystem: State of the Stocks Review. Report No. 2. Benguela Current Commission, Cape Town.

Kerley, G.I.H. 1983. Record for the Cape fur seal *Arctocephalus pusillus pusillus* from subantarctic Marion Island. S. Afr. J. Zool. 18: 139–140.

Kirkman, S.P., W.H. Oosthuizen, M.A. Meÿer, P.G.H. Kotze, J.-P. Roux and L.G. Underhill. 2007. Making sense of censuses and dealing with missing data: Trends in pup counts of Cape fur seal *Arctocephalus pusillus pusillus* for the period 1972–2004. Afr. J. Mar. Sci. 29: 161–176.

Kirkman, S.P., D. Yemane, W.H. Oosthuizen, M.A. Meÿer, P.G.H. Kotze, H. Skrypzeck et al. 2013. Spatio-temporal shifts of the dynamic Cape fur seal population in southern Africa, based on aerial censuses (1972–2009). Mar. Mamm. Sci. 29: 497–524.

Kooyman, G.L. and R.L. Gentry. 1986. Diving behaviour of South African fur seals. pp. 142–152. *In*: R.L. Gentry and G.L. Kooyman [eds.]. Fur Seals - Maternal Strategies on Land and at Sea. Princeton University Press, Princeton.

Kovacs, K.M., A. Aguilar, D. Aurioles, V. Burkanov, C. Campagna, N.J. Gales et al. 2012. Global threats to pinnipeds. Mar. Mamm. Sci. 28: 414–436.

Kuhn, B.F., I. Wiesel and J.D. Skinner. 2008. Diet of brown hyaenas (*Parahyaena brunnea*) on the Namibian coast. Trans. R. Soc. S. Afr. 63: 150–159.

Makhado, A.B., R.J.M. Crawford and L.G. Underhill. 2006. Impact of predation by Cape fur seals *Arctocephalus pusillus pusillus* on Cape gannets *Morus capensis* at Malgas Island, Western Cape, South Africa. S. Afr. J. Mar. Sci. 28: 681–687.

Martin, A.R., N. Hammerschlag, R.S. Collier and C. Fallows. 2005. Predatory behaviour of white sharks (*Carcharodon carcharias*) at seal Island, South Africa. J. Mar. Bio. Assoc. UK 85: 1121–1135.

Matthee, C.A., F. Fourie, W.H. Oosthuizen, M.A. Meÿer and K.A. Tolley. 2006. Mitochondrial DNA sequence data of the Cape fur seal (*Arctocephalus pusillus pusillus*) suggest that population numbers may be affected by climatic shifts. Mar. Biol. 148: 899–905.

Mecenero, S., S.P. Kirkman and J.-P. Roux. 2006. A dynamic fish consumption model for lactating Cape fur seals *Arctocephalus pusillus pusillus* based on scat analyses. J. Mar. Sci. 63: 1551–1566.

MLRA. 2007. Policy on the Management of Seals, Seabirds and Shorebirds, 2007. Marine Living Resources Act, 1998. Government Gazette No. 30534.

Moloney, C.L., S.T. Fennessy, M.J. Gibbons, A. Roychoudhury, F.A. Shillington, B.P. von der Heyden et al. 2013. Reviewing evidence of marine ecosystem change off South Africa. Afr. J. Mar. Sci. 35: 427–448.

O'Donoghue, S.H., P.A. Whittington, B.M. Dyer and V. Peddemors. 2010. Abundance and distribution of avian and marine mammal predators of sardine observed during the 2005 KwaZulu-Natal sardine run Survey. Afr. J. Mar. Sci. 32: 361–374.

Oosthuizen, W.H. 1991. General movements of South African (Cape) fur seals *Arctocephalus pusillus pusillus* from analysis of recoveries of tagged animals. S. Afr. J. Mar. Sci. 11: 21–29.

Oosthuizen, W.H., M.A. Meyer, J.H.M. David, N.M. Summers, P.G.H. Kotze, S.W. Swanson and P.D. Shaughnessy. 1997. Variation in jackal numbers at the Van Reenen Bay seal colony with comment on the likely importance of jackals as predators. S. Afr. J. Wildl. Res. 27: 26–29.

Punt, A.E. and D.S. Butterworth. 1995. The effects of future consumption by the Cape fur seal on catches and catch rates of the Cape hakes. 4. Modelling the biological interaction between Cape fur seals *Arctocephalus pusillus pusillus* and the Cape hakes *Merluccius capensis* and *M. paradoxus*. S. Afr. J. Mar. Sci. 16: 255–285.

Punt, A.E., J.H.M. David and R.M. Leslie. 1995. The effects of future consumption by the Cape fur seal on catches and catch rates of the Cape hakes. 2. Feeding and diet of the Cape fur seal *Arctocephalus pusillus pusillus*. S. Afr. J. Mar. Sci. 16: 85–99.

Rand, R.W. 1955. Reproduction in the female Cape fur seal, *Arctocephalus pusillus* (Schreber). Proc. Zool. Soc. Lond. 124: 717–740.

Rand, R.W. 1967. The Cape fur seal *Arctocephalus pusillus pusillus* 3. General behaviour on land and at sea. Sea Fish. Inst. Investgl. Rep. 60: 1–32.

Reijnders, P., S. Brasseur, J. van der Toorn, P. van der Wolf, I.L. Boyd, J. Harwood et al. 1993. Seals, fur seals, sea lions, and walrus. Status survey and conservation action plan. IUCN/SSC Seal Specialist Group. IUCN, Gland, Switzerland.

Repenning, C.A., R.S. Peterson and C.L. Hubbs. 1971. Contributions to the systematics of the southern fur seals, with particular reference to the Juan Fernández and Guadalupe species. pp. 1–34. In: W.H. Burt [ed.]. Antarctic Pinnipedia. Antarctic Research Series Vol. 18. American Geophysical Union, Washington.

Rice, F.H. and G.S. Saayman. 1987. Distribution and behaviour of killer whales (*Orcinus orca*) off the coasts of southern Africa. pp. 231–250. In: G. Pilleri [ed.]. Investigations on Cetacea. Vol. 20. Hirnanatomisches Institut der Universität, Berne.

Roux, J.P., C.D. van der Lingen, M.J. Gibbons, N.E. Moroff, L.J. Shannon, A.D. Smith et al. 2013. Jellyfication of marine ecosystems as a likely consequence of overfishing small pelagic fishes: Lessons from the Benguela. Bull. Mar. Sci. 89: 249–284.

Roy, C., C.D. Van Der Lingen, J.C. Coetzee and J.R.E. Lutjeharms. 2007. Abrupt environmental shift associated with changes in the distribution of Cape anchovy *Engraulis encrasicolus* spawners in the southern Benguela. Afr. J. Mar. Sci. 29: 309–319.

Shaughnessy, P.D. 1979. Cape (South African) fur seal. pp. 37–40. *In*: Mammals in the Seas, Vol. II: Pinniped Species Summaries and Report on Sirenians. FAO.

Shaughnessy, P.D. and G.J.B. Ross. 1980. Records of the sub-Antarctic fur seal (*Arctocephalus tropicalis*) from South Africa with notes on its biology and some observations of captive animals. Ann. S. Afr. Mus. 82: 71–89.

Skinner, J.D., R.J. van Aarde and R.A. Goss. 1995. Space and resource use by brown hyenas *Hyaena brunnea* in the Namib Desert. J. Zool. Lond. 237: 123–131.

Stewardson, C.L. 1999. The impact of the fur seal industry on the distribution and abundance of Cape fur seals *Arctocephalus pusillus pusillus* on the Eastern Cape Coast of South Africa. Trans. R. Soc. S. Afr. 54: 217–246.

Stewardson, C.L. 2001. Biology and conservation of the Cape (South African) fur seal *Arctocephalus pusillus pusillus* (Pinnipedia: Otariidae) from the Eastern Cape Coast of South Africa. Ph.D. Thesis, Australian National University, Canberra.

Thibault, M. 1999. Sighting of a South African fur seal on a beach in south-western Gabon. Afr. J. Ecol. 37: 119–120.

Warneke, R.M. and P. Shaughnessy. 1985. *Arctocephalus pusillus*, the South African and Australian fur seal: Taxonomy, evolution, biogeography, and life history. pp. 53–77. *In*: J.K. Ling and M.M. Bryden [eds.]. Studies of Sea Mammals in South Latitudes. South Australia Museum, Adelaide.

Wickens, P. and A.E. York. 1997. Comparative population dynamics of fur seals. Mar. Mamm. Sci. 13: 241–292.

Wickens, P.A., J.H.M. David, P.A. Shelton and J.G. Field. 1991. Trends in harvest and pup numbers of the South African fur seal: Implications for management. S. Afr. J. Mar. Sci. 11: 307–356.

Wickens, P.A., D.W. Japp, P.A. Shelton, F. Kriel, P.C. Goosen, B. Rose et al. 1992. Seals and fisheries in South Africa—competition and conflict. *In*: A.I.L. Payne, K.H. Brink, K.H. Mann and R. Hilborn [eds.]. Benguela Trophic Functioning. S. Afr. J. Mar. Sci. 12: 773–789.

15

Emerging Pathogens and Health Issues in the 21st Century
A Challenge for Tropical and Subtropical Pinnipeds

Karina Acevedo-Whitehouse* and Luis A. Soto-García

Introduction

Between 1985 and 2015, the number of published research on pinniped health, disease, or immune function has increased by roughly 1400%.[1] This trend might be due to the expansion and specialization of the marine mammalogy community. However, it is also likely to reflect growing understanding about the importance of pinniped health in the context of environmental changes. Despite mounting interest there is still a noticeable bias in our knowledge, as not all species are equally represented in publications. Roughly a third of the published studies on pinniped health, disease or immune function are focused on the California sea lion (*Zalophus californianus*) followed by harbor seals (*Phoca vitulina*), grey seals (*Halichoerus*

Autonomous University of Queretaro, Laboratory of Molecular Epidemiology and Eco-evolutionary Immunology.
Unit for Basic and Applied Microbiology, Av. de las Ciencias S/N., Queretaro, QRO, Mexico, 76230.
* Corresponding author: karina.acevedo.whitehouse@uaq.mx

[1] Unbiased search of published papers conducted on ISI Web of Knowledge and PubMed Feb 2016. Search string: (pinniped) AND ((disease) OR (pathogen) OR (health) OR (immune)).

grypus) and Northern elephant seals (*Mirounga angustirostris*). Pinniped species that reside in tropical and subtropical latitudes have received comparatively little (or no) attention on these topics, as shown in Table 1. This is unfortunate, as several of these species have limited distributions, small population sizes, low genetic diversity, and are already considered endangered.

In view of our limited knowledge, it is tempting to assume that (i) pathogens isolated from a species at a given location will be common across all areas of the species' distribution, (ii) they will affect phylogenetically-related or sympatric species, and (iii) their effects will be equal for all pinnipeds. However, it is clear that this pattern does not always hold true. For instance, the hematophagous hookworm *Uncinaria* (Nematoda) infects most otariid species (Nadler et al. 2013). However, hookworm-related pathology varies markedly among hosts, from mild infections with little or no associated mortality in the New Zealand sea lion (*Phocarctos hookeri*) (Castinel et al. 2007) and Galapagos sea lion (*Zalophus wollebaeki*) (Alejandra Herbert-Mainero, unpublished data), to severe anemia, hemorrhagic enteritis, and peritonitis in the California sea lion (Spraker et al. 2004, Lyons et al. 2011). Susceptibility can also vary for the same otariid host between locations. For example, hookworm-related mortality of California sea lion pups at San Miguel Island has reached 60% in some years, while California sea lion pups from breeding colonies in the Gulf of California are often infected, but associated pathology due to this parasite is minimal (K.A.-W., unpublished data).

The spirochete *Leptospira interrogans* also provides a good example of how the impact of a pathogen can vary across a species' distribution. Leptospirosis is endemic in the California sea lion population (Lloyd-Smith et al. 2007, Prager et al. 2013), but individuals off the coast of California periodically experience epidemic events, where adult and subadult sea lions often die due to renal failure, and pregnant females miscarry (Colagross-Schouten et al. 2002, Lloyd-Smith et al. 2007). Epidemics of leptospirosis have never been reported for California sea lions within the Gulf of California, although most of the common serovars typically detected in the US pacific coast are present (Godínez et al. 1999, Acevedo-Whitehouse et al. 2003, Avalos-Téllez et al. 2016), the species migrates seasonally (Zuerner et al. 2009), and there is gene transfer between regions (Maldonado et al. 1995, Schramm et al. 2009).

Thus, to assume that hookworm infections or leptospirosis are important source of neonate mortality for all otariid pinnipeds, or for all subpopulations, would be incorrect. For conservationists interested in disease, it is essential to comprehend that not all diseases constitute the same potential risk for all pinnipeds. To date, most disease surveillance efforts consist in attempting to detect pathogens that have previously

Table 1. Published studies on pathogens, disease or health status of free-ranging tropical and subtropical pinnipeds.

Pinniped species	Pathogen-oriented					Health or disease			
	Viruses	Bacteria	Helminth parasites	Protozoa	Pathology	Hematology	Physical examination	Serology	Immune potential
Galapagos sea lion [1-9]	1	-	2	-	1	2	1	3	3
Galapagos fur seal	-	-	-	-	-	-	-	-	-
Guadalupe fur seal	-	-	-	-	-	-	-	-	-
Juan Fernandez fur seal [10-12]	-	-	2	-	-	1	-	-	-
South American sea lion [13-21]	2	3	3	1	2	-	-	1	-
South American fur seal [18-28]	2	2	5	1	1	-	-	1	-
Cape fur seal [29, 30]	-	1	1	-	1	-	-	-	-
Hawaiian monk seal [31-37]	3	2	3	2	-	2	-	2	-
Mediterranean monk seal [38-41]	4	-	1	1	1	-	-	-	-

Unbiased search of peer-reviewed published papers conducted on ISI Web of Knowledge and cross-referenced with PubMed Feb 2016. Only those studies conducted on wild individuals were included. Studies that measured algal toxins and persistent organic pollutants without a health focus were not included here. References cited:

[1]Coria-Galindo et al. 2009, [2]Dailey et al. 2005, [3]Denkinger et al. 2015, [4]Rand 1975, [5]Meise and García-Parra 2015, [6]Brock et al. 2013a, [7]Páez-Rosas et al. 2016, [8]Brock et al. 2013b, [9]Brock et al. 2015, [10]Sepúlveda 1998, [11]Cattan et al. 1980, [12]Sepúlveda et al. 1999, [13]Jankowski et al. 2015, [14]Blanc et al 2009, [15]Goldman et al. 2008, [16]Bernardelli et al. 1996, [17]Bastida et al. 1999, [18]Silva et al. 2014, [19]Hernández-Orts et al. 2013, [20]Cattan et al. 1976, [21]George-Nascimento and Carvajal 1981, [22]Sepúlveda et al. 2015, [23]Wilson and Poglayen-Neuwall 1971, [24]de Amorim et al. 2014, [25]Pereira et al. 2013, [26]Hernández-Orts et al. 2012, [27]George-Nascimento and Llanos 1995, [28]Hernández-Orts et al. 2012, [29]Henton et al. 1999, [30]De Graaf et al. 1980, [31]Aguirre et al. 2007, [32]Goldstein et al. 2006, [33]Littnan et al. 2006, [34]Nielsen et al. 2005, [35]Dailey et al. 1988, [36]Reif et al. 2006, [37]Banish and Gilmartin 1988, [38]Toplu et al. 2007, [39]van de Bildt et al. 2000, [40]Osterhaus et al. 1997, [41]Papadopoulos et al. 2010.

been identified as agents of disease in other species. Various features influence intra- and inter-species variations in susceptibility or tolerance to a pathogen, and factors such as population size and density, overlapping distributions with other species, spatial and genetic structuring, and endemism, shape the risk of disease for each species (see Norman et al. 2008, Smith et al. 2009). In addition, environmental factors such as agro-industrial development, urbanization and encroachment of habitats by human settlements will also determine the possibility that diseases are introduced in a particular pinniped population. Having information on baseline health parameters, common pathogens and associated diseases of different stocks, subpopulations, and species is fundamental if we are to detect sudden changes in population health and predict likely outcomes in the face of different anthropogenic stressors.

In this chapter, we discuss some of the emerging conditions that could represent a health risk for tropical and subtropical pinnipeds. We highlight the importance of taking into account the environmental and demographic context of each population and incorporating health, disease and immune response markers in surveys of pinniped populations. This strategy could help us address some of the current conservation challenges that tropical and subtropical pinnipeds face in the 21st century.

Emerging Conditions that are Potential Threats to Tropical and Subtropical Pinnipeds

As the global human population expands, spillover of pathogens from domestic animals and humans to wildlife has become a conservation concern (Smith et al. 2009). Pinnipeds are not exempt from this risk, particularly coastal species that are exposed to industrial and urban runoff, and there is justified concern that they could impact their populations. Risk of disease introduction has even been used as a criterion for changing a species conservation status. Specifically, the Galapagos sea lion was listed as endangered in 2008 owing to the risk of disease introduction (Trillmich 2015). This was done despite the lack of evidence of any specific pathogen causing widespread disease or mortality in the population. The rationale was that in spite of the species' geographical isolation, tourism is significant in the Galapagos Archipelago, and human settlements have increased in some of the islands that are inhabited by sea lions. This has favored their exposure to domesticated animals and vermin associated with human settlements, and, potentially, increased their risk of pathogen spillover (Kilpatrick et al. 2006, Alava et al. 2014, Denkinger et al. 2015).

Based on the few published studies, we will now discuss the most relevant infectious and non-infectious diseases already detected in tropical and subtropical pinnipeds, as well as those that have been identified in other species, but that are likely to cause problems if they were to emerge to their populations.

Morbilliviruses

It is often quite difficult to determine whether a newly detected pathogen will pose a threat to the health of a given population. Some of the pathogens identified as causal agents for emerging and resurging diseases in pinnipeds more than a decade ago (see Miller et al. 2001) have not been proven to cause significant outbreaks, and often their detection in cases of illness or strandings have not been enough to unequivocally determine their clinical significance (e.g., Coronavirus and Influenza B virus; see Nollens et al. 2010, Bodewes et al. 2013). A few emerging pathogens, however, such as canine distemper virus (CDV) and other morbilliviruses can infect a large percentage of a susceptible population and cause high mortality rates. These pathogens certainly raise justifiable concern about the threat they pose to pinnipeds (Di Guardo et al. 2005, McCarthy et al. 2007, Wilson et al. 2014).

CDV was first identified in pinnipeds during an outbreak that caused the deaths of thousands of Siberian seals (*Phoca siberica*) in Lake Baikal between 1987 and 1988 (Kennedy 1990). The epidemic presumably began by transmission of the virus from domestic or feral dogs (Lyons et al. 1993) or other carnivores (Miller et al. 2001). In addition to CDV, other species within the morbillivirus genus, such as Phocine Distemper virus (PDV), have since been detected in pinnipeds and are known to be able to cause systemic disease in immunologically naïve or susceptible populations (see Duignan et al. 2014 for a recent review and discussion of this virus).

We are far from understanding morbillivirus transmission from terrestrial carnivores to pinnipeds, and even less is known about intra-species variations in susceptibility. However, their potential for causing widespread mortality has already been documented for pinnipeds. A well-known case is that of the Mediterranean monk seal (*Monachus monachus*), that in 1997 lost more than half of the total population. Evidence of morbillivirus-associated pathology was confirmed in some of the animals examined (Osterhaus et al. 1997, 1998; but see Hernández et al. 1998), and was consistent with past outbreaks (Barrett et al. 1995). There have not been more reported outbreaks since, but a dolphin morbillivirus was isolated a year after the epidemic from a stranded Mediterranean monk seal that showed neurological signs of disease, suggesting that viral transmission from dolphins to seals is possible (van de Bildt et al. 2001), and the possibility of entering the Mediterranean monk seal population once again is not inconceivable.

There have been relatively few recent disease surveys for morbilliviruses in other wild populations of pinnipeds inhabiting tropical and subtropical latitudes, and as far as we know, there are no published reports of exposure or disease of CDV or PDV in several species, such as the Guadalupe fur seal (*Arctocephalus townsendi*), Galapagos fur seal (*Arctocephalus galapagoensis*), Juan Fernandez fur seal (*Arctocephalys philippii*), or African Cape fur seal (*Arctocephalus pusillus*). Surveys show low levels of infection by CDV and PDV were recently detected in wild South American sea lions, *Otaria byronia* (Sepúlveda et al. 2015), but not in South American fur seals, *Arctocephalus australis* (Jankowski et al. 2015), suggesting that pinnipeds with overlapping distributions do not necessarily share the same risk of infection, despite the potential for transmission between closely related species. These results must, however, be considered with caution, as sample sizes reported are typically small, with limited power to detect seropositive animals.

Close contact with domestic dogs might not always derive in exposure. A good example of this are the suspected outbreaks of CDV in dogs from various inhabited islands of the Galapagos Archipelago (Alava and Salazar 2011), that did not lead to any observable CDV-associated disease or mortality of Galapagos sea lions, even at locations where dogs and sea lions have close contact rates. Serological surveys failed to detect antibodies against CDV in pup and adult sea lions sampled at those colonies in 2004, 2006 and 2007 (K.A.-W., unpublished data), suggesting that the dog epidemic might have not been CDV, or that despite the epidemic, sea lions were not exposed. More recent data suggests that Galapagos sea lions might have indeed been exposure to CDV, as antigens have been detected in the blood of dead Galapagos sea lion pups and apparently-healthy juveniles in San Cristobal Island in 2012, and PCR-based assays have detected CDV RNA in various tissues collected from dead pups (J.D., unpublished data). Despite these results, the lack of seroconversion in sea lions sampled at different islands in 2014 suggests that the CDV infection detected in 2012 was likely confined to few animals at that specific location (J.D., unpublished data), and that the general population is still naïve. From an epidemiological point of view, this implies a high risk of a future CDV outbreak for the population. The endangered Hawaiian monk seal (*Neomonachus schauinslandi*) also appears to be immunologically naïve to CDV and PDV (Aguirre et al. 2007), despite there being chances of encounter with domestic dogs at the Hawaiian main islands.

Regardless of the limited data available, considering the impact that morbilliviruses can have on susceptible pinnipeds, populations that have contact with domestic or feral dogs (e.g., Alava and Salazar 2011, García-Aguilar and Gallo-Reynoso 2012, Alava et al. 2014), or other terrestrial canids or procyonids, should be considered at risk; particularly those that appear to have no history of exposure to the virus (i.e., recently introduced dogs near their colonies). Both otariid species from the Galapagos Archipelago, the Hawaiian and Mediterranean monk seals, the South American sea lion and the Cape fur seal (*Arctocephalus pusillus*; see Lavigne and Schmitz 1990) fit that bill, and should be periodically monitored for neutralizing antibodies as a strategy for early detection of a potential epidemic that could cause high mortality of their populations.

Adenovirus, herpesvirus and other emerging viruses

As high-throughput sequencing and metagenomic approaches become more common, more viruses are being detected in pinnipeds. Their potential for causing disease remains unknown and, at least until now, there is little or no evidence that these newly-identified viruses cause significant effects on pinniped health. However, as diseased animals undergoing rehabilitation showed the highest diversity and intensity of eukaryotic virus (Li et al. 2011), it is possible that pinnipeds with weakened immune systems (see section below) or those experiencing severe disease could be less tolerant of enteric viruses. If this were proved to be true, it would suggest that even viruses that normally do not pose a risk could lead to disease under certain conditions.

Otarine Adenovirus type 1 (OtAdV-1; Goldstein et al. 2011) is commonly detected in stranded California sea lions, harbor seals and elephant seals (Wright et al. 2015). The virus, believed to have jumped from an unidentified host to otariids (Cortés-Hinojosa et al. 2015), appears to establish a "peaceful coexistence" that rarely derives in disease (see Virgin et al. 2009). In line with this hypothesis, OtAdV-1 has recently been detected in the enteric epithelium of more than 50% of pups sampled at different California sea lion breeding colonies along the Mexican coastline, but no evidence of associated disease was found (L.A. S.-G., unpublished data). If we generalize these findings, it would appear that OtAdV-1 does not pose a significant risk to the health of wild pinnipeds. However, recent reports of fatal acute hepatitis in three captive otariids (one California sea lion, one South American sea lion, and one Cape fur seal) that were caused by OtAdV-1 strain MJ12 (Inoshima et al. 2013) force us to consider this virus with caution, as it is possible that external factors can modify the normally host-virus relationship. Until this is understood, we suggest that long-term monitoring disease surveys of tropical and subtropical pinnipeds include this virus and search for changes in prevalence and intensity.

Pinniped herpesviruses are also becoming popular in the literature. Since their isolation in harbor seal pups three decades ago (Osterhaus et al. 1985), more than 15 herpesviruses have now been detected in otariids and phocids, and even in walrus (e.g., Dagleish et al. 2013, Melero et al. 2014, Jankowski et al. 2015, Wright et al. 2015, Bellehumeur et al. 2016). Judging by the high prevalence typically reported in for wild or stranded individuals, herpesviruses appear to be widespread in pinniped populations. The potential of some herpesvirus types to cause disease is known (e.g., Lipscomb et al. 2000, Martina et al. 2002, Himworth et al. 2010, Venn-Watson et al. 2012), or inferred (e.g., King et al. 2002, Buckles et al. 2006, Dagleish et al. 2013), but a search of the literature suggests that not all herpesviruses establish antagonistic relationships with their pinniped hosts. As discussed for adenoviruses, the fact that some herpesviruses are linked to pinniped diseases of population relevance, pathogen surveys of pinnipeds from tropical and subtropical latitudes would benefit from including herpesvirus as a virus of interest.

Toxoplasmosis

The apicomplexan parasite *Toxoplasma gondii* infects humans and endotherms in most parts of the world, including marine mammals. It is thought that marine mammals become infected by exposure to oocyst-contaminated run-off (Miller et al. 2002, Gaydos et al. 2007, Colegrove et al. 2011, VanWormer et al. 2014). This has been evidenced by cases of meningoencephalitis and myocarditis in species that reside near to inhabited coastal areas (Kreuder et al. 2003, Carlson-Bremer et al. 2015).

Visceral and cerebral toxoplasmosis was reported as the cause of death of one wild adult male Hawaiian monk seal (Honnold et al. 2005), but neutralizing antibodies against *T. gondii* appear to be uncommon in Hawaiian monk seals and, even when they are detected, titers are typically low (Aguirre et al. 2007; see Chapter 3). Relatively high antibody titers against *T. gondii* were also reported in a single adult male South American sea lion from Valdivia, Chile, possibly due to exposure to cat feces from a nearby river (Sepúlveda et al. 2015). Wild South American fur seals did not show evidence of exposure (Jankowski et al. 2015), and we do not know of studies conducted in other tropical and subtropical pinnipeds. Oocyst-like structures similar to *T. gondii* were detected in water samples from four islands of the Galapagos archipelago (Verant et al. 2013). Furthermore, antibodies against *T. gondii* were common in the cat population of Isabella, an inhabited island (Levy et al. 2008), and in Galapagos penguins and flightless cormorants from Isabella and Fernandina Islands (Deem et al. 2010). To our knowledge, evidence of exposure to *T. gondii* has not yet been assessed for either of the Galapagos pinnipeds, but as they inhabit these same islands, the potential risk of toxoplasmosis should not be ignored for these species.

In the absence of clinical signs or high antibody prevalence, it is reasonable to infer that *T. gondii* does not currently represent a health risk for the abovementioned species, be this by lack of exposure or by immune tolerance to the parasite (see Oksanen et al. 2014). However, host-pathogen relationships are rarely, if ever, static and present-day environmental changes could modify this situation. Diet composition and habitat use of marine mammals influence exposure to *T. gondii* (Conrad et al. 2005, Johnson et al. 2009) and shape risk of infection (see Johnson et al. 2009). Thus, environmentally-driven changes in the diet of pinnipeds could have consequences for their populations. If *T. gondii* were already present at a low level, prevalence could escalate, and depleted resources due to lesser availability of prey could impact immune responses, increasing the likelihood that toxoplasmosis emerges in the population.

Alopecia syndrome

Hair loss, termed alopecia, has been reported in various pinniped species. The cause of alopecia syndrome mostly remains unknown (Atwood et al. 2015), but proposed explanations are altered thyroid hormones caused by exposure to lipophilic persistent organic pollutants (POPs) (see Routti et al. 2010), as high concentrations of polychlorinated biphenyls (PCBs) have been found in animals with alopecia (e.g., Bergman and Olsson 1985, Beckmen et al. 1997). Fungi, viruses, bacteria and ectoparasites that could damage the hair shaft or root, have been suggested as causal agents in some (see Guillot et al. 1998, Dailey 2001, Dunn et al. 2001), but not all (see Lynch et al. 2012) cases, and nutritional deficiencies have also been invoked as potential causes (Trites and Donnelly 2003, Lynch et al. 2012).

Except for South American sea lions, alopecia has mostly been reported for species from higher latitudes (e.g., Bergman and Olsson 1985, Nettleton et al. 1995, Beckmen et al. 1997, Lynch et al. 2011, Pistorius and Baylis 2011, Lynch et al. 2012). There is some evidence that prevalence is increasing (Lynch et al. 2011). Individuals with alopecia syndrome have lower condition (Lynch et al. 2011), which might result in compromised health, and lower survival. There is evidence that alopecic patches have higher heat loss than areas of normal fur (Lynch et al. 2011). Energetic demand for thermoregulation is consequently higher and if this demand exceeds an individual's caloric intake, body condition will expectedly decrease (see Rutishauser et al. 2004, Rosen et al. 2007) while physiological stress increases (Acevedo-Whitehouse and Duffus 2009), consequently lowering survival probability.

In the context of environmental change it is possible that alopecia syndrome will emerge and become common in tropical and subtropical pinnipeds. Patchy loss of guard hair with a symmetrical bilateral distribution in the neck and rump has recently been observed in a number of Guadalupe fur seals in colonies from the Mexican Pacific (F. Elorriaga-Verplanken, personal communication). We do not yet know whether this condition is equal to the one described for Australian fur seals (see Lynch et al. 2012). However, guard hair is essential to protect the insulating layer of packed underfur of fur seals (Pabst et al. 1999), and its loss exposes the underfur to the environment and decreases the thermoregulatory ability of the skin and fur (see Boyd 2000). If alopecia syndrome is indeed related to micronutrient depletion, changes in prey distribution and shifts in ocean temperature could be extremely important in terms of its emergence for tropical and subtropical pinnipeds with specialist feeding habits, whose prey can decrease significantly with thermal shifts of the sea surface temperature (Trillmich and Limberger 1985). In the past years we have witnessed various oceanographic and climatic anomalies that lead to high sea surface temperatures (see Bond et al. 2014), and cyclical events such as El Niño-Southern Oscillation are projected to increase (Liu 2010). These events can deplete nutrient availability from the bottom up, increasing mortality and lowering reproductive success of pinnipeds (Trillmich and Limberger 1985). Future health surveys could easily include monitoring for evidence of patchy hair loss of fur seals from tropical and subtropical regions.

Ciguatera and other harmful algal blooms

Worldwide, coastal waters periodically experience harmful algal blooms (HABs), caused by algal toxins that are produced by unicellular algae when they proliferate in response to favorable environmental conditions (i.e., nutrient loads from continental sources and eutrophication). The toxins are normally present in low concentration and do not appear to impacts the ecosystem, but when the algal population increases and aggregates, toxin concentration increases (recently reviewed in Tan and Ransangan 2015). At high concentrations, the produced toxins can impact negatively on health and reproductive success of pinnipeds and other marine mammals when they are exposed through the food web (i.e., dietary intake) or by direct respiratory contact (i.e., inhalation). Biotransformation and bioaccumulation of biotoxins in high trophic level marine organisms make them susceptible to their detrimental effects (see Negri et al. 2007).

Four classes of HABs, namely saxitoxins, brevetoxins, domoic acid and ciguatoxins (also known as ciguatera), have been implied in events of mortality and widespread illness of a number of marine mammals. For pinnipeds, one of the most common HAB is linked to domoic acid, which has caused the intoxication and stranding of numerous California sea lions in the past decades, with cases of acute neurotoxicity and chronic illness following blooms of *Pseudo-nitzschia australis* (see Scholin et al. 2000, de la Riva et al. 2009, Goldstein et al. 2009).

In tropical latitudes around the globe, ciguatera is the most common algal toxin (Tester, 1994). To date, the only published report of ciguatera-associated mortality of Hawaiian monk seals (or indeed, any pinniped) was a mass die-off of seals at Laysan Island, in Northwestern Hawaii (Johnson and Johnson 1978). Although there is evidence that Hawaiian monk seals are exposed to toxic levels of the polyether ciguatoxins (Wilson and Jokiel 1986, Bottein et al. 2011), there have been no more reports of associated mortality that we are aware of (see Chapter 3). Saxitoxin is another marine algal toxin that can have negative effects in pinnipeds and other vertebrates. Known in humans as Paralytic Shellfish Poisoning (reviewed in James et al. 2010), there is some evidence that this water-soluble toxin was involved in the mass mortality of Mediterranean monk seals observed in 1997 (Hernández et al. 1998), and that has been attributed to morbillivirus (see section above). During the die-off, species of toxic dinoflagellates were found in water samples collected near monk seal colonies off the coast of the Western Sahara, and various tissues (brain, liver, kidney and muscle) contained saxitoxin at biologically relevant-levels (but see Osterhaus et al. 1998). We do not intend to rekindle a debate on the cause of the Mediterranean Monk seal mortality. Rather, we believe that it would be unwise to disregard HABs as a cause for morbidity and mortality of pinnipeds that feed in areas where saxitoxin blooms can occur.

Toxic algae are found in most parts of the world, and HABs are now common occurrences in different regions encompassing areas where some pinniped species are distributed. Risk of exposure is linked to increased eutrophication associated with agricultural run-off from coastal areas, increased marine transport and aquaculture activities (James et al. 2010) and overfishing (Jackson et al. 2001). For ciguatera, reef disturbance caused by human activities or storm damage (Kaly and Jones 1994) as well as coral bleaching (Chateau-Degat et al. 2007) often precedes blooms of ciguatera-producing algae. As yet, there is extremely limited data on genetic (and possibly, immune) factors that can influence resistance to algal toxins in pinnipeds (e.g., Acevedo-Whitehouse et al. 2003), and it is likely that complex interactions between stressors (such as pollutants, thermal stress and nutrient depletion due to climate change, and concurrent immune challenges) might increase susceptibility to HABs. This, together with the increased frequency of HABs registered over the past three decades (Smayda 1990, Hallegraeff 1993, Van Dolah 2000, Chateau-Degat et al. 2007), make it reasonable to believe that events of HAB associated-mortality or morbidity will be observed in tropical and subtropical pinniped populations.

Pinniped Immune Responses to Pathogens Under Environmental Stress

A healthy immune system is essential to mount proper responses to pathogens and emergent conditions. The pinniped immune system can be hampered by bioaccumulation and toxicity of POPs (i.e., immunotoxicity), including PCBs, polybrominated diphenyl ethers, (PBDEs) and polycyclic aromatic hydrocarbons (PAHs) (e.g., De Swart et al. 1996, Nakata et al. 2002, Hammond et al. 2005, Levin et al. 2005, Frouin et al. 2010, Dupont et al. 2013). Exposure to different POPs and heavy metals correlates with a range of systemic dysfunctions and reproductive failure (reviewed by O'Hara and O'Shea 2001), differences in transcription profiles for immune and metabolism-related genes (Brown et al. 2014), decreased thyroid activity (Sørmo et al. 2005, Tabuchi et al. 2006), and a wide range of effects pertaining to the immune system (recently reviewed by Desforges et al. 2016), including shifts in white blood cell populations, immunosuppression (but see Barron et al. 2003), decreased synthesis of immunoglobulin G (IgG) in response to a challenge, and raised haptoglobin levels (Zenteno-Savin et al. 1997). To our knowledge, there have been no studies on the effect that contaminants can exert on parasite susceptibility, but studies on harbor porpoises have shown increased parasite loads associated with high contaminant levels (Jepson et al. 2005), and it is likely that pinnipeds experience similar effects.

Expansion of human populations and increased agro-industrial activities (as is currently occurring along the Peruvian coast; see Jankowski et al. 2015) that result in runoff with high levels of persistent pollutants near pinniped habitats could impact the immune competence of their populations, as well as increase pathogen exposure from human, domestic, and wild animal sources. There is already some evidence of anthropogenic differences in pinniped immune activity. For instance, IgG concentration was found to be nearly $10 \text{ g} \cdot \text{L}^{-1}$ higher in Galapagos sea lions from a colony located in the midst of a rapidly growing town (Puerto Moreno Baquerizo, in San Cristobal Island), than in sea lions from a colony (Santa Fe) with no urban settlement and minimum human activity (Brock et al. 2013b). Serum albumin, inflammation-associated alpha globulins and swelling in response to phytohemagglutinin were also higher in the 'human-impacted colony' (Brock et al. 2013b). Sea lions did not show any obvious signs of disease in either colony (see Brock et al. 2013a), and there were no differences in the prevalence of selected pathogens (*Brucella* sp., CDV and leptospira; K.A.-W., unpublished data) between colonies, although sea lions from Puerto Moreno Baquerizo are exposed to sewage contaminated with fecal coliform bacteria (Cordoba et al. 2008, Alava et al. 2014). It is also possible that the sea lions at this location commonly experience stressful disturbances (e.g., harassment by dogs and noise from industrial machinery, car horns or fireworks) that contribute to over-reactive or pro-inflammatory states (see Pascuan et al. 2014, Denkinger et al. 2015, Recio et al. 2016). This is unfortunate; as a compromised immune system will affect an individual's ability of counteract disease. Hampered immune status is especially relevant during periods of nutritional stress and density-independent stressors in highly variable or unpredictable environments (e.g., El Niño-Southern Oscillation events) when mass mortality occurs and endangered populations frequently approach extinction thresholds (Alava et al. 2014). Exposure to immunotoxic contaminants may have significant consequences for pinniped populations as a contributing factor to increasing levels of anthropogenic stress, and can facilitate the emergence of infectious disease outbreaks (Ross 2002, Desforges et al. 2016).

Most of the pinnipeds that inhabit in tropical and subtropical latitudes are polygamous species, have small populations, marked site fidelity, and low overall genetic diversity, where high levels of homozygosity —and expression of deleterious recessive alleles—may occur due to consanguineous mating. Such genetic status is also relevant to disease risk, as limited variability and low levels of heterozygosity can decrease pinniped immune function and disease susceptibility (e.g., Acevedo-Whitehouse et al. 2003b, 2006, Rijks et al. 2008, de Assunção-Franco et al. 2012, Brock et al. 2015). Thus, populations with limited genetic diversity and those where inbreeding is known or expected, that reside in areas that are close to human settlements or agro-industrial activities, and that have high risk of contact with domestic or feral dogs, or other domesticated animals, should be given high priority in terms of systematic disease surveillance (see Fig. 1 for map of current and probable disease risk).

Final Thoughts about Assessing Pinniped Disease Risks

In this chapter, we have tried to highlight the most relevant diseases and health conditions that are known to affect pinnipeds and that could be relevant for those that inhabit tropical and subtropical latitudes. Our assessment has been hindered by limited (or lack of) reports on common microorganisms and their pathogenic potential for most of the pinnipeds covered here. We have also attempted to identify those populations that could be at risk of pathogen spillover and disease emergence. Nevertheless, the importance of specific pathogens or conditions for a given pinniped population is extremely difficult to predict under the current scenario of rapid environmental change. It is entirely possible that other, as yet unknown, conditions emerge. This means that we are in urgent need of tools that help us determine a population's response potential.

It is evident from the literature aforementioned in this chapter that most studies focus on pathogen detection and serological evidence of exposure

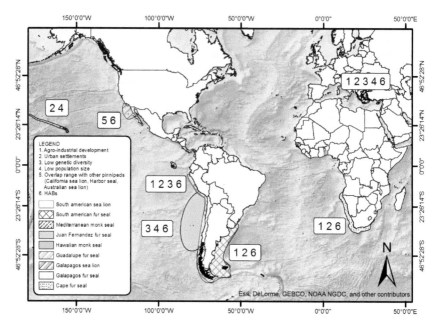

Figure 1. Map of potential disease risk for pinnipeds distributed in tropical and subtropical latitudes. Distributions for each species are indicated. Risk factors considered were agro-industrial development, proximity to human settlements, limited genetic diversity, low population size, overlapping range with other pinnipeds that are known to be hosts to various infectious diseases, and breeding or feeding ranges within zones where harmful algal blooms (HAB) are common.

to a pathogen. However, conceiving disease without assessing health biases our understanding of a population's status and risk. In other words, infection is not synonymous to disease, and the potential impacts of one and the other vary enormously. However, there are various indicators of health, disease and immune activity that can be, and are, used for pinnipeds. From basic hematology and serum chemistry (e.g., Bossart et al. 2001, Keogh et al. 2010, Páez-Rosas et al. 2016) to *ex vivo* or *in vitro* immune challenges (Hall et al. 1999, Drago et al. 2011, Brock et al. 2013a, Vera-Massieu et al. 2015); from quantitation of specific immune effectors to gene expression, immune transcriptomics and proteomics (e.g., Bowen et al. 2005, Mancia et al. 2012, McCarthy et al. 2011, Lehnert et al. 2014, Browning et al. 2015, Khudyakov et al. 2015), the literature is becoming enriched with tools that could be implemented in studies on disease of pinnipeds that inhabit areas of high pathogen risk, particularly those are near regions undergoing human development.

It is likely that modern "omics" approaches coupled with field assays and traditional measures of health and pathogen exposure will deepen our understanding of pinniped health. Evidently, protocols developed for one pinniped species may not necessarily be adequate for other pinnipeds, even those that are closely-related, and species-specific standardization of techniques will be essential; but, with some effort, this approach might help us to determine the immune response potential of a given species or population in its particular environmental context. If we chose to embark on this path, it is likely that our knowledge of disease risks and health threats to tropical and subtropical pinnipeds will greatly expand. Particularly, we may be able to understand why and how established or introduced pathogens can cause disease, and identify which populations are likely to be affected by specific disease conditions in terms of their viability and persistence. Given the rate of environmental change pinniped populations are currently facing, this information is urgent.

Acknowledgements

The authors thank Frances Gulland for reviewing the chapter and for her suggestions that greatly enriched the manuscript. Our sincere apologies to colleagues in the field whose work we were not able to mention because of space restrictions.

Keywords: Disease, emerging disease, health, immunity, pathogens, pinnipeds

References

Acevedo-Whitehouse, K., H. de la Cueva, F.M. Gulland, D. Aurioles-Gamboa, F. Arellano-Carbajal and F. Suarez-Güemes. 2003a. Evidence of *Leptospira interrogans* infection in California sea lion pups from the Gulf of California. J. Wildl. Dis. 39: 145–151.

Acevedo-Whitehouse, K., F.M. Gulland, D. Greig and W. Amos. 2003b. Inbreeding: Disease susceptibility in California sea lions. Nature 422: 435.

Acevedo-Whitehouse, K., T.R. Spraker, E. Lyons, S.R. Melin, F.M. Gulland, R.L. Delong et al. 2006. Contrasting effects of heterozygosity on survival and hookworm resistance in California sea lion pups. Mol. Ecol. 15: 1973–1982.

Acevedo-Whitehouse, K. and A.L. Duffus. 2009. Effects of environmental change on wildlife health. Phil. Trans. R. Soc. B 364: 3429–3438.

Aguirre, A.A., T.J. Keefe, J.S. Reif, L. Kashinsky, P.K. Yochem, J.T. Saliki et al. 2007. Infectious disease monitoring of the endangered Hawaiian monk seal. J. Wildl. Dis. 43: 229–241.

Alava, J.J. and S. Salazar. 2011. Status and conservation of Otariids in Ecuador and the Galapagos Islands. pp. 495–520. *In*: A.W. Trites, S.K. Atkinson, D.P. De Master, L.W. Fritz, T.S. Gelatt, L.D. Rea and K.M. Wynne [eds.]. Sea Lions of the World. Alaska Sea Grant College Program: Fairbanks, Alaska, USA.

Alava, J.J., C. Palomera, L. Bendell and P.S. Ross. 2014. Pollution as a threat for the conservation of the galapagos marine reserve: Environmental impacts and management perspectives. pp. 247–283. *In*: J. Denkinger and L. Vinueza [eds.]. The Galapagos Marine Reserve: A Dynamic Socio-ecological System. Springer Science+Business Media, New York.

Avalos-Téllez, R., E.M. Carrillo-Casas, D. Atilano-López, C.R. Godínez-Reyes, E. Díaz-Aparicio, D. Ramírez-Delgado et al. 2016. Pathogenic Leptospira serovars in free-Living sea lions in The Gulf of California and along the Baja California coast of Mexico. J. Wildl. Dis. 52: 199–208.

Atwood, T., E. Peacock, K. Burek-Huntington, V. Shearn-Bochsler, B. Bodenstein, K. Beckmen et al. 2015. Prevalence and spatio-temporal variation of an alopecia syndrome in polar bears (*Ursus maritimus*) of the southern Beaufort Sea. J. Wildl. Dis. 51: 48–59.

Banish, L.D. and W.G. Gilmartin. 1988. Hematology and serum chemistry of the young Hawaiian monk seal (*Monachus schauinslandi*). J. Wildl. Dis. 24: 225–230.

Barrett, T., M. Blixenkrone-Møller, G. Di Guardo, M. Domingo, P. Duignan, A. Hall et al. 1995. Morbilliviruses in aquatic mammals: Report on round table discussion. Vet. Microbiol. 44: 261–265.

Barron, M.G., R. Heintz and M.M. Krahn. 2003. Contaminant exposure and effects in pinnipeds: Implications for Steller sea lion declines in Alaska. Sci. Total. Environ. 311: 111–133.

Bastida, R., J. Loureiro, V. Quse, A. Bernardelli, D. Rodríguez and E. Costa. 1999. Tuberculosis in a wild subantarctic fur seal from Argentina. J. Wildl. Dis. 35: 796–798.

Beckmen, K.B., L.J. Lowenstine, J. Newman, J. Hill, K. Hanni and J. Gerber. 1997. Clinical and pathological characterization of northern elephant seal skin disease. J. Wildl. Dis. 33: 438–449.

Bellehumeur, C., O. Nielsen, L. Measures, L. Harwood, T. Goldstein, B. Boyle et al. 2016. Herpesviruses including novel gamma herpesviruses are widespread among phocid seal species in Canada. J. Wildl. Dis. 52: 70–81.

Bergman, A. and M. Olsson. 1985. Pathology of Baltic grey seal and ringed seal females with special reference to adrenocortical hyperplasia: Is environmental pollution the cause of widely distributed disease syndrome? Finn. Game Res. 44: 47–62.

Bernardelli, A., R. Bastida, J. Loureiro, H. Michelis, M.I. Romano, A.C. Ataldi et al. 1996. Tuberculosis in sea lions and fur seals from the south-western Atlantic coast. Revue. Scien. Tech. Off. Intern. Epizoot. 15: 985–1005.

Blanc, A., D. Ruchansky, M. Clara, F. Achaval, A. Le Bas and J. Arbiza. 2009. Serologic evidence of influenza A and B viruses in South American fur seals (*Arctocephalus australis*). J. Wildl. Dis. 45: 519–521.

Bodewes, R., D. Morick, G. de Mutsert, N. Osinga, T. Bestebroer, S. van der Vliet et al. 2013. Recurring influenza B virus infections in seals. Emerg. Infect. Dis. 19: 511–512.

Bond, N.A., M.F. Cronin, H. Freeland and N. Mantua. 2014. Causes and impacts of the 2014 warm anomaly in the NE Pacific. Geophys. Res. Lett. 42: 3414–3420.

Bossart, G.B., T.H. Reidarson, L.A. Dierauf and D.A. Duffield. 2001. Clinical pathology of marine mammals. pp. 383–430. *In*: L.A. Dierauf and M.D. Gulland [eds.]. Marine Mammal Medicine. CRC Press. Boca Raton, FL, USA.

Bottein, M.-Y.D., L. Kashinsky, Z. Wang, C. Littnan and J.S. Ramsdell. 2011. Identification of ciguatoxins in Hawaiian monk seals *Monachus schauinslandi* from the Northwestern and main Hawaiian islands. Environ. Sci. Technol. 45: 5403–5409.

Bowen, L., B.M. Aldridge, R. Delong, S. Melin, E.L. Buckles, F. Gulland et al. 2005. An immunogenetic basis for the high prevalence of urogenital cancer in a free-ranging population of California sea lions (*Zalophus californianus*). Immunogen. 56: 846–848.

Boyd, I.L. 2000. Skin temperatures during free-ranging swimming and diving in Antarctic fur seals. J. Exp. Biol. 203: 1907–1914.

Brock, P., A.J. Hall, S. Goodman, M. Cruz and K. Acevedo-Whitehouse. 2013a. Applying the tools of ecological immunology to conservation: A test case in the Galapagos sea lion. Anim. Conser. 16: 19–31.

Brock, P.M., A.J. Hall, S.J. Goodman, M. Cruz and K. Acevedo-Whitehouse. 2013b. Immune activity, body condition and human-associated environmental impacts in a wild marine mammal. PLoS ONE 8(6). e67132.

Brock, P.M., S.J. Goodman, A.J. Hall, M. Cruz and K. Acevedo-Whitehouse. 2015. Context-dependent associations between heterozygosity and immune variation in a wild carnivore. BMC Evol. Biol. 15: 242.
Brown, T.M., P.S. Ross, K.J. Reimer, N. Veldhoen, N.J. Dangerfield, A.T. Fisk et al. 2014. PCB related effects thresholds as derived through gene transcript profiles in locally contaminated ringed seals (*Pusa hispida*). Environ. Sci. Technol. 48: 12952–12961.
Browning, H.M., F.M. Gulland, J.A. Hammond, K.M. Colegrove and A.J. Hall. 2015. Common cancer in a wild animal: The California sea lion (*Zalophus californianus*) as an emerging model for carcinogenesis. Philos. Trans. R. Soc. B 370: 20140228.
Buckles, E.L., L.J. Lowenstine, C. Funke, R.K. Vittore, H.N. Wong, J.A. St Leger et al. 2006. Otarine Herpesvirus-1, not papillomavirus, is associated with endemic tumours in California sea lions (*Zalophus californianus*). J. Comp. Pathol. 135: 183–189.
Carlson-Bremer, D., K.M. Colegrove, F.M. Gulland, P.A. Conrad, J.A. Mazet and C.K. Johnson. 2015. Epidemiology and pathology of *Toxoplasma gondii* in free-ranging California sea lions (*Zalophus californianus*). J. Wildl. Dis. 51: 362–373.
Castinel, A., P.J. Duignan, E.T. Lyons, W.E. Pomroy, N. Gibbs and N. López-Villalobos. 2007. Epidemiology of hookworm (*Uncinaria* spp.) infection in New Zealand (Hooker's) sea lion (*Phocarctos hookeri*) pups on Enderby Island, Auckland Islands (New Zealand) during the breeding seasons from 1999/2000 to 2004/2005. Parasitol. Res. 101: 53–62.
Cattan, P.E., B.B. Babero and D. Torres. 1976. The helminth fauna of Chile: IV. Nematodes of the genera *Anisakis* Dujardin, 1845 and *Phocanema* Myers, 1954 in relation with gastric ulcers in a South American Sea Lion, *Otaria byronia*. J. Wildl. Dis. 12: 511–515.
Cattan, P.E., J.L. Yáñez and D. Torres. 1980. Helminth parasites of the Juan Fernández fur seal *Arctocephalus philippii* (Peters, 1866). Bol. Chil. Parasitol. 35: 73–75.
Chateau-Degat, M.L., E. Dewailly, N. Cerf, N.L. Nguyen, M.O. Huin-Blondey and B. Hubert. 2007. Temporal trends and epidemiological aspects of ciguatera in French Polynesia: A 10-year analysis. Trop. Med. Int. Health. 12: 485–492.
Colagross-Schouten, A.M., J.A. Mazet, F.M. Gulland, M.A. Miller and S. Hietala. 2002. Diagnosis and seroprevalence of leptospirosis in California sea lions from coastal California. J. Wildl. Dis. 38: 7–17.
Colegrove, K.M., M.E. Grigg, D. Carlson-Bremer, R.H. Miller, F.M. Gulland and D.J. Ferguson. 2011. Discovery of three novel coccidian parasites infecting California sea lions (*Zalophus californianus*), with evidence of sexual replication and interspecies pathogenicity. J. Parasitol. 97: 868–877.
Conrad, P.A., M.A. Miller, C. Kreuder, E.R. James, J. Mazet and H. Dabritz. 2005. Transmission of Toxoplasma: Clues from the study of sea otters as sentinels of *Toxoplasma gondii* flow into the marine environment. Int. J. Parasitol. 35: 1155–1168.
Cordoba, D.R., J.L. Medina and Y. Nagahama. 2008. Water Quality Monitoring on the Island of San Cristobal. Puerto Ayora, Ecuador: Galapagos National Park.
Coria-Galindo, E., E. Rangel-Huerta, A. Verdugo-Rodríguez, D. Brousset, S. Salazar and L. Padilla-Noriega. 2009. Rotavirus infections in Galapagos sea lions. J. Wildl. Dis. 45: 722–728.
Cortés-Hinojosa, G., F.M. Gulland, T. Goldstein, S. Venn-Watson, R. Rivera and T. Waltzek. 2015. Phylogenomic characterization of California sea lion adenovirus-1. Infect. Genet. Evol. 31: 270–276.
Dagleish, M.P., M. Barrows, M. Maley, R. Killick, J. Finlayson and R. Goodchild. 2013. The first report of otarine herpesvirus-1-associated urogenital carcinoma in a South American fur seal (*Arctocephalus australis*). J. Comp. Pathol. 149(1): 119–125.
Dailey, M.D., R.V. Santangelo and W.G. Gilmartin. 1988. A coprological survey of helminth parasites of the Hawaiian monk seal from the Northwestern Hawaiian Islands. Mar. Mamm. Sci. 4: 125–131.
Dailey, M.D. 2001. Parasitic diseases. pp. 383–430. *In*: L.A. Dierauf and M.D. Gulland [eds.]. Marine Mammal Medicine. CRC Press, Boca Raton, FL, USA.

Dailey, M., R. Ellin and A. Parás. 2005. First report of parasites from pinnipeds in the Galapagos Islands, Ecuador, with a description of a new species of *Philophthalmus* (Digenea: Philophthalmidae). J. Parasitol. 91: 614–617.
de Amorim, D.B., R.A. Casagrande, M.M. Alievi, F. Wouters, L.G. De Oliveira and D. Driemeier. 2014. *Mycobacterium pinnipedii* in a stranded South American sea lion (*Otaria byronia*) in Brazil. J. Wildl. Dis. 50: 419–422.
de Assunção-Franco, M., J.I. Hoffman, J. Harwood and W. Amos. 2012. MHC genotype and near-deterministic mortality in grey seals. Sci. Rep. 2: 659.
De Graaf, A.S., P.D. Shaughnessy, R.M. McCully and A. Verster. 1980. Occurrence of *Taenia solium* in a Cape fur seal (*Arctocephalus pusillus*). Onderstepoort J. Vet. Res. 47: 119–120.
de la Riva, G.T., C.K. Johnson, F.M. Gulland, G.W. Langlois, J.E. Heyning, T.K. Rowles et al. 2009. Association of an unusual marine mammal mortality event with *Pseudo-nitzschia* spp. Blooms along the southern California coastline. J. Wildl. Dis. 4: 109–121.
De Swart, R.L., P.S. Ross, J.G. Vos and A.D. Osterhaus. 1996. Impaired immunity in harbour seals (*Phoca vitulina*) fed environmentally contaminated herring. Vet. Q. 18 Suppl. 3: 127–128.
Deem, S.L., J. Merkel, L. Ballweber, F.H. Vargas, M.B. Cruz and P.G. Parker. 2010. Exposure to *Toxoplasma gondii* in Galapagos Penguins (*Spheniscus mendiculus*) and flightless cormorants (*Phalacrocorax harrisi*) in the Galapagos Islands, Ecuador. J. Wildl. Dis. 46: 1005–1011.
Denkinger, J., L. Gordillo, I. Montero-Serra, J.C. Murillo, N. Guevara, M. Hirschfeld et al. 2015. Urban life of Galápagos sea lions (*Zalophus wollebaeki*) on San Cristobal Island, Ecuador: Colony trends and threats. J. Sea Research 105: 10–14.
Desforges, J.P., C. Sonne, M. Levin, U. Siebert, S. De Guise and R. Dietz. 2016. Immunotoxic effects of environmental pollutants in marine mammals. Environ. Int. 86: 126–139.
Di Guardo, G., G. Marruchella, U. Agrimi and S. Kennedy. 2005. Morbillivirus infections in aquatic mammals: A brief overview. J. Vet. Med. A. Physiol. Pathol. Clin. Med. 52: 88–93.
Drago, M., L. Cardona, N. García, S. Ameghino and A. Aguilar. 2011. Influence of colony size on pup fitness and survival in South American sea lions. Mar. Mamm. Sci. 27: 167–181.
Duignan, P.J., M.F. Van Bressem, J.D. Baker, M. Barbieri, K.M. Colegrove, S. De Guise et al. 2014. Phocine distemper virus: Current knowledge and future directions. Viruses 6: 5093–5134.
Dunn, J.L., J.D. Buck and T.R. Robeck. 2001. Bacterial diseases of cetaceans and pinnipeds. pp. 309–335. In: L.A. Dierauf and M.D. Gulland [eds.]. Marine Mammal Medicine. CRC Press, Boca Raton, FL, USA.
Dupont, A., U. Siebert, A. Covaci, L. Weijs, G. Eppe, C. Debier et al. 2013. Relationships between *in vitro* lymphoproliferative responses and levels of contaminants in blood of free-ranging adult harbour seals (*Phoca vitulina*) from the North Sea. Aquat. Toxicol. 142-143: 210–220.
Frouin, H., M. Fortier and M. Fournier. 2010. Toxic effects of various pollutants in 11B7501 lymphoma B cell line from harbour seal (*Phoca vitulina*). Toxicol. 270(2-3): 66–76.
García-Aguilar, M.C. and J.P. Gallo-Reynoso. 2012. Perros ferales en la isla de Cedros, Baja California, México: una posible amenaza para los pinnípedos. Revista Mexicana de Biodiversidad 83: 785–789.
Gaydos, J.K., P.A. Conrad, K.V. Gilardi, G.M. Blundell and M. Ben-David. 2007. Does human proximity affect antibody prevalence in marine-foraging river otters (*Lontra canadensis*)? J. Wildl. Dis. 43: 116–123.
George-Nascimento, M. and A. Llanos. 1995. Micro-evolutionary implications of allozymic and morphometric variations in seal worms *Pseudoterranova* sp. (Ascaridoidea: Anisakidae) among sympatric hosts from the southeastern Pacific Ocean. Int. J. Parasitol. 25: 1163–1171.
George-Nascimento, M. and J. Carvajal. 1981. Helminth parasites of the South American sea lion *Otaria flavescens* from the Gulf of Arauco, Chile. Bol. Chil. Parasitol. 36: 72–73.
Godínez, C.R., B. Zelaya de Romillo, D. Aurioles-Gamboa, A. Verdugo-Rodríguez, E.A. Rodríguez-Reyes and A. De la Peña-Moctezuma. 1999. Antibodies against *Leptospira interrogans* in California sea lion pups from Gulf of California. J. Wildl. Dis. 35: 108–111.
Goldman, C., J.D. Loureiro, M.J. Matteo and G.A. Cremaschi. 2008. *Helicobacter* spp. from gastric biopsies of stranded South American fur seals (*Arctocephalus australis*). Res. Vet. Sci. 86: 18–21.

Goldstein, T., F.M.D. Gulland, R.G. Braun, G.A. Antonelis, L. Kashinski, T.K. Rowles et al. 2006. Molecular identification of a novel gamma herpesvirus in the endangered Hawaiian monk seal (*Monachus schauinslandi*). Mar. Mamm. Sci. 22: 465–471.

Goldstein, T., T.S. Zabka, R.L. Delong, E.A. Wheeler, G. Ylitalo, S. Bargu et al. 2009. The role of domoic acid in abortion and premature parturition of California sea lions (*Zalophus californianus*) on San Miguel Island, California. J. Wildl. Dis. 45: 91–108.

Goldstein, T., K.M. Colegrove, M. Hanson and F.M. Gulland. 2011. Isolation of a novel adenovirus from California sea lions *Zalophus californianus*. Dis. Aquat. Organ. 94: 243–248.

Guillot, J., T. Petit, F. Degorce-Rubiales, E. Gueho and R. Chermette. 1998. Dermatitis caused by *Malassezia pachydermatis* in a California sea lion (*Zalophus californianus*). Vet. Rec. 142: 311–312.

Hall, A.J., S.T. Licence and P.P. Pomeroy. 1999. The response of grey seal pups to intradermal phytahaemagglutinin injection. Aquat. Mamm. 25: 25–30.

Hallegraeff, G.M. 1993. A review of harmful algal blooms and their apparent global increase. Phycologia. 32: 79–99.

Hammond, J.A., A.J. Hall and E.A. Dyrynda. 2005. Comparison of polychlorinated biphenyl (PCB) induced effects on innate immune functions in harbour and grey seals. Aquat. Toxicol. 74: 126–138.

Henton, M.M., O. Zapke and P.A. Basson. 1999. *Streptococcus phocae* infections associated with starvation in Cape fur seals. J. S. Afr. Vet. Assoc. 70: 98–99.

Hernández, M., I. Robinson, A. Aguilar, L.M. González, L.F. López-Jurado, M.I. Reyero et al. 1998. Did algal toxins cause monk seal mortality? Nature 393: 28–29.

Hernández-Orts, J.S., F.E. Montero, E.A. Crespo, N.A. García, J.A. Raga and F.J. Aznar. 2012. A new species of Ascocotyle (Trematoda: Heterophyidae) from the South American sea lion, *Otaria flavescens*, off Patagonia, Argentina. J. Parasitol. 98: 810–816.

Hernández-Orts, J.S., F.E. Montero, A. Juan-García, N.A. García, E.A. Crespo, J.A. Raga et al. 2013. Intestinal helminth fauna of the South American sea lion *Otaria flavescens* and fur seal *Arctocephalus australis* from northern Patagonia, Argentina. J. Helminthol. 287: 336–347.

Himworth, C.G., M. Haulena, D.M. Lambourn, J.K. Gaydos, J. Huggins, J. Calambokidis et al. 2010. Pathology and epidemiology of phocid herpesvirus-1 in wild and rehabilitating harbor seals (*Phoca vitulina richardsi*) in the northeastern Pacific. J. Wildl. Dis. 46: 1046–1051.

Honnold, S.P., R. Braun, D.P. Scott, C. Sreekumar and J.P. Dubey. 2005. Toxoplasmosis in a Hawaiian monk seal (*Monachus schauinslandi*). J. Parasitol. 91: 695–697.

Inoshima, Y., T. Murakami, N. Ishiguro, K. Hasegawa and M. Kasamatsu. 2013. An outbreak of lethal adenovirus infection among different otariid species. Vet. Microbiol. 165: 455–459.

Jackson, J.B.C., M.X. Kirby, W.H. Berger, K.A. Bjordal, L.W. Botsford et al. 2001. Historical overfishing and the recent collapse of coastal ecosystems. Science 293: 629–637.

James, K.J., B. Carey, J. O'Halloran, F.N. van Pelt and Z. Skrabáková. 2010. Shellfish toxicity: Human health implications of marine algal toxins. Epidemiol. Infect. 138: 927–940.

Jankowski, G., M.J. Adkesson, J.T. Saliki, S. Cárdenas-Alayza and P. Majluf. 2015. Survey for infectious disease in the South American fur seal (*Arctocephalus australis*) population at Punta San Juan, Peru. J. Zoo. Wildl. Med. 46: 246–254.

Jepson, P.D., P.M. Bennett, R. Deaville, C.R. Allchin, J.R. Baker and R.J. Law. 2005. Relationships between polychlorinated biphenyls and health status in harbor porpoises (*Phocoena phocoena*) stranded in the United Kingdom. Environ. Toxicol. Chem. 24: 238–248.

Johnson, B.W. and P. A. Johnson. 1978. The Hawaiian monk seal on Laysan Island: 1977. U.S. Dep. of Commer. NTIS PB-285-428, 38 pp.

Johnson, C.K., M.T. Tinker, J.A. Estes, P.A. Conrad, M. Staedler, M.A. Miller et al. 2009. Prey choice and habitat use drive sea otter pathogen exposure in a resource-limited coastal system. Proc. Natl. Acad. Sci. USA 106: 2242–2247.

Kaly, U.L. and G.P. Jones. 1994. Test of the effect of disturbance on ciguatera in Tuvalu. Mem. Queensl. Mus. 34(3): 523–532.

Kennedy, S. 1990. A review of the 1988 European seal morbillivirus epizootic. Vet. Rec. 127: 563–567.

Keogh, M.J., J.M. Maniscalco and S. Atkinson. 2010. Steller sea lion (*Eumetopias jubatus*) pups undergo a decrease in circulating white blood cells and the ability of T cells to proliferate during early post-natal development. Vet. Immunol. Immunopathol. 137: 298–304.

Khudyakov, J.I., L. Preeyanon, C.D. Champagne, R.M. Ortiz and D.E. Crocker. 2015. Transcriptome analysis of northern elephant seal (*Mirounga angustirostris*) muscle tissue provides a novel molecular resource and physiological insights. BMC Genomics 16: 64.

Kilpatrick, A.M., P. Daszak, S.J. Goodman, H. Rogg, L.D. Kramer, V. Cedeño et al. 2006. Predicting pathogen introduction: West Nile virus spread to Galapagos. Conserv. Biol. 20: 1224–1231.

King, D.P., M.C. Hure, T. Goldstein, B.M. Aldridge, F.M. Gulland, J.T. Saliki et al. 2002. Otarine herpesvirus-1: A novel gamma herpesvirus associated with urogenital carcinoma in California sea lions (*Zalophus californianus*). Vet. Microbiol. 86(1-2): 131–137.

Kreuder, C., M.A. Miller, D.A. Jessup, L.J. Lowenstine, M.D. Harris, J.A. Ames et al. 2003. Patterns of mortality in southern sea otters (*Enhydra lutris nereis*) from 1998–2001. J. Wildl. Dis. 39: 495–509.

Lavigne, D.M. and O.J. Schmitz. 1990. Global warming and increasing population densities: A prescription for seal plagues. Mar. Pollut. Bull. 21: 280–284.

Lehnert, K., S. Müller, L. Weirup, K. Ronnenberg, I. Pawliczka, T. Rosenberger et al. 2014. Molecular biomarkers in grey seals (*Halichoerus grypus*) to evaluate pollutant exposure, health and immune status. Mar. Pollut. Bull. 88: 311–318.

Levin, M., S. de Guise and P.S. Ross. 2005. Association between lymphocyte proliferation and polychlorinated biphenyls in free-ranging harbor seal (*Phoca vitulina*) pups from British Columbia, Canada. Environ. Toxicol. Chem. 24: 1247–1252.

Levy, J.K., P.C. Crawford, M.R. Lappin, E.J. Dubovi, M.G. Levy, R. Alleman et al. 2008. Infectious diseases of dogs and cats on Isabela Island, Galapagos. J. Vet. Intern. Med. 22: 60–65.

Li, L., T. Shan, C. Wang, C. Côté, J. Kolman, D. Onions et al. 2011. The fecal viral flora of California sea lions. J. Virol. 85: 9909–9917.

Lipscomb, T.P., D.P. Scott, R.L. Garber, A.E. Krafft, M.M. Tsai, J.H. Lichy et al. 2000. Common metastatic carcinoma of California sea lions (*Zalophus californianus*): Evidence of genital origin and association with novel gamma herpesvirus. Vet. Pathol. 37: 609–617.

Littnan, C.L., B.S. Stewart, P.K. Yochem and R. Braun. 2006. Survey for selected pathogens and evaluation of disease risk factors for endangered Hawaiian monk seals in the main Hawaiian Islands. EcoHealth 3: 232–244.

Liu, Y. 2010. Simulation of ocean circulation around the Galapagos archipelago using a hybrid coordinate ocean model (HYCOM). Ph.D. Thesis, North Carolina State University, Raleigh NC.

Lloyd-Smith, J.O., D.J. Greig, S. Hietala, G.S. Ghneim, L. Palmer, J. St Leger et al. 2007. Cyclical changes in seroprevalence of leptospirosis in California sea lions: Endemic and epidemic disease in one host species? BMC Infect. Dis. 7: 125.

Lynch, M., R. Kirkwood, A. Mitchell, P. Duignan and J.P.Y. Arnould. 2011. Prevalence and significance of an alopecia syndrome in Australian fur seals (*Arctocephalus pusillus doriferus*). J. Mammal. 92: 342–351.

Lynch, M., R. Kirkwood, R. Gray, D. Robson, G. Burton, L. Jones et al. 2012. Characterization and causal investigations of an alopecia syndrome in Australian fur seals (*Arctocephalus pusillus doriferus*). J. Mammal. 93: 504–513.

Lyons, C., M.J. Welsh, J. Thorsen, K. Ronald and B.K. Rima. 1993. Canine distemper virus isolated from a captive seal. Vet. Rec. 132: 487–488.

Lyons, E.T., R.L. Delong, S.A. Nadler, J.L. Laake, A.J. Orr, B.L. Delong et al. 2011. Investigations of peritoneal and intestinal infections of adult hookworms (*Uncinaria* spp.) in northern fur seal (*Callorhinus ursinus*) and California sea lion (*Zalophus californianus*) pups on San Miguel Island, California (2003). Parasitol. Res. 109: 581–589.

Maldonado, J.E., F.O. Davila, B.S. Stewart, E. Geffen and R.K. Wayne. 1995. Intraspecific genetic differentiation in California sea lions (*Zalophus californianus*) from southern California and the Gulf of California. Mar. Mamm. Sci. 11: 46–58.

Mancia, A., J.C. Ryan, R.W. Chapman, Q. Wu, G.W. Warr, F.M. Gulland et al. 2012. Health status, infection and disease in California sea lions (Zalophus californianus) studied using a canine microarray platform and machine-learning approaches. Dev. Comp. Immunol. 36: 629–637.

Martina, B.E., T.H. Jensen, M.W. van de Bildt, T.C. Harder and A.D. Osterhaus. 2002. Variations in the severity of phocid herpesvirus type 1 infections with age in grey seals and harbour seals. Vet. Rec. 150: 572–575.

McCarthy, A.J., M.A. Shaw and S.J. Goodman. 2007. Pathogen evolution and disease emergence in carnivores. Proc. Biol. Sci. 274: 3165–3174.

McCarthy, A.J., M.A. Shaw, P.D. Jepson, S.M. Brasseur, P.J. Reijnders and S.J. Goodman. 2011. Variation in European harbour seal immune response genes and susceptibility to phocine distemper virus (PDV). Inf. Gen. Evol. 11: 1616–1623.

Meise, K. and C. García-Parra. 2015. Behavioural and environmental correlates of *Philophthalmus zalophi* infections and their impact on survival in juvenile Galapagos sea lions. Mar. Biol. 162: 2017–2117.

Melero, M., D. García-Párraga, J.M. Corpa, J. Ortega, C. Rubio-Guerri, J.L. Crespo et al. 2014. First molecular detection and characterization of herpesvirus and poxvirus in a Pacific walrus (*Odobenus rosmarus divergens*). BMC Vet. Res. 10: 968.

Miller, D.L., R.Y. Ewing and G.D. Bossart. 2001. Emerging and resurging diseases. pp. 15–30. *In*: L.A. Dierauf and M.D. Gulland [eds.]. Marine Mammal Medicine. CRC Press, Boca Raton, FL, USA.

Miller, M.A., I.A. Gardner, C. Kreuder, D.M. Paradies, K.R. Worcester, D.A. Jessup et al. 2002. Coastal freshwater runoff is a risk factor for *Toxoplasma gondii* infection of southern sea otters (*Enhydra lutris nereis*). Int. J. Parasitol. 32: 997–1006.

Nadler, S.A., E.T. Lyons, C. Pagan, D. Hyman, E.E. Lewis, K. Beckmen et al. 2013. Molecular systematics of pinniped hookworms (Nematoda: *Uncinaria*): Species delimitation, host associations and host-induced morphometric variation. Int. J. Parasitol. 43: 1119–1132.

Nakata, H., A. Sakakibara, M. Kanoh, S. Kudo, H. Watanabe, N. Nagai et al. 2002. Evaluation of mitogen-induced responses in marine mammal and human lymphocytes by *in-vitro* exposure of butyltins and non-ortho coplanar PCBs. Environ. Pollut. 120: 245–253.

Negri, A.P., C.J.S. Bolch, S. Geier, D.H. Green, T.G. Park and S.I. Blackburn. 2007. Widespread presence of hydrophobic paralytic shellfish toxins in *Gymnodinium catenatum*. Harmful Algae 6: 774–780.

Nettleton, P.F., R. Munro, I. Pow, J. Gilray, E.W. Gray and H.W. Reid. 1995. Isolation of parapoxvirus from a grey seal (*Halichoerus grypus*). Vet. Rec. 137: 562–564.

Nielsen, O., K. Nielsen, R. Braun and L. Kelly. 2005. A comparison of four serologic assays in screening for *Brucella* exposure in Hawaiian monk seals. J. Wildl. Dis. 41: 126–133.

Nollens, H.H., J.F. Wellehan, L. Archer, L.J. Lowenstine and F.M. Gulland. 2010. Detection of a respiratory coronavirus from tissues archived during a pneumonia epizootic in free-ranging Pacific harbor seals *Phoca vitulina richardsii*. Dis. Aquat. Organ. 90: 113–120.

Norman, S.A., R.F. Di Giacomo, F.M. Gulland, J.S. Meschke and M.S. Lowry. 2008. Risk factors for an outbreak of leptospirosis in California sea lions (*Zalophus californianus*) in California, 2004. J. Wildl. Dis. 44: 837–844.

O'Hara, T.M. and T.J. O'Shea. 2001. Toxicology. pp. 471–520. *In*: L.A. Dierauf and M.D. Gulland [eds.]. Marine Mammal Medicine. CRC Press, Boca Raton, FL, USA.

Oksanen, A., S. Aittomäki, D. Jankovic, Z. Ortutay, K. Pulkkinen, S. Hämäläinen et al. 2014. Proprotein convertase FURIN constrains Th2 differentiation and is critical for host resistance against Toxoplasma gondii. J. Immunol. 193(11): 5470–5479.

Osterhaus, A.D., H. Yang, H.E. Spijkers, J. Groen, J.S. Teppema and G. van Steenis. 1985. The isolation and partial characterization of a highly pathogenic herpesvirus from the harbor seal (*Phoca vitulina*). Arch. Virol. 86: 239–251.

Osterhaus, A., J. Groen, H. Niesters, M. van de Bildt, B. Martina, L. Vedder et al. 1997. Morbillivirus in monk seal mass mortality. Nature 388: 838–839.

Osterhaus, A., M. van de Bildt, L. Vedder, B. Martina, H. Niesters, J. Vos et al. 1998. Monk seal mortality: Virus or toxin? Vaccine 16: 979–981.

Pabst, D.A., S.A. Rommel and W.A. McLellan. 1999. The functional morphology of marine mammals. pp. 17–22. *In*: J.E. Reynolds and A.A. Rommel [eds.]. Biology of Marine Mammals. Smithsonian Institution Press, Washington, D.C.

Páez-Rosas, D., M. Hirschfeld, D. Deresienski and G.A. Lewbart. 2016. Health status of Galápagos sea lions (*Zalophus wollebaeki*) on San Cristóbal Island rookeries determined by hematology, biochemistry, blood gases, and physical examination. J. Wildl. Dis. 52: 100–105.

Papadopoulos, E., P. Loukopoulos, A. Komnenou, E. Androukaki and A.A. Karamanlidis. 2010. First report of *Acanthocheilonema spirocauda* in the Mediterranean monk seal (*Monachus monachus*). J. Wildl. Dis. 46: 570–573.

Pascuan, C.G., S.L. Uran, M.R. Gonzalez-Murano, M.R. Wald, L.R. Guelman and A.M. Genaro. 2014. Immune alterations induced by chronic noise exposure: Comparison with restraint stress in BALB/c and C57Bl/6 mice. J. Immunotoxicol. 11: 78–83.

Pereira, E.M., G. Müller, E. Secchi, J. Jr. Pereira and A.L. Valente. 2013. Digenetic trematodes in South American sea lions from southern Brazilian waters. J. Parasitol. 99: 910–913.

Prager, K.C., D.J. Greig, D.P. Alt, R.L. Galloway, R.L. Hornsby, L.J. Palmer et al. 2013. Asymptomatic and chronic carriage of *Leptospira interrogans* serovar Pomona in California sea lions (*Zalophus californianus*). Vet. Microbiol. 164: 177–183.

Pistorius, P.A. and A.M.M. Baylis. 2011. A bald encounter: Hairless southern sea lion at the Falkland Islands. Polar Biol. 34: 145–147.

Rand, C.S. 1975. Nodular suppurative cutaneous cellulitis in a Galapagos sea lion. J. Wildl. Dis. 11: 325–329.

Recio, A., C. Linares, J.R. Banegas and J. Díaz. 2016. Road traffic noise effects on cardiovascular, respiratory, and metabolic health: An integrative model of biological mechanisms. Environ. Res. 146: 359–370.

Reif, J.S., M.M. Kliks, A.A. Aguire, D.L. Borjesson, L. Kashinsky, R.C. Braun et al. 2006. Gastrointestinal helminths in the Hawaiian monk seal (*Monachus schauinslandi*): Associations with body size, hematology, and serum chemistry. Aquat. Mamm. 32: 157–167.

Rijks, J.M., J.I. Hoffman, T. Kuiken, A.D. Osterhaus and W. Amos. 2008. Heterozygosity and lungworm burden in harbour seals (*Phoca vitulina*). Heredity (Edinb) 100: 587–593.

Rosen, D.A.S., A.J. Winship and L.A. Hoopes. 2007. Thermal and digestive constraints to foraging behaviour in marine mammals. Phil. Trans. Royal Soc. Lon. B. Biol. Sciences 362: 2151–2168.

Ross, P.S. 2002. The role of immunotoxic environmental contaminants in facilitating the emergence of infectious diseases in marine mammals. Hum. Ecol. Risk. Assess. 8: 277–292.

Routti, H., B.M. Jenssen, C. Lydersen, C. Bäckman, A. Arukwe, M. Nyman et al. 2010. Hormone, vitamin and contaminant status during the moulting/fasting period in ringed seals (*Pusa* [*Phoca*] *hispida*) from Svalbard. Comp. Biochem. Physiol. A Mol. Integr. Physiol. 155: 70–76.

Rutishauser, M.R., D.P. Costa, T.E. Goebel and T.M. Williams. 2004. Ecological implications of body composition and thermal capabilities in young Antarctic fur seals (*Arctocephalus gazella*). Physiol. Biochem. Zool. 77: 669–681.

Scholin, C.A., F. Gulland, G.J. Doucette, S. Benson, M. Busman, F.P. Chavez et al. 2000. Mortality of sea lions along the central California coast linked to a toxic diatom bloom. Nature 403: 80–84.

Schramm, Y., S.L. Mesnick, J. de la Rosa, D.M. Palacios, M.S. Lowry, D. Aurioles-Gamboa et al. 2009. Phylogeography of California and Galápagos sea lions and population structure with in the California sea lion. Mar. Biol. 156: 1375–1387.

Sepúlveda, M.A., M. Seguel, M. Alvarado-Rybak, C. Verdugo, C. Muñoz-Zanzi and R. Tamayo. 2015. Postmortem findings in four South American sea lions (*Otaria byronia*) from an urban colony in Valdivia, Chile. J. Wildl. Dis. 51: 279–282.

Sepúlveda, M.S. 1998. Hookworms (*Uncinaria* sp.) in Juan Fernández fur seal pups (*Arctocephalus philippii*) from Alejandro Selkirk Island, Chile. J. Parasitol. 84: 1305–1307.
Sepúlveda, M.S., H. Ochoa-Acuña and B.L. Homer. 1999. Age-related changes in hematocrit, hemoglobin, and plasma protein in Juan Fernandez fur seals (*Arctocephalus philippii*). Mar. Mamm. Sci. 15: 575–581.
Silva, R.Z., J. Pereira, Jr. and J.C. Cousin. 2014. Histological patterns of the intestinal attachment of *Corynosoma australe* (Acanthocephala: Polymorphidae) in *Arctocephalus australis* (Mammalia: Pinnipedia). J. Parasit. Dis. 38: 410–416.
Smayda, T.J. 1990. Novel and nuisance phytoplankton blooms in the sea: Evidence for a global epidemic. pp. 24–90. *In*: Toxic Marine Phytoplankton. Proc. 4th Int. Conf. On Toxic Marine Phytoplankton. Elsevier, USA.
Smith, K.F., K. Acevedo-Whitehouse and A. Pederson. 2009. The role of infectious diseases in biological conservation. Anim. Cons. 12: 1–12.
Sørmo, E.G., I. Jüssi, M. Jüssi, M. Braathen, J.U. Skaare and B.M. Jenssen. 2005. Thyroid hormone status in gray seal (*Halichoerus grypus*) pups from the Baltic Sea and the Atlantic Ocean in relation to organochlorine pollutants. Environ. Toxicol. Chem. 24: 610–616.
Spraker, T.R., E.T. Lyons, R.L. DeLong and R.R. Zink. 2004. Penetration of the small intestine of a California sea lion (*Zalophus californianus*) pup by adult hookworms (*Uncinaria* spp). Parasitol. Res. 92: 436–438.
Tabuchi, M., N. Veldhoen, N. Dangerfield, S. Jeffries, C.C. Helbing and P.S. Ross. 2006. PCB-related alteration of thyroid hormones and thyroid hormone receptor gene expression in free-ranging harbor seals (*Phoca vitulina*). Environ. Health Perspect. 114: 1024–1031.
Tan, K.S. and J. Ransangan. 2015. Factors influencing the toxicity, detoxification and biotransformation of paralytic shellfish toxins. Rev. Environ. Contam. Toxicol. 235: 1–25.
Tester, P.A. 1994. Harmful marine phytoplankton and shellfish toxicity potential consequences of climate change. pp. 69–76. *In*: M.E. Wilson, R. Levins and A. Spielman [eds.]. Disease in Evolution: Global Changes and Emergence of Infectious Diseases. The New York Academy of Sciences, New York, USA.
Toplu, N.J., A. Aydogan and T.C. Oguzoglu. 2007. Visceral leishmaniosis and parapoxvirus infection in a Mediterranean monk seal (*Monachus monachus*). Comp. Pathol. 136: 283–287.
Trillmich, F. and D. Limberger. 1985. Drastic effects of El Niño on Galápagos pinnipeds. Oecologia. 67: 19–22.
Trillmich, F. 2015. *Zalophus wollebaeki* (Galapagos sea lion). The IUCN Red List of Threatened Species. 557 Version 2015.2. www.iucnredlist.org.
Trites, A.W. and C.P. Donnelly. 2003. The decline of Steller sea lions in Alaska: A review of the nutritional stress hypothesis. Mamm. Rev. 33: 3–28.
van de Bildt, M.W., B.E. Martina, E.J. Vedder, E. Androukaki, S. Kotomatas, A. Komnenou et al. 2000. Identification of morbilliviruses of probable cetacean origin in carcases of Mediterranean monk seals (*Monachus monachus*). Vet. Rec. 146: 691–694.
van de Bildt, M.W., B.E. Martina, B.A. Sidi and A.D. Osterhaus. 2001. Morbillivirus infection in a bottlenosed dolphin and a Mediterranean monk seal from the Atlantic coast of West Africa. Vet. Rec. 148: 210–211.
Van Dolah, F.M. 2000. Marine algal toxins: origins, health effects, and their increased occurrence. Environm. Health Persp. 108S1: 133–141.
Van Wormer, E., M.A. Miller, P.A. Conrad, M.E. Grigg, D. Rejmanek, T.E. Carpenter et al. 2014. Using molecular epidemiology to track *Toxoplasma gondii* from terrestrial carnivores to marine hosts: Implications for public health and conservation. PLoS Negl. Trop. Dis. 8(5): e2852.
Venn-Watson, S., C. Benham, F.M. Gulland, C.R. Smith, J. St Leger, P. Yochem et al. 2012. Clinical relevance of novel otarine herpesvirus-3 in California sea lions (*Zalophus californianus*): lymphoma, esophageal ulcers, and strandings. Vet. Res. 43: 85.
Vera-Massieu, C., P.M. Brock, C. Godínez-Reyes and K. Acevedo-Whitehouse. 2015. Activation of an inflammatory response is context-dependent during early development of the California sea lion. R. Soc. Open Sci. 2(4): 150108.

Verant, M.L., N. d'Ozouville, P.G. Parker, K. Shapiro, E. Van Wormer and S.L. Deem. 2013. Attempted detection of *Toxoplasma gondii* oocysts in environmental waters using a simple approach to evaluate the potential for waterborne transmission in the Galapagos Islands, Ecuador. Ecohealth 11: 207–214.

Virgin, H.W., E.J. Wherry and R. Ahmed. 2009. Redefining chronic viral infection. Cell 138: 30–50.

Wilson, M.T. and P.J. Jokiel. 1986. Ciguatera at Midway: An assessment using the Hokama "stick test" for ciguatoxin. NOAA Tech. Rep. NOAA-SWFSC-86. 23 pp.

Wilson, S.C., T.M. Eybatov, M. Amano, P.D. Jepson and S.J. Goodman. 2014. The role of canine distemper virus and persistent organic pollutants in mortality patterns of Caspian seals (*Pusa caspica*). PLoS One 9(7): e99265.

Wilson, T.M. and I. Poglayen-Neuwall. 1971. Pox in South American sea lions (*Otaria byronia*). Can. J. Comp. Med. 35: 174–177.

Wright, E.P., L.F. Waugh, T. Goldstein, K.S. Freeman, T.R. Kelly, E.A. Wheeler et al. 2015. Evaluation of viruses and their association with ocular lesions in pinnipeds in rehabilitation. Vet. Ophthalmol. 18 Suppl. 1: 148–159.

Zenteno-Savin, T., M.A. Castellini, L.D. Rea and B.S. Fadely. 1997. Plasma haptoglobin levels in threatened Alaskan pinniped populations. J. Wildl. Dis. 33: 64–71.

Zuerner, R.L., C.E. Cameron, S. Raverty, J. Robinson, K.M. Colegrove, S.A. Norman et al. 2009. Geographical dissemination of *Leptospira interrogans* serovar Pomona during seasonal migration of California sea lions. Vet. Microbiol. 137: 105–110.

16

Pathologies of Pinnipeds in Brazil

Paula Baldassin,[1,]* *Derek Blaese de Amorim,*[2] *Max R. Werneck*[1] *and Daniela Bueno Mariani*[3]

Introduction

In Brazil, there are no breeding colonies of tropical or subtropical pinnipeds and the presence of some species are observed more frequently in the southeastern and southern regions, from June to October, due to their post-reproductive movements, especially driven by the cold current of the Malvinas (Pinedo 1990, Simões-Lopes et al. 1995, Oliveira et al. 2006). Additionally, some species, such as the South American sea lion (*Otaria flavescens*), have marked seasonality (Simões-Lopes et al. 1995).

In these regions, there are rehabilitation centers specializing in marine animal veterinary care. In the Southeast, there are five rehabilitation centers belonging to Non-governmental organizations (NGOs), universities and private companies, which are contracted to monitor the coast for marine mammal strandings and rescue efforts due to environmental impacts resulting from conducting maritime exploration and production of oil and gas by the Oil and Gas General Coordination (CGPEG) of the Brazilian

---

[1] BW Consultoria Veterinária, Ubatuba, SP, Brazil.
[2] CECLIMAR - Centro de Estudos Costeiros, Limnológicos e Marinhos, Universidade Federal do Rio Grande do Sul, Imbé, RS, Brazil.
[3] Programa de Pós-graduação em Ciência Veterinária, Universidade Federal Rural de Pernambuco, Recife, Pernambuco, Brasil.
* Corresponding author: pauletsbj@gmail.com

Institute of Environment and Renewable Natural Resources (IBAMA). In addition to these centers, there are four centers from NGOs and Universities (Table 1) in the south.

The southern region has the highest incidence of pinnipeds in the country as most individuals are closer to some of the breeding colonies. However, it has been noticed that there has been an increase in the number of animals in both regions. It is believed that the increasing number of South American pinnipeds and even Antarctic and sub-Antarctic species along the Brazilian coast may be related to increased monitoring efforts by researchers and monitoring programs associated with the ban on hunting, thus, resulting in an increase in the number of individuals observed.

The species rescued and received by these institutions include tropical and subtropical species such as the South American sea lion (*O. flavescens*) and the South American fur seal (*Arctocephalus australis*) followed by the Sub antarctic fur seals (*Arctocephalus tropicalis*), Antarctic fur seal (*Arctocephalus gazella*), southern elephant seal (*Mirounga leonina*), crabeater seal (*Lobodon carcinophagus*) and the leopard seal (*Hydrurga leptonyx*).

This chapter addresses accounts for the two most frequent species of tropical pinnipeds in Brazil (i.e., *O. flavescens* and *A. australis*) with description of their main pathologies and threats currently found in this country.

Table 1. Marine animal rehabilitation centers of the southeast and southern regions of Brazil.

Regions	Rehabilitation centers	Management
Southeast region	CRAMAR	CTA Meio Ambiente (Company)
	CRETA – Centro de reabilitação e triagem de animais marinhos	Instituto Argonauta para a Conservação Costeira e Marinha (NGO)
	GREMAR	Instituto Gremar (NGO)
	IPEC – Instituto de Pesquisa Cananéia	IPEC – Instituto de Pesquisa Cananéia (NGO)
South region	PROAMAR – Projeto de Reabilitação e Estudos de Aves, Mamíferos e Répteis	Centro de Estudos do Mar da Universidade Federal do Paraná (CEM/UFPR)
	R3 Animal	R3 Animal (NGO)
	CERAM – Centro de reabilitação de animais silvestres e marinhos	Universidade Federal do Rio Grande do Sul
	CRAM – Centro de reabilitação de animais marinhos	Universidade Federal do Rio Grande

Government institutions and legislation

The Center for Aquatic Mammals (CMA) at the Chico Mendes Institute for Biodiversity Conservation (ICMBio) is the government agency responsible for coordinating, implementing and promoting studies, projects and research and management programs for the conservation of aquatic mammals, mainly focusing on endangered and migratory species. This agency developed the National Action Plan for Large Cetaceans and Pinnipeds, setting specific objectives, guidelines and priority actions for the conservation of marine mammals listed in the National List of Brazilian Fauna Threatened with Extinction (Normative Instruction Ministry of Environment No. 3, 26/05/2003), as well as for species jeopardized by anthropogenic threats. In Brazil, the fauna for both the rehabilitation centers and zoos and aquariums, came under the control of the States.

In Brazil, there are legal requirements on infrastructure, management and health aspects, to be followed by rehabilitation centers of marine mammals. These regulations include some aspects for licenses, facilities, management and release such as the Ordinance of the Ministry of Environment (MMA) No. 98, of April 14, 2000, IBAMA Normative Instruction No. 3 of February 8, 2002, IBAMA Normative Instruction 179 of June 25, 2008, IBAMA Normative Instruction No. 23, of December 31, 2014 and IBAMA Normative Instruction No. 07 of April 30, 2015 (IBAMA 2015).

First aid on the beach

As pinnipeds use both marine and terrestrial habitats, finding a specimen on the beach does not necessarily indicate that the animal is sick, thus it is important to assess the health and body conditions to differentiate whether it needs veterinary medical care or is just resting. Generally, the first step is to isolate the animal from the area to keep it away from other animals or environmental stressors. At this point, a trained veterinarian or technician should perform a routine checkup up to determine if the animal is alert, moving its head and body, and departing when something or someone approaches it. The approach should be gradual and careful, because some animals under high stress can become nervous and aggressive in self-defense. As protection, the veterinarian or rescue personnel should use shields and personal protective equipment.

If there is verification that the animal is just resting following the assessment, it should be kept under the shade if the day is warm and there is too much direct heat, in which case the animal should be moistened with water in a regular basis. If the event takes place during the tourist season or carnival in Brazil, the translocation of animals to a nearby, more quiet beach where the animal can rest without being harassed is recommended. In contrast to the situation described above, an animal exhibiting detrimental physical, serious behavioral conditions and/or presence of injuries or fracture must be referred to a rehabilitation center.

Trained personnel involved in the rescue and capture must always move around quietly, precluding sudden gestures, moving from the water toward the animal in order to avoid the animal escaping or returning to the water. With small animals, physical restraint can be performed using dip nets or blankets laced with hoops. For a large animal, wooden shields to corner and direct it to the transportation cage or kennel can be used.

To manage the beach commotion and visitors during the rescue operation, proper logistics are required in order to achieve the priority goal for safety of people and animals while avoiding accidents. This procedure requires at least two people, and everyone involved should use personal protective materials and protection of open wounds. The transportation kennel should have adequate size and ventilation. The transport should be done with the maximum of caution, requiring particular attention on the internal temperature of the animal so that it does not undergo hyperthermia or hypothermia.

Pathologies

A comprehensive knowledge of the general health of marine mammal populations can be acquired through multidisciplinary studies and necropsy is an essential tool. A necropsy can help to elucidate and define the cause of death of the animal, collect important biological data, determine direct human impacts, and establish health baselines of biological pathogens and disease. Information on morbidity and mortality in many wildlife populations is scarce and marine mammals are not the exception (Rowles et al. 2001).

A wide variety of infectious agents may affect pinnipeds, leading to morbidity or mortality. There are numerous viruses (e.g., pox viruses, papillomavirus, and morbillivirus), bacteria (such as *Salmonella* spp., *Corynebacterium* spp., *Pseudomonas* spp.) and fungi (like *Aspergillus fumigatus*, *Candida* spp., *Cryptococcus neoformans*) described for pinnipeds, causing lesions that lead to death (Stoskopf 2001, Lawrence et al. 2001, Reidarson et al. 2001). In addition, important zoonotic agents have also been isolated from pinnipeds, including Influenza A and B, *Brucella* spp., *Mycobacterium* spp., *Erysipelothrix* spp., *Leptospira* spp. (Lawrence et al. 2001, Cousins et al. 2003, Blanc et al. 2009). Knowledge of the causes of morbidity and mortality of pinnipeds in Brazil is scarce. This limitation is due to lack of resources and even local infrastructure for conducting more detailed studies. Unfortunately in other cases, the results are not reported at scientific meetings or journals to communicate findings. Information on the cause of death is an important element for developing preservation projects for the population health of marine mammals, including pinnipeds.

Main pathologies found in pinnipeds

Of 50 animals investigated during necropsies in a study (Amorim 2014), 19 cases (38%) were associated with severe trauma, seven cases (14%) with respiratory tract conditions, three cases (6%) showed gastrointestinal tract disorders, two cases (4%) in the muscular system, two cases (4%) in the central nervous system, one (2%) in the hepatic system and a single case (2%) with reproductive tract disorders. While mortality was associated with cachexia in nine cases (18%), the cause of death was not determinate in six cases (12%).

Of the 19 cases with severe trauma, tearing of muscles, extensive bleeding and multiple trauma were observed in 15 cases, while laceration of the musculature and extensive bleeding was found in four cases. In the seven animals showing pathologies in the respiratory tract, the cause of death was diagnosed as follows: three cases of pulmonary parasitosis, a case of purulent bronchopneumonia, two acute respiratory failures and a chronic pleurisy and peritonitis. Of the three cases with pathologies in the gastrointestinal tract, the cause of death included two cases of parasitic enteritis and a case of necrosis in the pyloric region associated with the presence of an external artefact. Of the two specimens that showed changes in the muscular system, one had muscle necrosis associated with bacterial myositis and one exhibited bacterial myositis. The two animal showing pathologies in the central nervous system exhibited cerebral hemorrhage. Furthermore, endometritis was observed in the specimen that had a lesion in the reproductive system; necrotic hepatitis was diagnosed in the animal that showed a pathology in the hepatic system. Finally, cachexia was diagnosed as the cause of death in six animals with poor body condition.

Tuberculosis

A key feature of tuberculosis is the clinically silent and opportunistic infection mechanism of this disease. Clinical signs are nonspecific and show slow evolution, depending on the extent and location of injuries. The affected specimen may show poor body condition, thinning of subcutaneous fat, exercise intolerance, open mouth breathing, progressive loss of body weight, anorexia, shortness of breath, weakness, depression, lethargy, and dysphagia. Coughing is not a common clinical sign reported in cases of tuberculosis in pinnipeds, even in cases where the lung injury is significant. Most often, the presumptive clinical diagnosis can be performed only when the disease is in a very advanced state. Due to the clinical signs, it is thought that the transmission of tuberculosis in aquatic carnivores occurs mainly via inhalation (i.e., airborne) during coughing and sneezing (Forshaw and Phelps 1991, Bernardelli et al. 1996, Bastida et al. 1999, Kiers et al. 2008, Lefèvre et al. 2010, Arbiza et al. 2012, Jurczynski et al. 2012).

During necropsies, macroscopic changes were frequently observed in cases of tuberculosis in aquatic carnivores suffering of calcified granulomatous lesions in thoracic, mediastinal lymph node and lungs with large quantity of fluid in the thoracic cavity. Other changes are accentuated such as diffuse thickening of the parietal and visceral pleura, papillary and proliferative lesions in the parietal and visceral pleura, pulmonary emphysema, granulomatous lesions on liver, spleen, pleura, peritoneum and lymph nodes, exudate in the tracheal lumen, lymph nodes enlargement and calcified foci on lymph nodes cut surfaces (Forshaw and Phelps 1991, Cousins et al. 1993, Bernadelli et al. 1996, Bastida et al. 1999, Cousins et al. 2003, Kiers et al. 2008, Kriz et al. 2011, Jurczynski et al. 2011, 2012, Amorim et al. 2014, Boardman et al. 2014).

Histopathologically, granulomas characterized by caseous necrosis with inflammatory infiltrate composed of lymphocytes, neutrophils, macrophages and epithelioid cells are observed. The lymph nodes usually are markedly hyperplastic with necrotic area, with possible central calcification. On pleural surface, fibroplasia with marked lymphocytic infiltration is observed. There are mononuclear exudates in the bronchi and bronchioles' lumen and aggregates composed by lymphocytes and macrophage below the cartilage (Forshaw and Phelps 1991, Bernadelli et al. 1996, Arbiza et al. 2012, Amorim et al. 2014, Boardman et al. 2014). In the Ziehl-Neelsen staining, alcohol-acid resistant bacilli in the focus of the organ affected can be observed (Forshaw and Phelps 1991, Cousins et al. 1993, Bernadelli et al. 1996).

In cases of tuberculosis in aquatic carnivores, clinical evaluation is not characterized as an effective diagnostic procedure method, since the clinical symptoms and lesions are usually observed in cases where the disease is already at an advanced stage (Forshaw and Phelps 1991, Cousins et al. 1993, Jurczynski et al. 2012, Arbiza et al. 2012). Different methods for antemortem diagnosis of the disease were tested in pinniped specimens and compared with the results of *Mycobacterium* isolation, PCR, Ziehl-Neelsen stain and/or pathological changes after death or euthanasia of positive cases (Forshaw and Phelps 1991, Kiers et al. 2008, Jurczynski et al. 2011, Jurczynski et al. 2012). Testing using bovine and avian tuberculin has shown that this technique is sensitive, economical and easy to perform, but there is a risk of false positives to occur (i.e., TB-positive in cases where the animal had contact with other non-Mycobacteria belonging to the *Mycobacterium tuberculosis* Complex or MTBC), as well as false negatives in advanced cases of infection with specimen anergy or recent infections, in which the animal immune system has not yet generated an adequate response (Forshaw and Phelps 1991, Arbiza et al. 2012).

From the sputum of South American sea lions, acid-fast staining, PCR (DNeasy Blood and Tissue Kit, Qiagen) and culture were performed. The bacterial isolation is considered the gold standard for the confirmation of *Mycobacterium*, despite the long time for confirmation because of the slow growth of bacteria. PCR was used in the bacterial isolation to identify the etiologic agent belonging to the MTBC and it should be noted that the test is not validated for tuberculosis in pinnipeds. Similarly, the acid-fast staining can give a false-positive result because the isolated bacillus can be a *Mycobacterium* not belonging to the MTBC. The use of sputum has shown low sensitivity for the diagnosis of infections caused by *Mycobacterium*, as well as in other species (Cousins et al. 1993, Dierauf and Gulland 2001, Jurczynski et al. 2011, Arbiza et al. 2012, Jurczynski et al. 2012). Radiographic examination has not been proven to be a good method of diagnostic imaging in sea lions due to the size of the evaluated animals and the thickness of the blubber, features interfering with the penetration of X-rays (Forshaw and Phelps 1991, Dierauf and Gulland 2001, Jurczynski et al. 2011, 2012).

No antemortem method of tuberculosis detection in aquatic carnivores is completely accurate when used alone. However, combining the results from the sputum, serology and computerized tomography (CT) scan has proved to be a useful tool for the diagnosis of tuberculosis in captive South American sea lions. The application of these tests in free-ranging animals and reproductive colonies is difficult to implement, especially for otariids due to the biological characteristics of species and distribution of geographical colonies. In wild animals, possible diagnostic methods have been suggested, including serological tests (i.e., ElephantTB STAT-PAK® assay, multi-antigen print immunoassay [MAPIA], and dual path platform assay [DPP]), culture of sputum samples and PCR (Arbiza et al. 2012, Jurczynski et al. 2012). Connecting the results obtained from the study on tuberculosis in Uruguay to the high prevalence of the disease in animals found in captivity from South American colonies, Arbiza et al. (2012) was able to provide an alert concerning the wide distribution of this disease in colonies of the country (Uruguay). Even so, the authors described the isolation of *Mycobacterium pinnipedii* in offspring of *O. flavescens* and *A. australis* and juvenile specimens of *O. flavescens*, *A. australis* and *M. leonina*, indicating an early transmission of the pathogen.

Amorim et al. (2014) described the first report of *M. pinnipedii* in a South American sea lion from Brazil. They reported a carcass of an adult male in poor physical condition with accentuated muscular atrophy and thinning of subcutaneous fat. The necropsy indicated a large quantity of reddish serous fluid (i.e., exudate) in the thoracic cavity along with thickening of the parietal and visceral pleura (Fig. 1A), a 3 cm diameter firm nodule in the right lung and no lymph node chain enlargement. The histology revealed parietal pleura with diffuse caseous necrosis with areas of calcification, lungs with accentuated calcification of the alveolar septum and visceral pleura, and the mesenteric lymph node had an extensive area of caseous necrosis with moderate central calcification (Fig. 1B). The kidneys and stomach had metastatic calcification of the intima layer of the arteries. The cause of the death was granulomatous pleuropneumonia and granulomatous mesenteric lymphadenitis caused by *M. pinnipedii*. Following 18 days under these pathological findings, acid-fast bacilli were isolated from frozen lungs and mesenteric lymph nodes fragments, as shown in Fig. 1C.

The prevalence of tuberculosis disease within a pinniped colony can be quite high due to the gregarious behavior of otariids, mainly during the period when South American sea lion breeding concur with other pinnipeds,

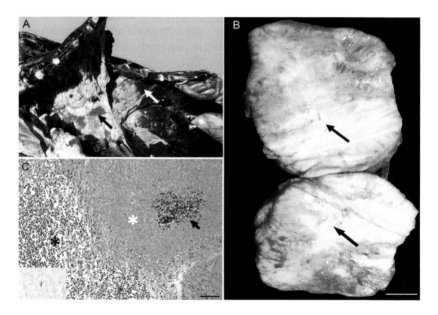

Figure 1. Tuberculosis caused by *Mycobacterium pinnipedii* in a South American sea lion (*Otaria flavescens*). (A) Thoracic cavity with a large quantity of reddish serous fluid (black and white arrows) with accentuated and diffuse thickening of visceral and parietal pleura (asterisk). (B) Transversal cut of mesenteric lymph node, diffusely whitish and with central calcification (arrows); Bar 51.5 cm. (C) Mesenteric lymph node with extensive area of caseous necrosis (white asterisk) with moderate central calcification (arrow) surrounded by infiltrate predominantly of lymphocytes and plasmocytes with few macrophages (black asterisk). H&E stain; Bar 5100 mm. Alcohol-acid resistant bacillus in detail (see inset pale square at the inferior margin to the left in Fig. 1C), Ziehl-Neelsen staining (Amorim et al. 2014).

highlighting the high risk of transmission to other species (Riedman 1990, Bernadelli et al. 1996, Bastida et al. 1999, Arbiza et al. 2012). The information gaps on tuberculosis caused by *M. pinnipedii* in *O. flavescens* and *A. australis* in South America countries where both species occur highlights the need for further studies to understand the epidemiology of this disease.

Campylobacter sp.

There is only one report describing an abortion caused by *Campylobacter* sp. in a female specimen of *A. australis* (Machado et al. 2007) in the north of the Rio Grande do Sul state, Brazil. The animal was resting on the beach and beside her was a new fetus aborted, with 17 cm in total length and a mass of 130 g. Following transportation of the animal to the rehabilitation center, nasal, oral and anal swabs were performed, and, in the latter, the presence of *Campylobacter* sp. was detected.

Helminths

There is limited data about intestinal parasites and helminthiasis and their health impact on the following pinniped hosts, *A. australis* and *O. flavescens*, along the Brazilian coast. Even the few available reports present specific findings and basically restrict these to the southern region of Brazil, especially along the coast of the state of Rio Grande do Sul. Terms such as prevalence (P), average intensity (IM) and average abundance (AM) are reported in several studies, following the definitions by Bush et al. (1997).

The first report was performed by Andrade et al. (1998) through the gastrointestinal tract analysis of *O. flavescens* adults found dead in the southern coast of Rio Grande do Sul, reporting for the first time the occurrence of *Corynosoma australe* Johnston, 1937 (Acanthocephala: Polymorphidae) in gut, as well as specimens of *Contracaecum* sp. (Nematoda: Anisakidae) in stomach.

Subsequently, Marigo (2003) analyzed stranded animals from the Southern Region beaches (State of Rio Grande do Sul) of Brazil, evaluating two males and a young specimen of *A. australis* and a young male of *O. flavescens*. The death of one *A. australis* was related to the presence of metastrongylid nematodes in lungs causing bronchopneumonia and acute-moderate multifocal interstitial pneumonia. Although the cause of death of the other specimen was due to a physical trauma, it also showed pulmonary congestion associated with the presence of parasites in the alveolar space. Furthermore, in one of the deceased *A. australis*, the author reports the occurrence of *Corynosoma* sp. (Acanthocephala: Polymorphidae), a parasite that has not apparently been observed and associated to these kind of lesions in the host. Necropsy results of this animal revealed the presence of helminth parasites in the pulmonary system of *O. flavescens*, leading to moderate to severe fibrinous parasitic pneumonia, with congestion, edema and hemorrhage.

Silva et al. (2013) analyzed the infection by *Corynosoma cetaceum* Johnston and Best, 1942 (Acanthocephala: Polymorphidae) in 32 specimens of *A. australis* found dead on the beach of Cassino, in Rio Grande do Sul, between 2008 and 2010. The hosts were divided into 26 pups (12 males and 14 females), 4 prepuberal males and 2 reproductive/breeding females. These groups were seasonally evaluated (i.e., fall and winter, spring and summer). The overall prevalence (P = 34.4%) by *C. cetaceum* was 34.4%, with an average intensity (MI) of 2.9 and average abundance (AM) of 1.0. Histologic analysis of stomach samples indicated that the infection by *C. cetaceum* caused changes in the gastric glands, exhibiting areas of necrosis, erosion and size reduction of glands. No significant differences for average prevalence and intensity were observed among host genders or group inter-seasonality during the study. Due to some factors (i.e., low parasitological indexes, severity of inflammatory reactions and behavioral changed in diet) the infection of *C. cetaceum* was considered accidental in the host groups analyzed.

Through the analysis of 29 specimens of *O. flavescens* (13 subadults and 16 adults, with a total length ranging from 1.58 to 2.64 m), found dead in beach strandings in the state of Rio Grande do Sul between June 2010 and September 2011, Pereira (2012) identified six species of parasites: *Contracaecum ogmorhini* Johnston and Mawson, 1941 (P = 10.3%, MI = 332, AM = 34.3), *C. australe* (P = 100%, IM and AM = 1,756), *Bolbosoma turbinella* (Diesing 1851) (P = 50%, IM = 42.7, AM = 21.3), *Diphyllobothrium* sp. (P = 4.2%), *Stephanoprora uruguayense* Holeman et Olagüe 1989 (P = 17.4%, IM = 202 and AM = 8.4) and *Ascocotyle (Phagicola) longa* Ransom, 1920 (P = 33.3%, IM = 248.5 ± 539.1 [Mean ± SD], AM = 83.0 ± 3201 [Mean ± SD]), being the last two parasite species the first digenetic records for *O. flavescens* in Brazil (Pereira et al. 2013).

More recently, Silva et al. (2014) evaluated 30 (15 males and 15 females) *A. australis* from Cassino Beach, in Rio Grande do Sul state and evaluated different degrees of intestinal disorders resulting from *C. australes* infection. The lesions observed ranged from edema of the submucosa, with dilatation of the lymphatic vessels in the superficial areas of parasite insertion in the gut of the host to total destruction of the mucosa and the muscle layer at those points where the insertion of the parasite was most profound in the host gastrointestinal wall. The tissue alterations were generally found to be restricted to the parasites' insertion points, causing different local reactions, but no changes were observed in other regions of the host guts.

At present, the analysis of helminth parasites of *A. australis* in the Brazilian region is restricted to the genus *Corynosoma*, which is represented by the following species: *C. australes* and *C. cetaceum*. The helminth parasites of *O. flavescens* in the same territory include *C. ogmorhini*, *C. australes*, *B. turbinella*, *S. uruguayense*, *A. (Phagicola) longa* and *Diphyllobothrium* sp. These data are presented along with the sites of infection and their references in Table 2.

Table 2. Helminth parasite species reported in tropical pinnipeds from the Brazilian coast.

Host	Species	Infection site	Reference
Arctocephalus australis	*Corynosoma* sp.	Intestine	Marigo (2003)
	Corynosoma cetaceum	Stomach	Silva et al. (2013)
	Corynosoma australe	Small intestine	Silva et al. (2014)
Otaria flavescens	*Contracaecum* sp.	Stomach	Andrade et al. (1998)
	Contracaecum ogmorhini	Stomach	Pereira (2012)
	Corynosoma australe	Intestine, Small intestine	Andrade et al. (1998), Pereira (2012)
	Bolbosoma turbinella	Small intestine	Pereira (2012)
	Stephanoprora uruguayense	Small intestine	Pereira (2012)
	Ascocotyle (Phagicola) longa	Small and large intestine	Pereira (2012)
	Diphyllobothrium sp.	Intestine	Pereira (2012)

Threats

Fisheries interaction

According to Rigon et al. (2014), in a survey conducted between 2011 and 2012 in southern Brazil, there were records of 6 entanglements of *A. australis*, in which the length of the animals ranged from 88 to 130 cm (i.e., 2 males, 2 females and 2 indeterminate). The entanglements of pinnipeds have increased worldwide and have been associated with declining populations in some cases (e.g., Steller sea lion, *Eumetopias jubatus*; Hawaiian monk seal, *Neomonachus schauinslandi*, and Northern fur seal, *Callorhinus ursinus*). The interaction resulting from the fishing artifacts is a potential threat to all marine mammals, since the fishing effort in southern Brazil has increased.

Pollution

Records of non-infectious diseases were also reported in aquatic mammals such as congenital defects, neoplasia and trauma (Gulland et al. 2001).

In addition, studies on heavy metals and persistent organic pollutants (POPs) found in high concentrations in various organs and tissues can affect the homeostasis of the animal (Fillmann et al. 2007, Baraj et al. 2009). Environmental pollutants, including organochlorines and heavy metals have also been reported to contribute to modulating the pathogenic and pathogenetic activity primarily displayed by marine mammal morbilliviruses, associated to death in several free-living pinniped and cetacean populations around the world (Di Guardo et al. 2005).

Dog attack

A study on the cause of death of *A. australis* in the southern coast of Brazil conducted in 2011 and 2012 by Amorim (2014) reports that of 50 necropsied animals, 19 cases (38%) were associated with dog attacks.

In the 19 cases, lacerations of skeletal muscle associated with extensive areas of hemorrhage were observed. In addition to the aforementioned injury, 15 cases (15/19) showed multiple trauma observed mainly in cervical vertebrae and ribs (Fig. 2A). Other changes observed in lower frequency were pneumothorax, avulsion of salivary glands, disruption of the trachea, presence of blood in the trachea and pulmonary hemorrhage, as shown in Figs. 2B and 2C. The specimens usually showed no skin lesions suggestive of attack or aggression. In some cases, the gross examination revealed infrequent, circular lesions suggestive of a puncture wound or holes by dog bites when the skin tissue was thoroughly evaluated. This type of attack that canids inflict on their prey causes minor injuries to the skin that can cover avulsions, severe lacerations and muscle, lesions in the vascular system and internal organs. The low amount of perforating injuries into the skin observed in South American sea lions can also be related to the tough and robust skin and connective tissues of the species (Duignan et al. 2003, Munro and Munro 2008, Pough et al. 2008). Histopathologically, there were multifocal hemorrhages mainly in skeletal muscle and lungs (Fig. 2).

Similar lesions found in specimens of *A. australis* at the Rio Grande do Sul coast were also observed in animals from Cape Polonius and Isla de Lobos, Uruguay (Katz et al. 2012). During the breeding season in rookeries of South American fur seal from Uruguay and Peru, there is often observations of dogs attacking fur seal pups (Katz et al. 2012).

Three direct cases of interaction between South American fur seal and dogs were witnessed during beach monitoring along the Rio Grande do Sul coast. A juvenile was observed entering the sea after chasing and barkings from a dog pack; another animal was trying to enter the sea to find protection and defend itself from the attack of three dogs (Fig. 3A); and a third animal was found dead after being carried by one of the dogs from the pack (Fig. 3B).

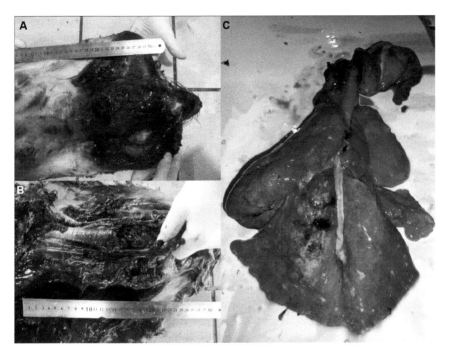

Figure 2. Pathological lesion caused by dog attacks in South American fur seal (*A. australis*). Extensive hemorrhage area observed in the ventral cervical region; decubitus dorsal (A). Extensive muscle laceration and hemorrhagic area, with disruption of the trachea; decubitus dorsal (B). Lungs showing a pulmonary hemorrhagic area observed during the gross analyses of a dead specimen (C).

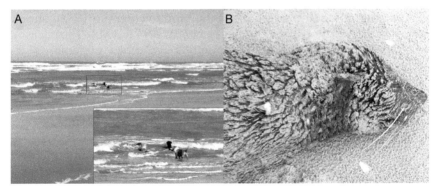

Figure 3. Dogs attacking South American fur seals in a beach from the southern coast of Brazil (A). Dead South American fur seal after a dog attack (B).

Public health policies for reducing the presence of dogs on the beaches are required to prevent transmission of diseases to and mortality of pinnipeds inhabiting the coast and natural area of occurrence and resting. Moreover, the contact between dogs and fur seals can transmit infectious diseases (e.g., morbilliviruses, including canine distemper virusor CDV) that can affect breeding colonies of tropical pinnipeds, and potentially cause mass mortality events.

Marking and release

To optimize the success of a release, the animal must spend as little time as possible in captivity and showing acceptable health body condition. Before releasing the animal, the medical evaluation, blood tests, microbiological and parasitological analyses are clinical aspects of paramount importance to be conducted. These procedures helps to ensure that the released animal is able to perform all their activities (e.g., feeding, migration and reproduction) and reduce the maximum risk of carrying diseases to breeding colonies.

Marking procedures are important to assess the success of the release and migration of species. Ear tags used for cows and sheep are the marking devices most commonly used on pinnipeds (colored, with sequential numbers and contacts information for the home institution) in Brazil. The two species studied in this chapter are marked according to Lander et al. (2001), using the skinfold thickness of the axillary region of the pinniped for marking. No standardized Brazilian system of pinnipeds marking exists at the Center for Aquatic Mammals (CMA), thus each rehabilitation center has its own markup protocols in Brazil.

For the release of animals, rehabilitation centers in Brazil should follow the Instruction No. 23, of December 31, 2014 IBAMA. The releasing of rehabilitated animals should be carefully performed, preferably at the location where the specimen was found to avoid the introduction or spread of any subclinical diseases for breeding colonies. If this is not feasible, the animal should be relocated to another nearby beach for later release.

Acknowledgements

The authors thank Dr. Patricia Fair for reviewing the chapter and providing insights.

Keywords: Rehabilitation center, bacterial, viral and parasitis diseases, pinnipeds, Brazil

References

Amorim, D.B. 2014. Estudo de causa Mortis de *Arctocephalus australis* (Zimmermann, 1783) (Lobo-marinho-Sul-Americano) no litoral norte do Rio Grande do Sul. Brasil. MSc. Dissertation, Universidade Federal do Rio Grande do Sul, Brazil.

Amorim, D.B., R.A. Casagrande, M.M. Alievi, F. Wouters, L.G.S. De Oliveira, D. Driemeier et al. 2014. *Mycobacterium pinnipedii* in a Stranded South American Sea Lion (*Otaria byronia*) in Brazil. J. Wildl. Dis. 50: 419–422.

Andrade, A.L., M.C. Pinedo and J. Pereira, Jr. 1998. Parasitos bioindicadores dos habitats dos mamíferos aquáticos? Brasil. RT/ SOLAMAC. Brazil 8: 7.

Arbiza, J., A. Blanc, M. Castro-Ramos, H. Katz, A. Ponce de León and M. Clara. 2012. Uruguayan pinnipeds (*Arctocephalus australis* and *Otaria flavescens*): Evidence of influenza virus and *Mycobacterium pinnipedii* infections. pp. 151–182. *In*: A. Romero and O.E. Keith [eds.]. New Approaches to the Study of Marine Mammals. Intech, Rijeka, Croatia.

Baraj, B., L.F. Niecheski, H. Windom and L. Hermanns. 2009. Trace metal concentration in liver, kidney and heart in South American fur seal (*Arctocephalus australis*) from Southern Brazil. Mar. Pollut. Bull. 58: 1922–1952.

Bastida, R., J. Loureiro, V. Quse, A. Bernardelli, D. Rodriguez and E. Costa. 1999. Tuberculosis in a wild sub Antarctic fur seal from Argentina. J. Wildl. Dis. 35: 796–798.

Bernardelli, A., R. Bastida, J. Loureiro, H. Michelis, M.I. Romano, A. Cataldi et al. 1996. Tuberculosis in sea lions and fur seals from the south-western Atlantic coast. Rev. Sci. Tech. Off. Int. Epiz. 15: 985–1005.

Blanc, A., D. Ruchansky, M. Clara, F. Achaval, A. Le Bas and J. Arbiza. 2009. Serologic evidence of influenza A and B viruses in South American fur feals (*Arctocephalus australis*). J. Wildl. Dis. 45: 519–521.

Boardman, W.S., L. Shephard, I. Bastian, M. Globan, J.A. Fyfe, D.V. Cousins et al. 2014. *Mycobacterium pinnipedii* tuberculosis in a free-ranging Australian fur seal (*Arctocephalus pusillus doriferus*) in South Australia. J. Zoo. Wildl. Med. 45: 970–972.

Bush, A.O., K.D. Lafferty, J.M. Lotz and A.W. Shostak. 1997. Parasitology meets ecology on its own terms: Margolis et al. revisited. J. Parasitol. 83: 575–583.

Cousins, D.V., S.N. Williams, R. Reuter, D. Forshaw, B. Chadwick, D. Coughran et al. 1993. Tuberculosis in wild seals and characterization of the seal bacillus. Aust. Vet. J. 70: 92–97.

Cousins, D.V., R. Bastida, A. Cataldi, V. Quse, S. Redrobe, S. Dow et al. 2003. Tuberculosis in seals caused by a novel member of the *Mycobacterium tuberculosis* complex: *Mycobacterium pinnipedii* sp. nov. Int. J. Syst. Evol. Microbiol. 53: 1305–1314.

Di Guardo, G., G. Marruchella, U. Agrimi and S. Kennedy. 2005. Morbillivirus infection in aquatic mammals: A brief overview. J. Vet. Med. A 52: 88–93.

Dierauf, L.A. and F.M.D. Gulland. 2001. CRC Handbook of Marine Mammal Medicine, 2nd Ed. CRC Press, Boca Raton, USA.

Duignan, P.J., N.J. Gibbs and G.W. Jones. 2003. Autopsy of pinnipeds incidentally caught in commercial fisheries, 2001/02. DOC Science international series 131. Department of Conservation, Wellington.

Fillmann, G., L. Hermanns, T.W. Fileman and J.W. Readman. 2007. Accumulation patters of organoclhorines in juveniles of *Arctocephalus australis* found stranded along the coast of Southern Brazil. Environ. Pollut. 146: 262–267.

Forshaw, D. and G.R. Phelps. 1991. Tuberculosis in a captive colony of pinnipeds. J. Wildl. Dis. 27: 288–295.

Gulland, F.M.D., L.J. Lowenstine and R.T. Spraker. 2001. Noninfectious diseases. pp. 521–547. *In*: L.A. Dierauf and F.M.D. Gulland [eds.]. CRC Handbook of Marine Mammal Medicine, 2nd Ed. CRC Press, Boca Raton, USA.

IBAMA (Instituto Brasileiro de Meio Ambiente) 2015. Available from: http://www.ibama.gov.br/ (September 30, 2015).

Jurczynski, K., J. Scharpegge, J. Ley-Zaporozhan, S. Ley, J. Cracknell, K. Lyashchenko et al. 2011. Computed tomographic examination of South American sea lions (*Otaria flavescens*) with suspected *Mycobacterium pinnipedii* infection. Vet. Rec. 169: 608–613.
Jurczynski, K., J.K.P. Lyashchenko, J. Scharpegge, M. Fluegger, G. Lacave, I. Moser et al. 2012. Use of multiple diagnostic tests to detect *Mycobacterium pinnipedii* infections in a large group of South American Sea Lions (*Otaria flavescens*). Aquat. Toxicol. 38: 43–55.
Katz, H., D. Morgades and M. Castro-Ramos. 2012. Pathological and parasitological findings in South American fur seal pups (*Arctocephalus australis*) in Uruguay. ISRN Zoology ID 586079.
Kiers, A., A. Klarenbeek, B. Mandelts, D. Van Soolingen and G. Koeter. 2008. Transmission of *Mycobacterium pinnipedii* to humans in a zoo with marine mammals. Int. J. Tuberc. Lung. D 12: 1469–1473.
Kriz, P., P. Kralik, M. Slany, I. Slana, J. Svobodova, I. Parmova et al. 2011. *Mycobacterium pinnipedii* in captive Southern sea lion (*Otaria flavescens*): A case report. Vet. Med-Czech. 56: 307–313.
Lander, M.E., A.J. Westgate, R.K. Bonde and M.J. Murray. 2001. Tagging and tracking. pp. 851–880. *In*: L.A. Dierauf and F.M.D. Gulland [eds.]. CRC Handbook of Marine Mammal Medicine, 2nd Ed. CRC Press, Boca Raton, USA.
Lawrence, D.L.J., J.D. Buck and T.R. Robeck. 2001. Bacterial diseases of cetaceans and pinnipeds. pp. 309–335. *In*: L.A. Dierauf and F.M.D. Gulland [eds.]. CRC Handbook of Marine Mammal Medicine, 2nd Ed. CRC Press, Boca Raton, USA.
Lefèvre, P., J. Blancou, R. Chermette and G. Uilenberg. 2010. Infectious and Parasitic Diseases of Livestock. Lavoisier, France.
Machado, R., J. Barreto, S. Siciliano, A.L.L. Filgueiras, L.R. Oliveira, J.P.A.V. Rocha et al. 2007. Evidência de aborto em lobo-marinho-sul-americano, *Arctocephalus australis* (Carnivora: Otariidae), possivelmente causado por *Campylobacter* sp. no litoral norte do Rio Grande do Sul. SIC UFRGS, Brazil 19: 358.
Marigo, J. 2003. Patologia Comparada das Principais Enfermidades Parasitárias de Mamíferos Marinhos encontrados na Costa Sudeste e Sul do Brasil. MsC. Dissertation, Universidade de São Paulo, São Paulo, Brazil.
Munro, R. and H.M.C. Munro. 2008. Animal Abuse and Unlawful Killing. Forensic Veterinary Pathology. Saunders Elsevier, London.
Oliveira, L.R., M. Arias-Schreiber, D. Meyer and J.S. Morgante. 2006. Effective population size in a bottlenecked fur seal population. Biol. Cons. 131: 505–509.
Pereira, E.M. 2012. Identificação da comunidade componente de helmintos gastrointestinais, hepáticos, pulmonares, cardíacos e renais de *Otaria flavescens* (Shaw, 1800) Leão-marinho-do-sul, no litoral sul do Brasil. MsC. Dissertation, Universidade Federal de Pelotas, Pelotas, Brazil.
Pereira, E.M., G. Müller, E. Secchi, J. Pereira, Jr. and A.L.S. Valente. 2013. Digenetic trematodes in South American sea lions from Southern Brazilian waters. J. Parasitol. 99: 910–913.
Pinedo, M.C. 1990. Ocorrência de pinípedes na costa brasileira. Garcia de Orta, Sér. Zool. 15: 37–48.
Pough, F.H., C.M. Janis and J.B. Heinser. 2008. A vida dos vertebrados. Atheneu, São Paulo.
Reidarson, T.H., J.F. McBain, L.M. Dalton and M.G. Rinaldi. 2001. Mycotic diseases. pp. 337–356. *In*: L.A. Dierauf and F.M.D. Gulland [eds.]. CRC Handbook of Marine Mammal Medicine, 2nd Ed. CRC Press, Boca Raton, USA.
Riedman, M. 1990. The Pinnipeds: Seals, Sea Lions and Walruses. University of California Press, Los Angeles.
Rigon, C.T., D.B. Amorim, I.R. Machado and L.R. Oliveira. 2014. Registros de interação de *Arctocephalus australis* e *Arctocephalus tropicalis* com a pesca no sul do Brasil. RT/SOLAMAC. Colombia 10: 786.
Rowles, T.K., M.F. Van Dolah and A.A. Hohn. 2001. Gross necropsy and specimen collection protocols. pp. 449–470. *In*: L.A. Dierauf and F.M.D. Gulland [eds.]. CRC Handbook of Marine Mammal Medicine, 2nd Ed. CRC Press, Boca Raton, USA.

Silva, R.Z., J.C.B. Cousin and J. Pereira, Jr. 2013. *Corynosoma cetaceum* Johnston & Best, 1942 (Acanthocephala, Polymorphidae) in *Arctocephalus australis* Zimmermann, 1783 (Mammalia: Pinnipedia): Histopathology, parasitological indices, seasonality and host gender influences. Estud. Biol. 35: 121–134.

Silva, R.Z., J. Pereira, Jr. and J.C.B. Cousin. 2014. Histological patterns of the intestinal attachment of *Corynosoma australe* (Acanthocephala: Polymorphidae) in *Arctocephalus australis* (Mammalia: Pinnipedia). J. Parasit. Dis. 38: 410–416.

Simões-Lopes, P.C., C.J. Drehmer and P.H. Ott. 1995. Nota sobre os Otariidae e Phocidae (Mammalia: Carnivora) da costa norte do Rio Grande do Sul e Santa Catarina, Brasil. Rev. Bras. Biociênc. 3: 173–181.

Stoskopf, S.K. 2001. Viral diseases. pp. 285–308. *In*: L.A. Dierauf and F.M.D. Gulland [eds.]. CRC Handbook of Marine Mammal Medicine, 2nd Ed. CRC Press, Boca Raton, USA.

Index

A

Africa 234–240
anthropogenic interactions 239
anthropogenic threats 69, 128, 164
Arctocephalus australis 211
Atlantic Ocean 220

B

bacterial 273, 275
basicranium 39
biological pollution 123, 124
Brazil 269–272, 275, 277–282
breeding 234, 236, 237, 239

C

Cabo Blanco 221, 223–229
California sea lion 23, 24, 26, 29, 43–45
canine distemper virus 124
Cape fur seal 234–241
charismatic species 159, 160
Chile 176–179
chlorophyll "a" 93, 100
climate change 241
conservation 69–75, 86–88, 120, 122, 123, 126–128, 133, 135, 139, 140, 151, 219–221, 227–230
conservation strategies 170

D

DDTs 125
Disease 245–252, 256–259
distribution 234, 238–241
diving 198, 199, 201
diving behavior 134, 135, 138, 141–143, 147, 148
diving physiology 132, 141

E

eastern Mediterranean Sea 221, 225, 227, 229

ecology 50, 51, 57
endangered 51, 219, 220, 226, 230
endangered species 160, 161
Endangered Species Act 70
evolution 22, 23, 38
expansion 91
exploitation 178, 179, 186, 188

F

fisheries interaction 85, 227, 201
foraging behavior 134, 134, 137, 138, 140, 146, 148, 149, 197
foraging ecology 133, 134, 237
foraging niche overlap 146, 147, 149, 150, 151
fur seals 1, 4–6

G

Galapagos fur seal 120, 122, 123, 125, 127
Galapagos Islands 161, 162, 167, 170, 172, 174
Galapagos sea lion 26, 29, 32, 36, 37, 38, 42, 43, 132, 133, 137, 142, 150, 159–162, 167, 168, 170, 171, 173, 174
genetic inbreeding 226, 228, 229
Guadalupe fur seals 102, 105, 112, 113
Gulf of Ulloa 91, 93, 100, 101, 110, 116

H

habitat deterioration 226
harvesting 194, 196, 204
Hawaiian monk seal 50–52, 59, 62, 63, 65, 69–76, 81–83, 86, 87
health 245, 247–249, 251, 253, 254, 257–259
Humboldt squid 93, 95, 96, 102, 104

I

infectious disease 81
Influenza 82
intrinsic growth rate 191
Island of Gyaros 221, 229

J

Juan Fernandez Archipelago 177–179, 184, 189, 191
Juan Fernández fur seal 176, 177, 179

L

Leptospira 124
life-history 51, 52, 57
Lobodontini 15, 16, 18, 19

M

Management Plan 160, 163, 167, 171
marine mammal 50, 51, 63, 64
Marine Mammal Protection Act 70
maternal behaviour 212
Mediterranean monk seal 219–230
Miocene 13, 15–19
Monachinae 13, 15, 19
Monachini 13, 15, 18–20
Morbillivirus 81, 82
morphology 22, 23, 29, 30, 32, 33, 41, 43, 44

N

Namibia 234, 236, 238–241
Neomonachus schauinslandi 50

O

Otaria flavescens 194, 195 ,199
Otariidae 12, 13, 17–20
otariids 3, 8
oxygen stores 141–143, 148

P

pathogens 245–249, 256, 257, 259
PCBs 125
persistent organic pollutants 123, 125
Phocidae 13, 17–19
phocids 3
physiological ecology 141
pinniped evolution 2, 13
pinnipeds 1, 3–9, 245–249, 269–276, 279, 282

Pliocene 13, 16–20
pollution 226, 229
population 91–93, 95, 96, 103, 104, 113, 114, 116, 120–123, 126–128, 219–223, 226–230
population dynamics 176, 181

R

recovery 179
Rehabilitation center 269–272, 277, 282
reproduction 146
reproductive behaviour 196, 214

S

sagittal crest 32, 38, 41, 43, 44
sea lions 1, 3, 8
sea surface temperature 93, 101, 113, 116
seals 1, 3–6
shark predation 77, 78
South African fur seal 234
South American fur seal 211, 213, 216
South American sea lion 194, 195, 197, 199
southern Angola 234

T

Tropic of Cancer 3
Tropic of Capricorn 3
tropical 1–9

U

Uruguay 194–204, 211–213, 216

V

viruses 127

Q

West Nile virus 81

Z

Zalophus wollebaeki 132–134
zygomatic arches 32, 33, 38, 41–44

Alejandro Selkirk de mi corazón

Alejada de mí en la distancia, lates entre nubes de agua y de sal. Te azota el viento y la marea.
Tu costa tenue es bañada por horizontes de espuma. La isla de la abundancia te llaman, la isla de mi corazón te nombro.
Majestuosamente naces de brazos de gigantes desde fondo del mar.
La cordillera submarina te alza hacia el firmamento y deslumbras con tu pureza, tu esplendor, tu magnificencia.
La más bella de todas. Tu geografía furtiva me conmueve. Eres una sola y parte de todas "Masafuera" querida.
Navegué un océano de tiempo encontrándote. Viví un océano de tiempo recorriéndote.
Marcado por fuego volcánico y voluntad marina el sendero de lobos respira. En el camino a "Lobería Vieja" se siente el latido. Doblando por "Vicente Porras" el mundo se descubre y la mirada se llena, se condensa,
se sacia de peludos habitantes de esta tierra lejana.
Lobos por millares, altura de "Los inocentes". La tierra se expande poderosa y te sientes en tierra de leyendas. Perdida, inalcanzable, apartada de toda ley excepto por el susurro del viento, el vuelo del Blindado, la paz de la distancia, el rugido del lobo.
Ahí, en pleno campamento lobero, con el castigo de la gran ola, de día, de noche, de luna, los lobos existen, persisten, se encuentran.
El Blindado rasga el cielo en silencio. Observa tranquilo el transcurso de una vida.
Quebrada de las casas y hombres y mujeres y niños. Corazones marinos. Voluntades pesqueras.
Los botes pintados dan vueltas en la circunferencia isleña. Langostas, vidriolas, bacalaos, erizos.
Las fardelas llegan de noche a veces encandiladas con las luces del pueblo.
Pescadores luchando, viviendo, heroicos habitantes del Pacifico. Historias sublimes del esfuerzo humano.
Han seguido la estrella por salvar a sus amados.
Isla querida que estas en la distancia. Cuanto murmullo has traído.
El estero baja raudo al compás de la lluvia. Te escucho aun en mi guarida.
Selkirk, endémica tierra, mi corazón late en tus entrañas.

For Product Safety Concerns and Information please contact our EU representative GPSR@taylorandfrancis.com Taylor & Francis Verlag GmbH, Kaufingerstraße 24, 80331 München, Germany